The views expressed in this article are those of the author and do not reflect the official policy or position of the Air Force, Department of Defense, or the U.S. government.

This page intentionally blank.

Preface

This book grew out of several courses on atmospheric entry offered at the Air Force Institute of Technology. I was a student in the first course when it was created in the mid-1980s. The course was taught from an analytical perspective, with finding closed-form solutions being the preferred approach. The professor's goal was to get us to understand how "families of solutions" behaved as well as the general trends, tradeoffs, and (above all else) the nature of atmospheric entry before picking point designs to study in detail. That analytical mindset stuck with me and carried over to how I approached engineering tasks for the next 20+ years.

This book uses the same approach to "get back to the basics" for a new generation of students who've become more comfortable with numerical solutions than analytical ones. You'll find the pages are loaded with equations. That's because I've tried to include the details of many of the derivations so you won't have to spend inordinate amounts of time recreating something that "is easily shown."

Does this book present anything new? Well, yes and no. Many of the fundamental areas in atmospheric entry have studied by very capable people already and it would be a disservice to gloss over that fact. This book pulls together many classical analyses and presents them in a consistent notation for the first time. It provides a convenient starting point for an analytical understanding of atmospheric entry, with plenty of references to those original works. It ties together results that were originally published years apart by different authors. And, peppered throughout, you'll find some new approaches and results.

How is this book different? This book approaches solutions with one thought always in the back of our minds: "How little can I know about the vehicle and still study its atmospheric entry?" Current books tend to quickly turn to numerical solutions. Many of the numerical approaches require you to know the mass, surface area, etc. of the vehicle before beginning. While that approach certainly has its place, what happens early in the design when you don't even know if the vehicle should be a cone or a sphere? The approaches in this book help you make those early decisions by avoiding "point designs."

Why is there so much emphasis on analytical formulations? There's definitely a time to run screaming to the computer for a solution. But, in doing so, it's easy to lose sight of the "big picture" or miss "general trends" that could lead you to a better vehicle design. By thoroughly understanding the classic analytic analyses first, we can better use the computer to solve the hard problems. We'll use easily visualized variables to solve the analytical problems and keep them (more-or-less) consistent as we move to the computer. As you'll see, the consistency will enable us to "know very little about the vehicle" and still study its entry!

Who should use this book? It was designed for senior undergraduate and graduate engineering students since a basic understanding of differential equations is required. Students should also be currently enrolled in or have taken a "basic orbital mechanics" class to get the most out of it.

I've tried to catch and correct all of the mistakes in this book. But, I realize errors will "magically" appear. If you find any, or you have any other comments, I encourage you to contact me by email at the address below. All comments are important!

spam4dayton-reentrybook@yahoo.com

ACKNOWLEDGMENTS

I'd like to thank the 55+ graduate students that helped me tailor this book to meet their needs. Their comments and feedback helped me immeasurably in crafting a stack of lecture notes into a book.

Lt Col Kerry D. Hicks

This page intentionally blank.

Contents

This page intentionally blank.

Table of Figures

This page intentionally blank.

Nomenclature

Latin Symbols:

a	semi-major axis
a_{decel}	total deceleration
A	surface area for heating
$B = \dfrac{C_D S}{2m}$	ballistic coefficient
c	speed of sound
c_f	average skin friction coefficient
C_D	coefficient of drag
$\bar{C}_D = \dfrac{C_D}{C_{D*}}$	"normalized" coefficient of lift
C_{D*}	coefficient of drag at the maximum lift-to-drag ratio
C_{D_0}	zero-lift coefficient of drag
C_L	coefficient of lift
$\bar{C}_L = \dfrac{C_L}{C_{L*}}$	"normalized" coefficient of lift
C_{L*}	coefficient of lift at the maximum lift-to-drag ratio
$\dfrac{C_L}{C_D}$	lift-to-drag ratio
$\left(\dfrac{C_L}{C_D}\right)_* = \dfrac{C_{L*}}{C_{D*}}$	maximum lift-to-drag ratio
C_P	fluid heat capacity at constant pressure
$D = \dfrac{\rho C_D S}{2}{}^R V^2$	drag

e	eccentricity
E	eccentric anomaly
$Ei(\alpha) = \int_{-\infty}^{\alpha} \dfrac{e^x}{x}\, dx$	exponential integral function
f	atmospheric deceleration sufficient to assume entry is occurring
$f = \left(1 + 2e\cos v + e^2\right)^{\frac{1}{2}}$	shorthand term used when relating the ORSW and ONTW frames
$f(\lambda) = \overline{C}_D$	"normalized" coefficient of drag
$f(\lambda) = \dfrac{(n-1) + \lambda^n}{n}$	"normalized" coefficient of drag (expressed as a function of the lift parameter or "normalized" coefficient of lift)
$\mathcal{F}(W, v, \gamma) = 0$	constraint relation
$\vec{F}$	force vector
$F = \left(1 - \dfrac{r_{p0}\,\omega_\oplus \cos i_0}{^{I}V_{p0}}\right)^2$	constant term used to relate inertial and relative velocity in drag-perturbed orbits
$F_p = \dfrac{\rho_p S C_D}{2m}\sqrt{\dfrac{r_p}{\beta}}$	Chapman's periapsis parameter
g	acceleration due to gravity
$G = \dfrac{-1}{\beta r_0 \eta}\left(1 - \dfrac{1}{2T}\right)\xi$	Loh's "constant"
$G = \dfrac{-\cos\gamma}{\sqrt{\beta r}\,Z}\left(1 - \dfrac{\cos^2\gamma}{u}\right)$	Loh's "constant" written in universal formulation
$h = rV\cos\gamma$	orbital specific momentum
$h = r - R_\oplus$	altitude above planet surface

$h = \dfrac{1}{W}$ nondimensional "height" in universal formulation

$h = C_p \mathcal{T}$ specific enthalpy

h^* heat transfer coefficient of fluid

$H = \dfrac{V^2}{2} + C_p \mathcal{T}$ total enthalpy

i inclination

$I_n(x)$ modified Bessel function of the first kind and of order n

ℓ specific heat ratio

k auxiliary variable used in derivations

K auxiliary variable used in derivations

$K = \dfrac{C_{D_0}}{(n-1)C_{L*}{}^n}$ constant used to define the drag polar

L reference length

$L = \dfrac{\rho C_L S}{2}\,{}^R V^2$ lift

$L = \dfrac{C_L}{C_D}$ shorthand used for lift-to-drag ratio

$L_* = \dfrac{C_{L*}}{C_{D*}}$ shorthand used for maximum lift-to-drag ratio

m vehicle mass

$M = \dfrac{{}^R V}{c}$ mach number

n constant used in various equations, including the drag polar and series solutions

$n = \sqrt{\dfrac{\mu}{a^3}}$ mean motion

$$Nu_L = \frac{h^*L}{\kappa}$$ — Nusselt number

p — semilatus rectum

$$Pr = \frac{\mu C_P}{\kappa}$$ — Prandtl number

$\dot{q}_s$ — stagnation heat flux

$\dot{q}_w$ — average wall heat flux

$\bar{\dot{q}}_s$ — non-dimensional stagnation heat flux

$\bar{\dot{q}}_w$ — non-dimensional average wall heat flux

Q — total heat transfer

$\bar{Q}$ — non-dimensional total heat transfer

r — radius from the center of the planet to the vehicle

r_a — apoapsis radius

r_p — periapsis radius

R — nose radius of curvature

$\mathcal{R}$ — gas constant per unit weight

$R_\oplus$ — radius of the planet

$$Re_L = \frac{\rho V L}{\mu}$$ — Reynold's number

RMS — root mean square

s — arc-length

$$s = \int_0^t \frac{RV}{r}\cos\gamma\, dt$$ — independent variable used universal formulations

S — reference area for drag

$$St = \frac{Nu_L}{Pr \cdot Re_L}$$ — Stanton number

t	time
T	thrust
T	component of drag acceleration perturbation tangent to orbit
$T = \dfrac{1}{2}\left(\dfrac{{}^{R}V^2}{g_0 r_0}\right)$	non-dimensional kinetic energy
$\mathcal{T}$	absolute temperature
$u = \dfrac{{}^{R}V^2 \cos^2 \gamma}{gr}$	nondimensional kinetic energy perpendicular to radius in universal formulation
$u = \omega + v$	argument of latitude
$v = \dfrac{u}{\cos^2 \gamma} = \dfrac{{}^{R}V^2}{gr}$	a nondimensional kinetic energy in universal formulation (used when lift and drag coefficients are not constant)
V	velocity (no specific reference)
$\bar{V} = \dfrac{{}^{R}V}{\sqrt{g_\oplus R_\oplus}}$	nondimensional velocity in Yaroshevskii's formulation
${}^{I}V$	inertial velocity of the vehicle
${}^{R}V$	velocity of the vehicle relative to the atmosphere
V_A	velocity of the atmosphere
W	component of drag acceleration perturbation perpendicular to the orbit plane
$W = \dfrac{\rho C_{L*} S}{2m}\sqrt{\dfrac{r}{\beta}}$	a nondimensional altitude in universal formulation (used when lift and drag coefficients are not constant)
x	auxiliary variable used in derivation

$$x = -\ln\left(\frac{^{R}V}{\sqrt{g_{\oplus}R_{\oplus}}}\right)$$ independent variable in Yaroshevskii's formulation

X horizontal distance traveled

y distance from vehicle wall in boundary layer

$$y = \frac{\rho C_{D} S}{2m}\sqrt{\frac{R_{\oplus}}{\beta}}$$ nondimensional altitude in Yaroshevskii's formulation

$$z = \frac{r_{0}}{r}$$ a nondimensional distance

$$Z = \frac{\rho C_{D} S}{2m}\sqrt{\frac{r}{\beta}}$$ a nondimensional altitude in universal formulation

$Z_{p} = F_{p}$ a nondimensional altitude of periapsis in universal formulation (equal to Chapman's periapsis parameter)

Greek Symbols:

$\overline{\beta r}$ average value of βr for an atmosphere

β^{-1} scale height for atmospheric density

γ flight-path angle

γ_{E} Euler constant

$$\varepsilon = \frac{V^{2}}{2} - \frac{\mu}{r}$$ specific mechanical energy

ε one of the angles describing the thrust vector relative to the velocity vector

ζ one of the angles describing the thrust vector relative to the velocity vector

$\eta = \dfrac{\rho S C_D}{2m\beta}$	a non-dimensional altitude
θ	longitude
κ	coefficient of thermal conductivity of atmosphere
$\lambda = \bar{C}_L$	normalized coefficient of lift (sometimes referred to as "lift parameter")
μ	fluid viscosity of atmosphere; gravitational constant
ν	true anomaly
$\xi = \cos\gamma$	cosine of the flight-path angle
ρ	atmospheric density
$\rho_{\oplus}$	atmospheric density at the planet's surface
ρ_s	atmospheric density at the planet's surface
σ	bank angle
$\bar{\sigma}$	optimal (constant) bank angle (Egger's approach)
$\bar{\sigma}_*$	optimal (constant) bank angle (Gell's approach)
τ	auxiliary variable used in derivations
ϕ	latitude
ψ	heading angle
ω	argument of periapsis
$\omega_{\oplus}$	rotation rate of the planet
Ω	rotation rate between reference frames
Ω	right-ascension of the ascending node

Subscripts, Superscripts, Overscripts:

$(\ )_*$	conditions at some "critical" point
$(\ )_\infty$	conditions far away from the vehicle

$(\ )_\phi$	component in the direction of increasing latitude
$(\ )_0$	conditions at a reference radius
$(\ )_e$	conditions at entry
$(\ )_f$	final conditions
$(\ )_{gd}$	geodetic value
$(\ )_L$	component in lift direction
$(\ )_{oe}$	conditions at the "outer edge" of the boundary layer
$(\ )_{over}$	conditions on the overshoot boundary
$(\ )_r$	component in radial direction
$(\ )_S$	conditions at planet surface
$(\ )_{under}$	conditions on the undershoot boundary
$(\ )_v$	component in velocity direction
$(\ )_w$	conditions at the wall (surface) of the vehicle
$^I(\ )$	measured with respect to an inertial frame (e.g., inertial velocity ^IV)
$^R(\ )$	measured with respect to a rotating frame (e.g., velocity relative to the rotating atmosphere is ^RV)
$\overrightarrow{(\ )}$	vector
$\widehat{(\ )}$	unit vector
$\langle\ \rangle$	values "averaged over an orbit"

Chapter 1

Introduction

1.1 Applicable Missions

Not all space missions end with a satellite, vehicle, or payload entering a planet's atmosphere; in fact, most do not. And, of those that do end with a plunge through the atmosphere, only some of those require the object to actually *survive* the trip! At least three missions require surviving the entry: ballistic-missile warheads, planetary probes, and manned missions.

Ballistic missiles provided the first real experiences with atmospheric entry (or "reentry"). The German V-2 program was the first to encounter reentry problems. (The rockets had a nasty habit of exploding when they hit the Earth's atmosphere.) Later, the warheads of ballistic missiles became the first reentry "vehicles" to be tested as the United States and the Soviet Union sought to develop better and more accurate ways to deliver nuclear weapons to each other's cities.

Planetary probes developed to explore the solar system are similar in many ways to ballistic-missile warheads. However, the probes usually contain more delicate sensors so they might be designed to reduce the deceleration forces.

Similarly, the probes may intend to *gently* land on the planet's surface whereas the warhead doesn't mind plowing into the surface at hypersonic speeds. The National Reconnaissance Office's Corona spy satellites (1959-1972) used a type of "film bucket" comparable to a planetary probe. After taking pictures in orbit, the satellite deorbited a small reentry capsule (shown in Figure 1-1) with the film. The capsule was recovered and the film developed.

Manned missions absolutely require surviving the atmospheric entry. There are even more constraints on the vehicle (and its trajectory) when humans are onboard. At the very least, there is a relatively small maximum deceleration limit and a greater need to control the interior temperature.

1.2 Phases of Flight

Typical ballistic missile trajectories are composed of three parts: the *powered boost phase* from launch until burnout, the *free-flight* (or *orbital*) phase, and the *reentry phase* that begins at the somewhat arbitrary point where the atmosphere "begins." (Perhaps more correctly, we should say it is at the point

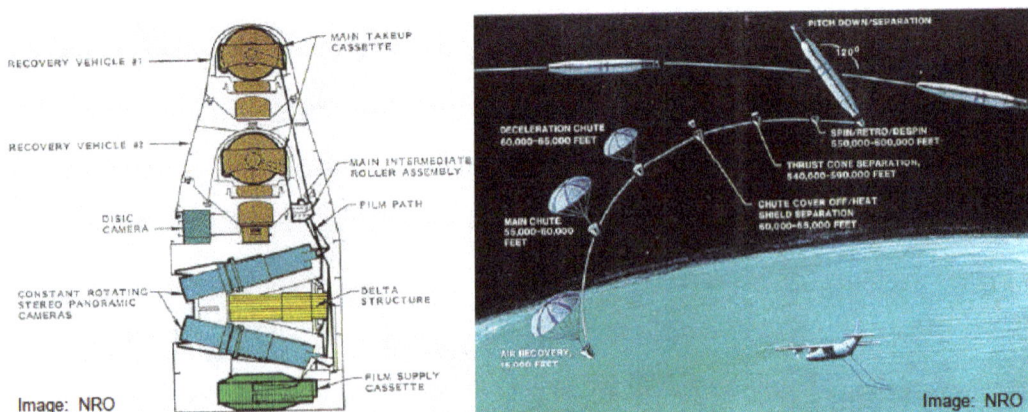

Figure 1-1: Corona Reentry Capsules and Their Recovery

where the atmospheric drag becomes significant in some sense.) Figure 1-2 shows a typical ballistic missile trajectory.

During the powered boost phase, the propulsion system (rocket motor, hypersonic scramjet, etc.) is continuously adding energy to the vehicle. The on-board guidance and navigation system largely determines the trajectory followed, so gravity and atmospheric effects take a secondary role. There may be a period of time between when powered flight ends and the vehicle "clears" the atmosphere. During that "gap," the vehicle is "coasting" unpowered, but might still be influenced by atmospheric drag. We will, however, restrict our study to assuming the boost phase ends after the vehicle has "cleared" the atmosphere. Further, since the guidance and navigation system plays the dominant role in setting the trajectory during boost, we will not study that motion in detail here.

Once the engine cuts off and the vehicle has cleared the atmosphere, the free-flight phase begins. In this region, the trajectory is governed by orbital motion. For our purposes, we will assume the only gravity source is the planet we

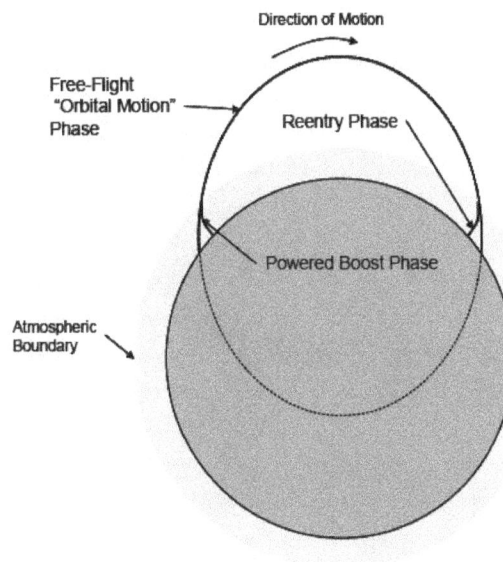

Figure 1-2: Geometry of a Typical Ballistic Missile Trajectory

are near and that the planet is perfectly spherical. (In-other-words, simple two-body orbital mechanics is sufficient to describe the motion.) This motion will be addressed in Chapter 2.

When the vehicle (or its payload) returns to the atmosphere, drag dramatically influences the trajectory and the reentry phase begins. Chapter 3 derives the general equations of motion for this phase. Chapter 7 addresses the aerodynamic heating encountered during this reentry.

For certain missions, the boost phase may be absent (or so long ago as to be irrelevant). Figure 1-3 illustrates this. Examples include the space shuttle's return to Earth as well as an Apollo capsule return from the moon. The free-flight/orbital motion phase is identical that in the ballistic missile discussion above while the portion in the atmosphere is simply called "entry" instead of "reentry." For consistency, "entry" will be used for all situations where a vehicle is entering a planet's atmosphere in the remainder of this text.

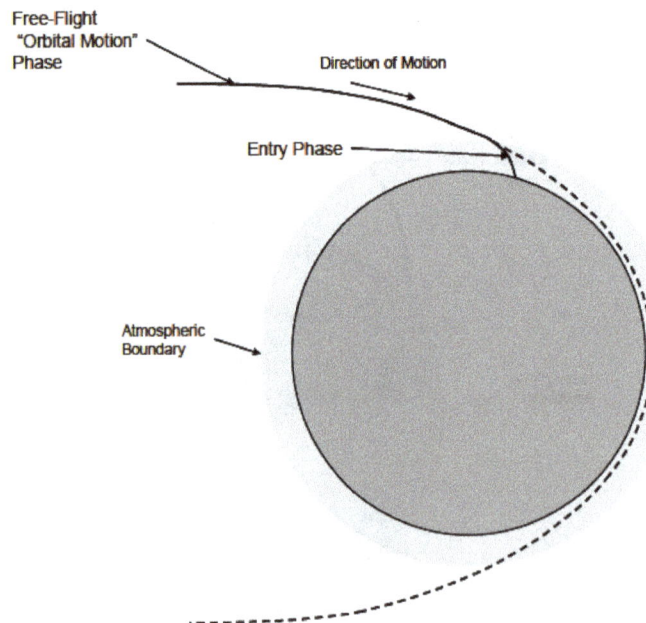

Figure 1-3: Geometry of a Typical Atmospheric Entry Trajectory

1.3 Tradeoffs for Atmospheric Entry Planning

Designing a reentry trajectory requires carefully balancing three competing requirements: deceleration limits, heating limits, and impact/landing accuracy. A large portion of this text is dedicated to qualitatively comparing entry profiles so that we can make *initial* design choices to satisfy these requirements. Chapter 5 looks at some broad categories of entry profiles and (among other things) examines the deceleration during the entry. Chapter 7 revisits some of those same entry profiles and examines the total heating a vehicle might experience as well as its instantaneous heating rates. The entry theories are expanded and generalized in Chapter 9 and Chapter 10. In Chapter 11, orbits perturbed by drag are studied.

This page intentionally blank.

Chapter 2

Orbital Motion Near a Planet

2.1 Introduction

In this text, any motion outside an atmosphere and "near" a planet will be referred to as "orbital motion." In terms of the flight phases introduced in Section 1.2, this is the "free-flight" or "orbital" phase after burnout (if applicable) and before atmospheric entry. We will assume the only gravitational force encountered by the vehicle (or satellite) will be from that nearby spherical planet. (In other words, the motion can be modeled as a simple two-body problem.)

Simple two-body motion is well-defined and well-derived in a multitude of other texts, including References 5, 20, 24, 63, and 65. Since this is not a study in orbital motion, we'll limit our effort to presenting and understanding the equations to be used rather than deriving them.

2.2 Equations of Motion

This is the simplest gravitational problem: two point masses where one mass (the satellite) is negligibly small compared to the other (the planet). (Perfectly spherical planets behave as if they are point masses, hence the assumption our planet is spherical!) If the center of mass is used as a reference point (Figure 2-1), then the equation of motion for the (small) satellite about the planet's center is

$$\ddot{\vec{r}} = -\frac{\mu \vec{r}}{r^3}$$

2.1

where $\mu = Gm_{\oplus}$. G is the universal gravitational constant and $m_{\oplus}$ is the mass of the planet. For our purposes, we can assume a coordinate system at the planet's center is fixed and inertial. We could write the acceleration in Eq. (2.1) as

$$\ddot{\vec{r}} = \frac{{}^{I}d^2 \vec{r}}{dt^2}$$

2.2

to help distinguish it as "inertial" and different from the "relative" acceleration that will be introduced in the next chapter. At this point, however, we won't introduce this new (more explicit) notation because all of the velocities and

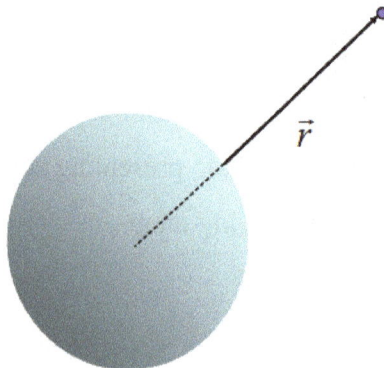

Figure 2-1: Simple Two-Body Orbital Motion

accelerations in this chapter are inertial and there is no danger of confusion. The solution to Eq. (2.1) provides us with expressions we will use in other chapters.

2.3 Constants of the Motion

With simple two-body motion, specific momentum is conserved in magnitude and in direction:

$$\vec{r} \times \dot{\vec{r}} = \vec{r} \times \vec{V} = \vec{h} = \text{a constant vector} \qquad \textbf{2.3}$$

In Eq. (2.3), $\vec{V} = \dot{\vec{r}}$ has been introduced as the inertial velocity. A constant momentum vector $\vec{h}$ requires that the orbital plane is *fixed in space*. Thus, the motion in this phase is restricted to a fixed plane that passes through the center of the planet and contains the radius and velocity vectors. The orbital plane the satellite "flies" during this phase will be the initial "entry plane" when atmospheric entry begins. (Once the atmosphere begins to affect the motion, Eq. (2.3) will no longer hold true.)

If we introduce the "local horizontal plane" as the plane perpendicular to the radius at any given instant (Figure 2-2), then the (constant) magnitude of the

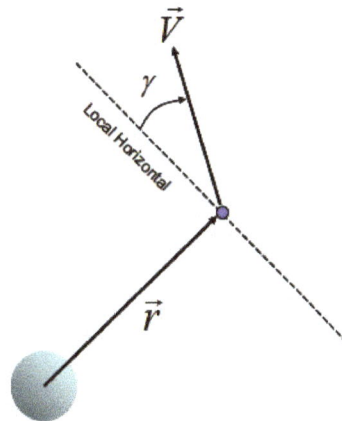

Figure 2-2: Flight-Path Angle Defined

specific momentum will give us another expression

$$h = rV \sin\left(90° - \gamma\right) = rV \cos\gamma$$ **2.4**

where γ is the "flight-path angle" and is defined to be positive when the velocity is above the local horizontal plane. (When $\gamma > 0$, the satellite is moving away from the planet; i.e., it is gaining altitude.)

Other constants related to the energy and momentum are the specific mechanical energy ε

$$\varepsilon = \frac{V^2}{2} - \frac{\mu}{r} = \text{constant}$$ **2.5**

and the semilatus rectum p:

$$p = \frac{h^2}{\mu} = \text{constant}$$ **2.6**

The geometric interpretation of p is shown in Figure 2-3 for the various orbit types.

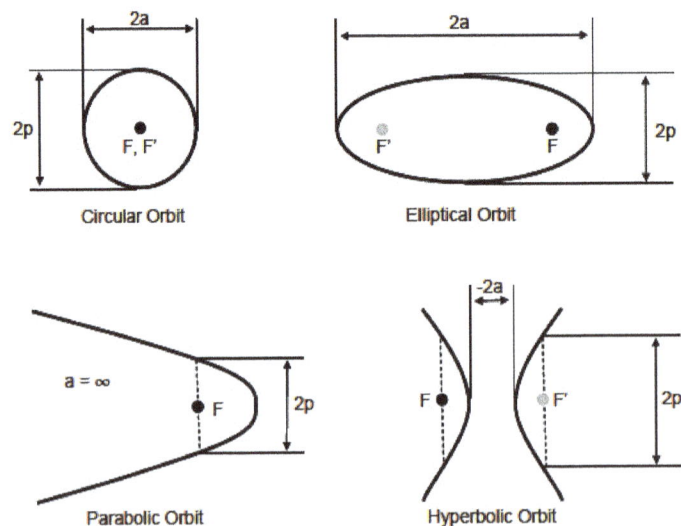

Figure 2-3: *Geometrical Dimensions in Conic Sections (Orbital Types)*

Finally, there are the constants describing the orbit itself. The classical orbital elements are well-known and easily visualized, so they will be used in most instances throughout this text. Two elements (semimajor axis a and eccentricity e) describe the size and shape of the orbit. Figure 2-3 already shows how the semimajor axis defines the *size* or the orbit. The eccentricity describes the *shape*: $e = 0$ is circular, $0 < e < 1$ is elliptical, $e = 1$ is parabolic, and $e > 1$ is hyperbolic. Orbital inclination i and right ascension of the ascending node Ω define how the orbit plane is oriented relative to the planet. Figure 2-4 shows these angles for an elliptical orbit, but they are defined identically for parabolic and hyperbolic orbits. The argument of periapsis ω (also shown in Figure 2-4 and similarly defined for parabolic and hyperbolic orbits) completes the list of constants of the motion. (Note, the Cartesian coordinate system formed from the $\hat{e}_x$, $\hat{e}_y$, and $\hat{e}_z$ directions is often called the "geocentric-equatorial coordinate system.")

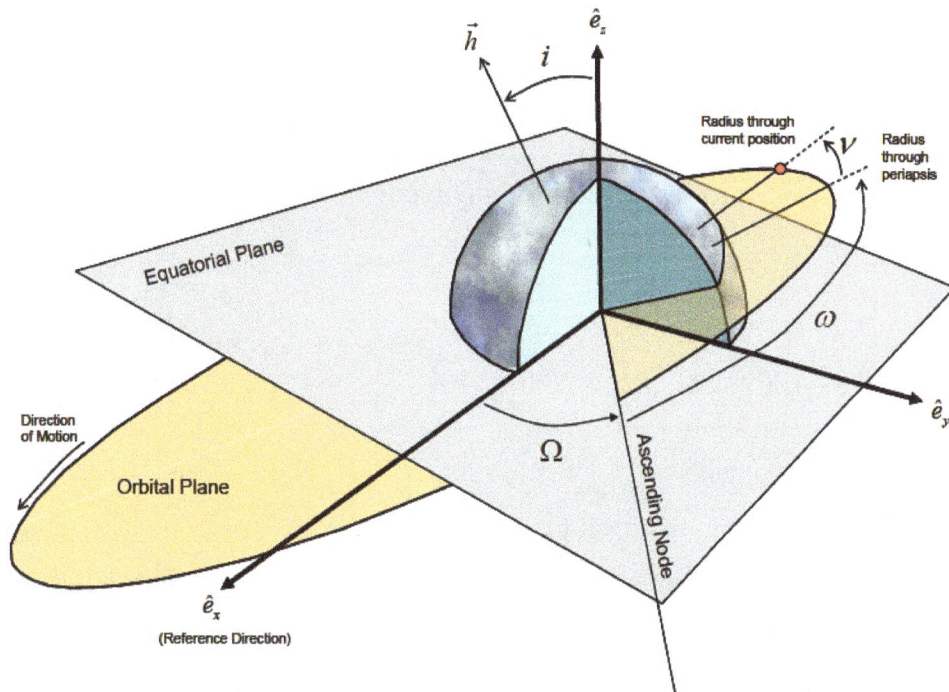

Figure 2-4: Angles Used in Classical Orbital Elements

2.4 Motion in the Orbit

For completeness, Figure 2-4 shows the true anomaly v of the orbit at any given time. It is *not* a constant of the motion, but rather fixes the position of the satellite at a specific point along the trajectory at any given time. The true anomaly and the radius between the planet center and satellite are related by:

$$r = \frac{p}{1 + e\cos v} \qquad\qquad 2.7$$

For the mathematician, Eq. (2.7) represents the polar form of a conic section with the origin at one focus. This equation explains why orbits take the shape of a conic section (circle, ellipse, parabola, or hyperbola).

To find an expression for the velocity corresponding to any radius in the orbit, the energy equation (Eq. (2.5)) can be solved by evaluating the constant at the periapsis (the point of closest approach) and rearranging:

$$V^2 = \mu\left(\frac{2}{r} - \frac{1}{a}\right) \qquad\qquad 2.8$$

Equation (2.8) is called the vis-viva equation in some texts and is valid for all of the orbit types. Equations specific to orbital types are given in the following sections.

2.4.1 Elliptical Orbits

Satellite and ballistic missile trajectories are elliptical orbits. Since an ellipse is a closed curve, an object in an elliptical orbit will repeat its path around the planet over and over again. In the case of ballistic missiles, however, the (theoretical) orbit happens to pass *through* the Earth so it never completes the first orbit.

For *any non-parabolic* orbit, the semilatus rectum is related to the semimajor axis and eccentricity by:

$$p = a\left(1 - e^2\right) \tag{2.9}$$

Then, for ellipses, Eq. (2.7) can be rewritten as:

$$r = \frac{a\left(1 - e^2\right)}{1 + e\cos v} \tag{2.10}$$

From this, it is clear the periapsis (closest approach) r_p is at $v = 0$ and:

$$r_p = a\left(1 - e\right) \tag{2.11}$$

Again, for ballistic missiles, the satellite (or, more correctly, reentry vehicle) may never reach this point because it impacted the planet's surface. The apoapsis (furthest orbital point) r_a is at $v = \pi$ and:

$$r_a = a\left(1 + e\right) \tag{2.12}$$

For ellipses, the true anomaly is sometimes replaced by the eccentric anomaly E. Figure 2-5 shows the geometric relationship between v and E. Mathematically, they are related by:

$$\cos E = \frac{e + \cos v}{1 + e\cos v} \tag{2.13}$$

$$\sin E = \frac{\left(1 - e^2\right)^{\frac{1}{2}} \sin v}{1 + e\cos v} \tag{2.14}$$

$$\tan\left(\frac{v}{2}\right) = \left(\frac{1 + e}{1 - e}\right)^{\frac{1}{2}} \tan\left(\frac{E}{2}\right) \tag{2.15}$$

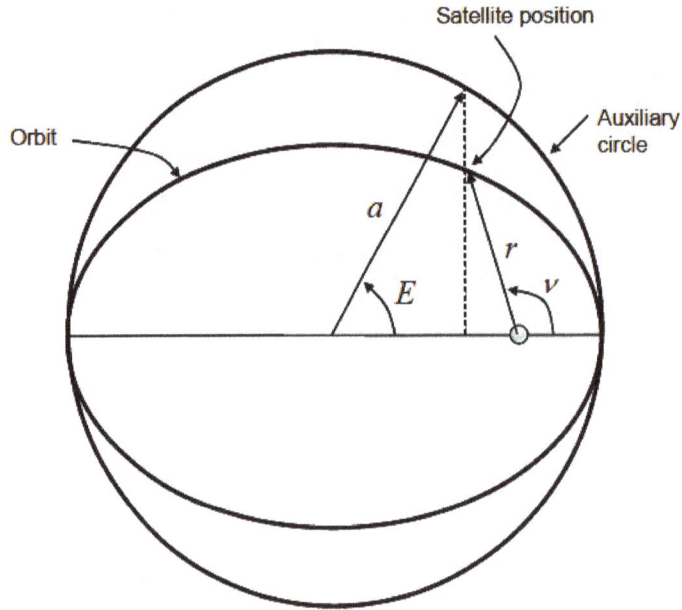

Figure 2-5: *The Geometric Relationship between True Anomaly v and Eccentric Anomaly E*

In terms of E, Eq. (2.8) for the velocity becomes:

$$V^2 = \left(\frac{\mu}{a}\right)\left(\frac{1 + e\cos E}{1 - e\cos E}\right)$$

2.16

and Eq. (2.10) for the radius:

$$r = a(1 - e\cos E)$$

2.17

The orbital period $\mathcal{P}$ is the time it takes to travel completely around the ellipse once (barring impact with the planet, of course). The period is given by:

$$\mathcal{P} = 2\pi\sqrt{\frac{a^3}{\mu}}$$

2.18

The mean motion n is the "average angular rate" of the satellite in the orbit. Thus, to travel 2π in true anomaly in one period:

$$n = \frac{2\pi}{\mathcal{P}} = \sqrt{\frac{\mu}{a^3}}$$

2.19

2.4.2 Circular Orbits

Circular orbits are really just a special case of elliptical orbits. An important exception is that circular orbits cannot impact the planet. For circular orbits, Eq. (2.7) becomes trivial:

$$r = a$$

2.20

Often, this radius (or, equivalently, semimajor axis) is written as r_c, with the "c" signifying "circular." Obviously, a periapsis and apoapsis do not exist for circular orbits and the true anomaly and eccentric anomaly are equal.

The orbital period $\mathcal{P}$ is given by

$$\mathcal{P} = 2\pi \sqrt{\frac{r_c^3}{\mu}}$$

2.21

and the mean motion by:

$$n = \sqrt{\frac{\mu}{r_c^3}}$$

2.22

The vis-viva equation reduces somewhat to give a simplified expression for the constant velocity found with circular orbits:

$$V_c = \sqrt{\frac{\mu}{r_c}}$$

2.23

2.4.3 Parabolic Orbits

Parabolic orbits might be encountered when examining asteroids or interplanetary probes. Such objects might be in elliptical orbits about the sun, but as they approach a planet, they are moving fast enough to be in parabolic (or hyperbolic) orbits with respect to the planet.

For parabolic orbits, Eq. (2.7) can be written to relate radius, momentum, and true anomaly:

$$r = \frac{\left(\dfrac{h^2}{\mu}\right)}{1 + \cos v} \qquad\qquad 2.24$$

The periapsis is found from Eq. (2.24) by substituting $v = 0$:

$$r_p = \frac{h^2}{2\mu} \qquad\qquad 2.25$$

Since this is an "escape" orbit, the apoapsis is undefined. Similarly, the period and mean motion are meaningless. Eccentric anomaly is undefined.

The vis-viva equation can be used to find the velocity "dividing line" between closed (circular/elliptical) orbits and hyperbolic orbits. At speeds above the "escape velocity," the satellite is on a hyperbolic orbit. At speeds below the escape velocity, the satellite is on a closed orbit. If the velocity is exactly the escape speed, then the orbit is parabolic. Escape velocity is calculated by evaluating:

$$V_{esc} = \sqrt{\frac{2\mu}{r}} \qquad\qquad 2.26$$

With parabolic orbits, the satellite has just enough energy to reach "infinity," where both the velocity and gravitational attraction drop to exactly zero.

2.4.4 Hyperbolic Orbits

If an orbit isn't elliptical it is much more likely to be hyperbolic than parabolic. This is because a parabolic orbit requires an exact speed at a given radius $\left(V = V_{esc}\right)$ while a hyperbolic orbit only requires that the speed is greater than minimum speed at that radius $\left(V > V_{esc}\right)$.

For hyperbolic orbits, the radius and true anomaly are related by

$$r = \frac{a\left(1-e^2\right)}{1+e\cos\nu} \tag{2.27}$$

where $a < 0$ and $e > 1$. The periapsis r_p is at $\nu = 0$ and:

$$r_p = a\left(1-e\right) \tag{2.28}$$

As with parabolic orbits, the eccentric anomaly, apoapsis, period, and mean motion are not defined.

The specific mechanical energy can be evaluated at two different points along the orbit and set equal:

$$\varepsilon = \frac{V_a^2}{2} - \frac{\mu}{r_a} = \frac{V_b^2}{2} - \frac{\mu}{r_b} \tag{2.29}$$

If the first point is at a distance r, and the other is at infinity, we can get an expression for the "left over" speed when the satellite reaches infinity:

$$V_\infty^2 = V^2 - \frac{2\mu}{r}$$
$$= V^2 - V_{esc}^2 \tag{2.30}$$

V_∞ is termed the "hyperbolic excess speed" because it represents the speed beyond the minimum required to escape the gravity field of the planet. Notice that for a parabolic orbit, $a = \infty$ and the hyperbolic excess speed is zero.

2.5 Determining the Orbital Elements from Position and Velocity

If you have the position and velocity for a satellite at any given time, you can calculate the corresponding orbital elements. This section is intended to give you a "cookbook" approach to getting the classical orbital elements when you start with position and velocity measured in the coordinate frame shown in Figure 2-4. (We will ignore the situations where the classical orbital elements fail to be defined. Any of the orbital mechanics books in the references will provide solutions using other orbital element sets.)

Step 1: Form the three fundamental vectors: $\vec{h}$ (specific momentum), $\vec{n}$ (a vector along the line of nodes in the direction of the ascending node), and $\vec{e}$ (a vector pointing at periapsis and having a magnitude equal to the eccentricity of the orbit).

$$\boxed{\vec{h} = \vec{r} \times \vec{V}} \qquad \text{2.31}$$

$$\boxed{\vec{n} = \hat{e}_z \times \vec{h}} \qquad \text{2.32}$$

$$\boxed{\vec{e} = \frac{1}{\mu}\left[\left(V^2 - \frac{\mu}{r}\right)\vec{r} - \left(\vec{r} \cdot \vec{V}\right)\vec{V}\right]} \qquad \text{2.33}$$

Step 2:

$$\varepsilon = \frac{V^2}{2} - \frac{\mu}{r}$$

2.34

and

$$a = \begin{cases} -\dfrac{\mu}{2\varepsilon} & \text{for } \varepsilon \neq 0 \\ \infty & \text{for } \varepsilon = 0 \end{cases}$$

2.35

Step 3:

$$e = |\vec{e}|$$

2.36

Step 4: The inclination i is the angle between $\vec{h}$ and $\hat{e}_z$:

$$i = \cos^{-1}\left(\frac{\vec{h} \cdot \hat{e}_z}{h}\right)$$

2.37

Since the $i \leq 180°$, there is no quadrant ambiguity in Eq. (2.37).

Step 5: The $\vec{n}$ vector lies in the equatorial plane, so it must have the form:

$$\vec{n} = n_x \hat{e}_x + n_y \hat{e}_y$$

2.38

Or, when converted to a unit vector $\hat{n} = \dfrac{\vec{n}}{n}$, this is simply

$$\hat{n} = \cos\Omega\, \hat{e}_x + \sin\Omega\, \hat{e}_y$$

2.39

since Ω is the angle between $\hat{e}_x$ and the line of nodes. With both the sine and cosine of Ω known in Eq. (2.39), there isn't any quadrant ambiguity in the angle. (There is a case, however, where Ω is undefined. When the inclination is exactly zero, $\hat{n} = \vec{0}$ and the orbit plane does not "break" the equatorial plane, so Ω does not exist.)

Step 6: The angle between $\vec{n}$ and $\vec{e}$ is the argument of periapsis ω. Its cosine can be found by taking the dot product:

$$\omega = \begin{cases} \cos^{-1}\left(\dfrac{\hat{n} \bullet \vec{e}}{e}\right) & for\ \ \vec{e} \bullet \hat{e}_z > 0 \\[4mm] \cos^{-1}\left(\dfrac{\hat{n} \bullet \vec{e}}{e}\right) + 180° & for\ \ \vec{e} \bullet \hat{e}_z < 0 \end{cases}$$

2.40

Notice the solution for ω when $\vec{e} \bullet \hat{e}_z = 0$ is not listed in Eq. (2.40). This case is more complicated because it can occur in multiple situations: either the line of nodes contains the periapsis ($\omega = 0°$ or $\omega = 180°$) or the orbit is circular ($\vec{e} = \vec{0}$ and ω is undefined). Working through this special case is left as an exercise for the reader.

Step 7: Similarly, the angle between $\vec{r}$ and $\vec{e}$ is the true anomaly (at the time of interest) v. Its cosine can be found by taking the dot product:

$$v = \begin{cases} \cos^{-1}\left(\dfrac{\vec{e} \bullet \vec{r}}{er}\right) & for\ \ \vec{r} \bullet \vec{V} > 0 \ (\text{moving towards apocenter}) \\[4mm] \cos^{-1}\left(\dfrac{\vec{e} \bullet \vec{r}}{er}\right) + 180° & for\ \ \vec{r} \bullet \vec{V} < 0 \ (\text{moving towards pericenter}) \end{cases}$$

2.41

Once again, this solution isn't all-inclusive. It does not give a solution for when $\vec{r} \bullet \vec{V} = 0$. When $\vec{r} \bullet \vec{V} = 0$, the vehicle is at periapsis or apoapsis ($v = 0°$ or

$\nu = 180°$, respectively). A less obvious "feature" of Eq. (2.41) is that it does not work for circular ($e = 0$) orbits because ν is undefined!

Similar algorithms exist for other orbital element sets. This one will suffice for most of our work.

2.6 Determining Position and Velocity from the Orbital Elements

Similar to the last section, we can create a cookbook algorithm for finding the position and velocity vectors from the classical orbital elements at any specific time. The process is simplified somewhat by introducing the *perifocal coordinate system* shown in Figure 2-6. The unit vector $\hat{e}_P$ points from the planet center to the periapsis, $\hat{e}_Q$ is out the semilatus rectum (i.e., in the orbit plane and rotated 90° from $\hat{e}_P$), and $\hat{e}_W$ is along the momentum vector $\vec{h}$.

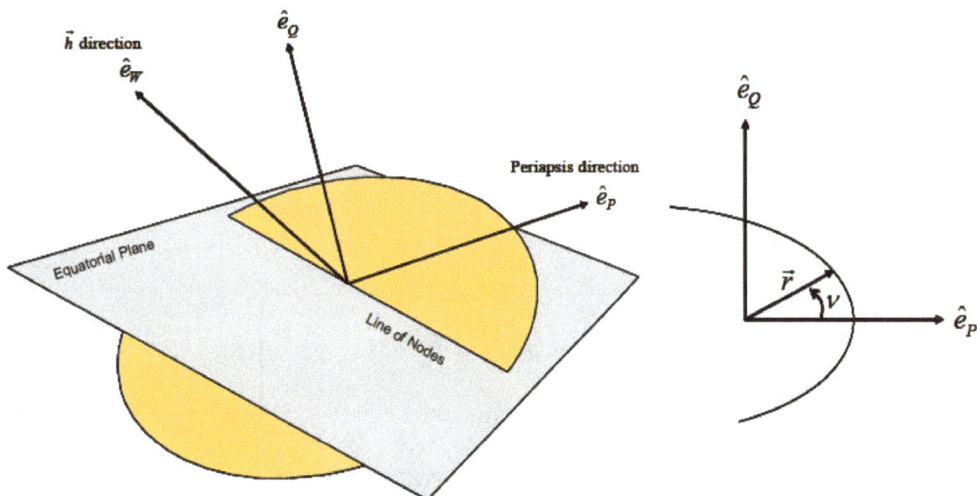

Figure 2-6: Perifocal Coordinate System

Step 1: We can immediately write the position vector

$$\vec{r} = r\cos v\ \hat{e}_P + r\sin v\ \hat{e}_Q$$

2.42

where, for *non-parabolic* orbits the scalar radius is calculated with:

$$r = \frac{a(1-e^2)}{1+e\cos v}$$

2.43

For the odd case when the orbit turns out to be parabolic, Eq. (2.43) can replaced with

$$r = \frac{\left(\dfrac{h^2}{\mu}\right)}{1+\cos v}$$

2.44

and we are required to have additional information that allows us to evaluate h.

Step 2: The velocity is found by simply differentiating Eq. (2.42) and simplifying:

$$\vec{V} = \sqrt{\frac{\mu}{p}}\left[-\sin v\ \hat{e}_P +(e+\cos v)\hat{e}_Q\right]$$

2.45

p is found from:

$$p = \begin{cases} \left(\dfrac{h^2}{\mu}\right) & \textit{for all orbits} \\ & \textit{or} \\ a(1-e^2) & \textit{for non-parabolic orbits} \end{cases}$$

2.46

Step 3: The unit vectors of the perifocal system ($\hat{e}_P$, $\hat{e}_Q$, and $\hat{e}_W$) are replaced to put the position (Eq. (2.42)) and velocity (Eq. (2.45)) in terms of the geocentric-equatorial coordinates introduced back in Figure 2-4. This involves a simple rotation from one coordinate system to another. The substitutions for the perifocal unit vectors are given below (5:80-83).

$$\hat{e}_P = \left(\cos\Omega\cos\omega - \sin\Omega\sin\omega\cos i\right)\hat{e}_x \\ + \left(\sin\Omega\cos\omega + \cos\Omega\sin\omega\cos i\right)\hat{e}_y + \left(\sin\omega\sin i\right)\hat{e}_z$$

2.47

$$\hat{e}_Q = \left(-\cos\Omega\sin\omega - \sin\Omega\cos\omega\cos i\right)\hat{e}_x \\ + \left(-\sin\Omega\sin\omega + \cos\Omega\cos\omega\cos i\right)\hat{e}_y + \left(\cos\omega\sin i\right)\hat{e}_z$$

2.48

$$\hat{e}_w = \left(\sin\Omega\sin i\right)\hat{e}_x + \left(-\cos\Omega\sin i\right)\hat{e}_y + \left(\cos i\right)\hat{e}_z$$

2.49

2.7 Problems

1. At a given moment in time (the "epoch" time), a satellite's orbital elements are:

$$a = 2.21 \text{ DU}$$
$$e = 0.870$$
$$i = 132.9°$$
$$\Omega = 21.8°$$
$$\omega = 203°$$
$$v = 157°$$

The units are canonical units, where unit length = 1 DU, unit time = 1 TU, and the gravitational parameter, μ, is given by $\mu \equiv 1 \ DU^3/TU^2$.

   a. Calculate the position and velocity vectors in perifocal coordinates.

   b. Calculate the position and velocity vectors in geocentric equatorial coordinates.

2. The orbit given in Problem 1 above is representative of a ballistic missile trajectory.

   a. Calculate the periapsis radius of the orbit in the Problem 1 above. If the radius of the planet is 1 DU, is the periapsis you found realistic? Why or why not?

   b. If the atmosphere of the planet extends to a radius of 1.2 DU, what is the flight-path angle γ_e and the magnitude of the velocity $V_e = \left| \vec{V}_e \right|$ when atmospheric reentry begins?

3. At a given moment in time, a satellite's position and velocity are given by:

$$\vec{r} = 2\hat{e}_x \text{ DU}$$
$$\vec{V} = 0.6422\hat{e}_y + 0.3708\hat{e}_z \text{ DU/TU}$$

Find the orbital elements $(a, e, i, \Omega, \omega, \nu)$ at that moment in time. (Assume $\mu \equiv 1 \ DU^3/TU^2$.)

4. For the orbit in Problem 3, calculate the flight-path angle γ at the time the position and velocity were measured.

This page intentionally blank.

Chapter 3

Equations of Flight

3.1 Introduction

In this chapter, we'll derive the equations of motion for a point mass flying in the atmosphere of a rotating planet. The position and velocity will be given by vector equations $\vec{r}(t)$ and $\vec{v}(t)$, respectively. We begin by defining the coordinate systems to be used in the derivations.

3.2 Reference Frames

A variety of coordinate systems and reference frames are needed. These are the inertial geocentric-equatorial system, a planet-fixed rotating system, and a quickly rotating, vehicle-pointing frame.

3.2.1 Geocentric-Equatorial Coordinate Frame

The geocentric-equatorial system (OXYZ) is an inertial reference frame with its origin at the center of the planet. The x-axis, $\hat{e}_x$, points in the direction of the vernal equinox and the z-axis, $\hat{e}_z$, passes out the North Pole. The y-axis, $\hat{e}_y$, completes the system such that it lies in the planet's equatorial plane and $\hat{e}_z = \hat{e}_x \times \hat{e}_y$.

3.2.2 Planet-Fixed Coordinate Frame

Choose the $OX_1Y_1Z_1$ frame to be a planet-fixed system originating at the planet's center with the z_1-axis pointing out the North Pole and the x_1-y_1 plane in the plane of the equator. Further, assume the planet rotates with constant velocity given by $\overline{\omega}_\oplus = \omega_\oplus \hat{e}_{z_1} = \omega_\oplus \hat{e}_z$. Figure 3-1 shows the relationship of the $OX_1Y_1Z_1$ frame to the geocentric-equatorial system where it has been arbitrarily assumed the axes align when $\Delta t = 0$.

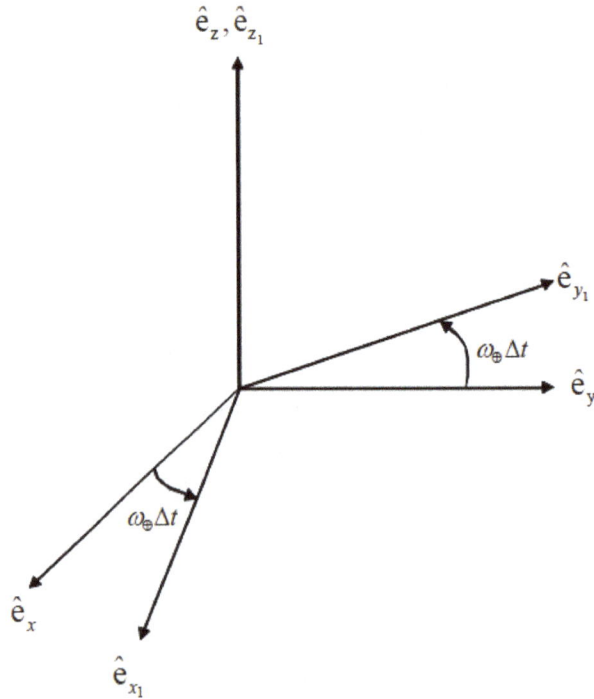

Figure 3-1: Inertial (OXYZ) and Planet-Fixed (OX₁Y₁Z₁) Reference Systems

28

A coordinate transformation from the OXYZ frame to the $OX_1Y_1Z_1$ frame is given by

$$[\hat{\mathbf{e}}_1] = \begin{bmatrix} \cos(\omega_\oplus \Delta t) & \sin(\omega_\oplus \Delta t) & 0 \\ -\sin(\omega_\oplus \Delta t) & \cos(\omega_\oplus \Delta t) & 0 \\ 0 & 0 & 1 \end{bmatrix} [\hat{\mathbf{e}}] \qquad \textbf{3.1}$$

where $[\hat{\mathbf{e}}]$ and $[\hat{\mathbf{e}}_1]$ represent the unit vectors of the OXYZ and $OX_1Y_1Z_1$ frames, respectively. A convenient way to express the "transformation" matrix (or "rotation" matrix) in Eq. (3.1) is to rewrite the equation as

$$[\hat{\mathbf{e}}_1] = \mathbf{R}_z(\omega_\oplus \Delta t) \, [\hat{\mathbf{e}}] \qquad \textbf{3.2}$$

where the subscript on the **R** matrix indicates the axis of rotation and the argument indicates the angle of the rotation.

3.2.3 Vehicle-Pointing System

As shown in Figure 3-2, let the frame $OX_2Y_2Z_2$ originate at the planet's center and be aligned with the x_2-axis pointing along the vehicle's position vector, the y_2-axis parallel to the equatorial plane, and the z_2-axis completing a right-handed system. θ and ϕ represent the longitude and latitude of the vehicle's position, respectively.

This frame can be created by a rotation θ around the z_1-axis followed by a $-\phi$ rotation about the y_2-axis:

$$[\hat{\mathbf{e}}_2] = \begin{bmatrix} \cos(-\phi) & 0 & -\sin(-\phi) \\ 0 & 1 & 0 \\ \sin(-\phi) & 0 & \cos(-\phi) \end{bmatrix} \begin{bmatrix} \cos\theta & \sin\theta & 0 \\ -\sin\theta & \cos\theta & 0 \\ 0 & 0 & 1 \end{bmatrix} [\hat{\mathbf{e}}_1] \qquad \textbf{3.3}$$

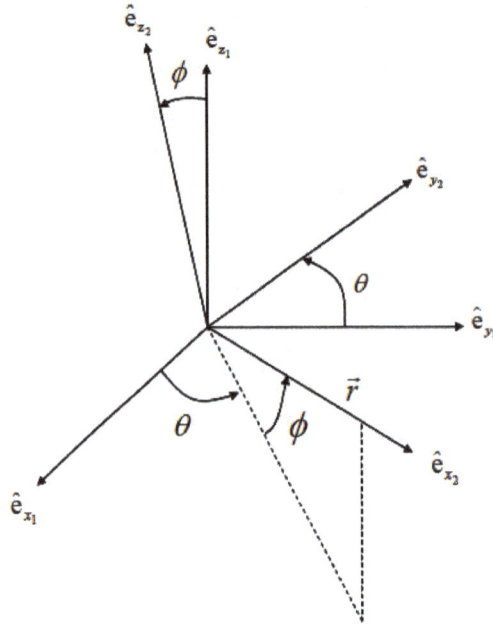

Figure 3-2: Planet-Fixed (OX₁Y₁Z₁) and Vehicle-Pointing (OX₂Y₂Z₂) Systems

Using notation similar to that in Eq. (3.2), we could write:

$$[\hat{\mathbf{e}}_2] = \mathbf{R}_{y_2}(-\phi)\, \mathbf{R}_{z_1}(\theta)\, [\hat{\mathbf{e}}_1] \qquad\qquad \textbf{3.4}$$

Multiplying the matrices gives us a single matrix for relating the unit vectors of the OX₂Y₂Z₂ and OX₁Y₁Z₁ systems:

$$[\hat{\mathbf{e}}_2] = \begin{bmatrix} \cos\phi\cos\theta & \cos\phi\sin\theta & \sin\phi \\ -\sin\theta & \cos\theta & 0 \\ -\sin\phi\cos\theta & -\sin\phi\sin\theta & \cos\phi \end{bmatrix} [\hat{\mathbf{e}}_1] \qquad \textbf{3.5}$$

Substituting for $\left[\hat{\mathbf{e}}_1\right]$ from Eq. (3.2), provides the relationship between the OX$_2$Y$_2$Z$_2$ and OXYZ systems as well:

$$\left[\hat{\mathbf{e}}_2\right] = \mathbf{R}_{y_2}\left(-\phi\right)\,\mathbf{R}_{z_1}\left(\theta\right)\,\mathbf{R}_z\left(\omega\Delta t\right)\,\left[\hat{\mathbf{e}}\right] \qquad \textbf{3.6}$$

Multiplying out $\mathbf{R}_{y_2}\left(-\phi\right)\,\mathbf{R}_{z_1}\left(\theta\right)\,\mathbf{R}_z\left(\omega_\oplus\Delta t\right)$ to form a single transition matrix is left as an exercise.

Figure 3-3 illustrates two more angles we'll need to understand and use. The flight-path angle γ is defined as the angle between the local horizontal plane (the plane passing through the vehicle and orthogonal to $\vec{r}$) and the velocity $\vec{v}$.

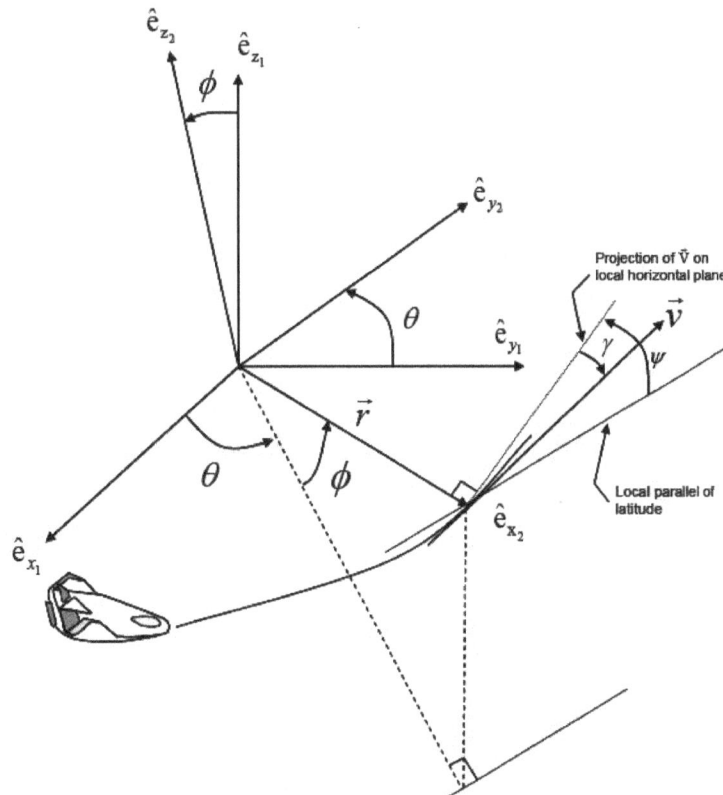

Figure 3-3: Flight-path and heading angles in relation to Planet-Fixed (OX$_1$Y$_1$Z$_1$) and Vehicle-Pointing (OX$_2$Y$_2$Z$_2$) Systems

By convention, γ is positive when $\vec{v}$ is above the local horizontal plane. The heading angle, ψ, is the angle between the local parallel of latitude and the projection of $\vec{v}$ on the local horizontal plane. ψ is measured positively in the right-handed direction about the x_2-axis. (Another way to visualize these two angles is to think of their effects on the motion. γ describes how much the velocity vector contributes to moving "in and out" along the radius while ψ describes how much the velocity vector contributes to moving "toward and away" from the planetary equator.)

3.2.4   A Velocity-Referenced Coordinate System

Before moving too far from the introduction of the vehicle-pointing $OX_2Y_2Z_2$ system, it is helpful to develop another coordinate system for use later in the derivations involving aerodynamic and thrusting forces.

Consider a rotation of the $OX_2Y_2Z_2$ axes about the x_2-axis by ψ so that the "new" y-axis, y', is aligned with the projection of $\vec{v}$ on the local horizontal plane. This is shown in Figure 3-4. The unit vectors in the new $OX'Y'Z'$ system are given by the rotation:

$$\begin{bmatrix} \hat{e}'_x \\ \hat{e}'_y \\ \hat{e}'_z \end{bmatrix} = \begin{bmatrix} \hat{e}_{x_2} \\ \hat{e}'_y \\ \hat{e}'_z \end{bmatrix} = \begin{bmatrix} 1 & 0 & 0 \\ 0 & \cos\psi & \sin\psi \\ 0 & -\sin\psi & \cos\psi \end{bmatrix} \begin{bmatrix} \hat{e}_{x_2} \\ \hat{e}_{y_2} \\ \hat{e}_{z_2} \end{bmatrix} \qquad 3.7$$

Or, more simply,

$$[\hat{e}'] = \mathbf{R}_{x_2}(\psi)[\hat{e}_2] \qquad 3.8$$

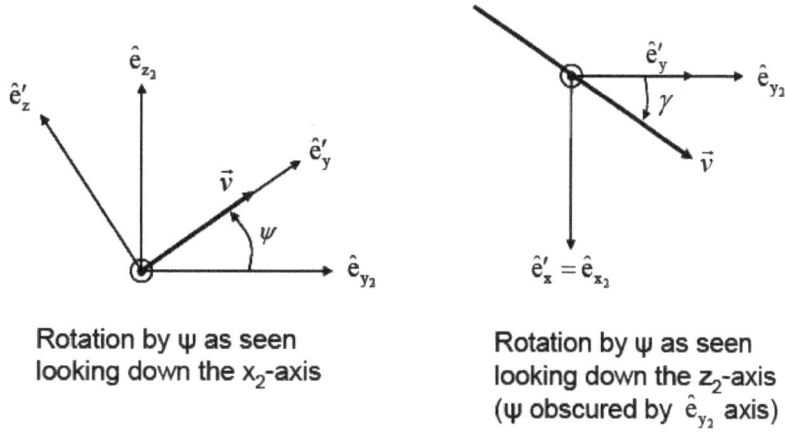

Figure 3-4: Rotation of OX₂Y₂Z₂ System About x₂-axis

where the notation the subscript and argument on the rotation matrix indicate the axis and angle of rotation, respectively.

Next, consider a *negative* rotation of the OX'Y'Z' system by γ about the z'-axis. This aligns the "new" y-axis, y″, with the velocity vector $\vec{v}$ as shown in Figure 3-5. Expressions similar to those in Eqs. (3.7) and (3.8) are readily deduced as

$$\begin{bmatrix} \hat{e}''_x \\ \hat{e}''_y \\ \hat{e}''_z \end{bmatrix} = \begin{bmatrix} \hat{e}''_x \\ \hat{e}''_y \\ \hat{e}''_z \end{bmatrix} = \begin{bmatrix} \cos(-\gamma) & \sin(-\gamma) & 0 \\ -\sin(-\gamma) & \cos(-\gamma) & 0 \\ 0 & 0 & 1 \end{bmatrix} \begin{bmatrix} \hat{e}'_x \\ \hat{e}'_y \\ \hat{e}'_z \end{bmatrix} \qquad \textbf{3.9}$$

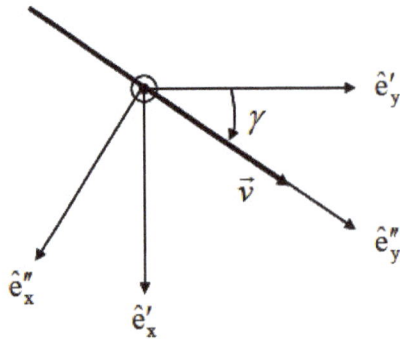

Negative rotation by γ as
seen looking down the z'-axis

Figure 3-5: Rotation of the OX'Y'Z' System About the z'-axis

or, after using trigonometric identities,

$$\begin{bmatrix} \hat{e}_x'' \\ \hat{e}_y'' \\ \hat{e}_z'' \end{bmatrix} = \begin{bmatrix} \hat{e}_x'' \\ \hat{e}_y'' \\ \hat{e}_z' \end{bmatrix} = \begin{bmatrix} \cos\gamma & -\sin\gamma & 0 \\ \sin\gamma & \cos\gamma & 0 \\ 0 & 0 & 1 \end{bmatrix} \begin{bmatrix} \hat{e}_x' \\ \hat{e}_y' \\ \hat{e}_z' \end{bmatrix}$$ **3.10**

and

$$[\hat{e}''] = \mathbf{R}_{\underline{z}'}(-\gamma)[\hat{e}']$$ **3.11**

where the subscript and argument on the rotation matrix have again been used to designate the axis of rotation and the angle. Combining these two rotations relates the final OX"Y"Z" system to the original vehicle-pointing system $OX_2Y_2Z_2$:

$$[\hat{e}''] = \mathbf{R}_{\underline{z}'}(-\gamma)\mathbf{R}_{x_2}(\psi)[\hat{e}_2]$$ **3.12**

34

After multiplying out the matrices, the relationship becomes

$$\begin{bmatrix} \hat{e}_x'' \\ \hat{e}_y'' \\ \hat{e}_z'' \end{bmatrix} = \begin{bmatrix} \cos\gamma & -\sin\gamma\cos\psi & -\sin\gamma\sin\psi \\ \sin\gamma & \cos\gamma\cos\psi & \cos\gamma\sin\psi \\ 0 & -\sin\psi & \cos\psi \end{bmatrix} \begin{bmatrix} \hat{e}_{x_2} \\ \hat{e}_{y_2} \\ \hat{e}_{z_2} \end{bmatrix} \qquad \textbf{3.13}$$

We'll return to these relationships later in the chapter. For now, just note the following relationships:

- $\hat{e}_y''$ points along the velocity vector $\vec{v}$ and can be equivalently called $\hat{e}_v$ to clarify its direction

- $\hat{e}_z''$ is perpendicular to both the velocity $\vec{v}$ and radius $\vec{r}$ vectors and lies in the local horizontal plane

3.3 Relative Angular Motion

As we work towards deriving the equations of motion, the relative angular motion between several of the coordinate frames will become important. So, while it may seem like a diversion, it is probably best to approach the topic now, while all of the coordinate transformations are still fresh in our minds.

We have already seen an angular rotation between two frames when Eq. (3.1) was written. In that case, the angular rotation of the $OX_1Y_1Z_1$ frame relative to the OXYZ (and written in terms of the planet-fixed $OX_1Y_1Z_1$ frame) is simply:

$$\vec{\omega} = \omega_\oplus\, \hat{e}_{z_1} \qquad \textbf{3.14}$$

Eq. (3.5) lets us convert coordinates in a third frame – the $OX_2Y_2Z_2$ frame:

$$\left[\vec{\omega}\right]_{1/0}^{\hat{\mathbf{e}}_2} = \begin{bmatrix} \cos\phi\cos\theta & \cos\phi\sin\theta & \sin\phi \\ -\sin\theta & \cos\theta & 0 \\ -\sin\phi\cos\theta & -\sin\phi\sin\theta & \cos\phi \end{bmatrix} \begin{bmatrix} 0 \\ 0 \\ \omega_\oplus \end{bmatrix}$$

$$= \omega_\oplus \begin{bmatrix} \sin\phi \\ 0 \\ \cos\phi \end{bmatrix}$$

$$= \left(\omega_\oplus \sin\phi\right)\hat{e}_{x_2} + \left(\omega_\oplus \cos\phi\right)\hat{e}_{z_2} \qquad \textbf{3.15}$$

The notation $\left[\vec{\omega}\right]_{1/0}^{\hat{\mathbf{e}}_2}$ is a somewhat cumbersome way to explicitly remind us we are writing the angular rate of Frame 1 ($OX_1Y_1Z_1$) relative to Frame 0 ($OXYZ$) written in terms of coordinates along the $\left[\hat{\mathbf{e}}_2\right]$ unit vectors. This notation is shown to reinforce the fact that, in spite of using $\left[\hat{\mathbf{e}}_2\right]$ coordinates, the rotation rate given by Eq. (3.14) is, indeed, that of $\left[\hat{\mathbf{e}}_1\right]$ unit vectors relative to the inertial $\left[\hat{\mathbf{e}}\right]$ unit vectors (or, equivalently, the $OX_1Y_1Z_1$ system relative to the $OXYZ$ system). We normally will not require (or use) such explicit notation.

Another relative angular motion relationship that will be needed is the one between the planet-fixed $OX_1Y_1Z_1$ system and the quickly rotating, vehicle-pointing $OX_2Y_2Z_2$ system. The overall angular rate can be assembled by remembering how we created the $\left[\hat{\mathbf{e}}_2\right]$ vectors from the $\left[\hat{\mathbf{e}}_1\right]$ vectors: a rotation θ around the z_1-axis followed by a $-\phi$ rotation about the y_2-axis. In terms of angular *rates*:

$$\vec{\Omega} = \dot{\theta}\,\hat{e}_{z_1} - \dot{\phi}\,\hat{e}_{y_2} \qquad \textbf{3.16}$$

Once again, Eq. (3.5) lets us write this entirely in terms of the vehicle-pointing frame:

$$\vec{\Omega} = \left(\dot{\theta}\sin\phi\right)\hat{e}_{x_2} - \dot{\phi}\,\hat{e}_{y_2} + (\dot{\theta}\cos\phi)\,\hat{e}_{z_2} \qquad \textbf{3.17}$$

For completeness, this could be expressed as

$$\left[\vec{\Omega}\right]_{2/1}^{\hat{\mathbf{e}}_2} = \left(\dot{\theta}\sin\phi\right)\hat{e}_{x_2} - \dot{\phi}\,\hat{e}_{y_2} + (\dot{\theta}\cos\phi)\,\hat{e}_{z_2} \qquad \textbf{3.18}$$

if we wanted to use the more explicit notation analogous to that in Eq. (3.15).

3.4 Equations of Motion

It requires six independent quantities to define the motion of a point mass at any time. Normally, we think of these as being the three components of a position and three components of a velocity. However, that is not always the most convenient description. As an example, the motion of a satellite orbiting a planet is often given in "classical orbital elements," none of which independently equate to a component of position or velocity in the traditional (e.g., Cartesian or polar) sense. We will be finding three position quantities, a velocity magnitude, and two angles to describe the direction of the velocity. With that in mind, we'll go back to the basics to begin finding our equations of motion.

At any instant, the total force, $\vec{F}$, acting on the point mass is given by

$$\vec{F} = \vec{T} + \vec{A} + m\vec{g} \qquad \textbf{3.19}$$

where $\vec{T}$ is the force from thrust, $\vec{A}$ is the aerodynamic forces (lift and drag), m is the mass, and $\vec{g}$ is the force of gravity. Newton's Second Law of Motion can be applied when the mass is constant (or "constant enough") and the reference frame is inertial:

$$m\frac{^I d\vec{v}}{dt} = \vec{F}$$

$$= \vec{T} + \vec{A} + m\vec{g} \qquad\qquad \textbf{3.20}$$

The notation "I" superscript has been used to reinforce the fact that the derivative is with respect to an inertial frame.

Much of the motion we wish to describe is more conveniently measured relative to the planet-fixed (i.e. rotating) reference frame. In order to accomplish this, recall from dynamics that an inertial derivative can be written in terms of a rotating reference frame by

$$\frac{^I d\vec{r}}{dt} = \frac{^R d\vec{r}}{dt} + \vec{\omega} \times \vec{r} \qquad\qquad \textbf{3.21}$$

and

$$\frac{^I d\vec{v}}{dt} = \frac{^R d}{dt}\left[\frac{^R d\vec{r}}{dt} + \vec{\omega} \times \vec{r} \right] + \vec{\omega} \times \left[\frac{^R d\vec{r}}{dt} + \vec{\omega} \times \vec{r} \right] \qquad\qquad \textbf{3.22}$$

where the "R" superscript has been introduced to indicate a derivative in a rotating frame. $\vec{\omega}$ is the angular velocity of that rotating frame relative to the inertial frame. Take note: The $\vec{\omega}$ in Eqs. (3.21) and (3.22) is general and should not *automatically* be assumed to be the same as the $\vec{\omega} = \vec{\omega}_\oplus$ developed in Section 3.3!

If the rotating frame is assumed to be our planet-fixed $OX_1Y_1Z_1$ system, then Eq. (3.22) can be simplified since $\vec{\omega} = \vec{\omega}_\oplus$ is a constant vector:

$$\frac{^I d\vec{v}}{dt} = \frac{^R d^2\vec{r}}{dt^2} + 2\vec{\omega}_\oplus \times \frac{^R d\vec{r}}{dt} + \vec{\omega}_\oplus \times (\vec{\omega}_\oplus \times \vec{r}) \qquad\qquad \textbf{3.23}$$

The terms $\dfrac{{}^{R}d^2\vec{r}}{dt^2}$ and $\dfrac{{}^{R}d\vec{r}}{dt}$ should be recognized as the "relative acceleration" and "relative velocity" (i.e., as seen in the rotating frame), respectively. In addition, the superscript "R" on these terms could be replaced with a superscript "1" to indicate they are derivatives *taken with respect to the $OX_1Y_1Z_1$ planet-fixed frame*; however, we will retain the "R" notation for the time being to avoid any confusion between "I" and "1." The term $2\vec{\omega}_{\oplus} \times \dfrac{{}^{R}d\vec{r}}{dt}$ is the Coriolis acceleration and $\vec{\omega}_{\oplus} \times (\vec{\omega}_{\oplus} \times \vec{r})$ is the centripetal acceleration term. Introducing Newton's Second Law from Eq. (3.20) into this gives:

$$m\frac{{}^{R}d^2\vec{r}}{dt^2} = \vec{F} - 2m\vec{\omega}_{\oplus} \times \frac{{}^{R}d\vec{r}}{dt} - m\vec{\omega}_{\oplus} \times (\vec{\omega}_{\oplus} \times \vec{r}) \qquad \textbf{3.24}$$

The last term on the right is what an observer in the rotating frame would call "centrifugal force." In reality it is just an acceleration term taken to the "wrong side of the equation."

For convenience, we can introduce ${}^{R}\vec{V}$ as the velocity *relative to the planet*

$$ {}^{R}\vec{V} = \frac{{}^{R}d\vec{r}}{dt} \qquad \textbf{3.25}$$

and substitute it into Eq. (3.24)

$$m\frac{{}^{R}d\left({}^{R}\vec{V}\right)}{dt} = m\frac{{}^{1}d\left({}^{R}\vec{V}\right)}{dt} = \vec{F} - 2m\vec{\omega}_{\oplus} \times {}^{R}\vec{V} - m\vec{\omega}_{\oplus} \times (\vec{\omega}_{\oplus} \times \vec{r}) \qquad \textbf{3.26}$$

where we have, for completeness, noted that the rotating ("R") frame is actually the $OX_1Y_1Z_1$ ("1") frame in this situation by showing both corresponding superscripts on the derivative. At this point, we turn our attention to deriving expressions for the various vector quantities on the right-hand side of Eq. (3.26).

Once that is done, we can substitute into Eqs. (3.25) and (3.26) and equate components on the left and right sides to find the equations of motion.

3.4.1   Kinematic Equations

The position vector $\vec{r}$ is easily expressed in terms of the vehicle-pointing system $OX_2Y_2Z_2$:

$$\vec{r} = r\hat{e}_{x_2} \qquad\qquad \textbf{3.27}$$

The (relative) velocity vector is more complicated, but we have already done the hardest part of the derivation! In Eq. (3.13), the unit vector $\hat{e}_y''$ points in the direction of the velocity, so:

$$
\begin{aligned}
{}^R\vec{V} &= {}^RV\hat{e}_y'' \\
&= {}^RV\hat{e}_v
\end{aligned}
\qquad\qquad \textbf{3.28}
$$

The second line of Eq. (3.28) simply uses notation to reinforce the visualization of the unit vector's direction. Replacing that unit vector with what is found in Eq. (3.13), we get the velocity in terms of the vehicle-pointing system:

$$
{}^R\vec{V} = \left({}^RV \sin \gamma\right)\hat{e}_{x_2} + \left({}^RV \cos \gamma \cos \psi\right)\hat{e}_{y_2} + \left({}^RV \cos \gamma \sin \psi\right)\hat{e}_{z_2} \qquad \textbf{3.29}
$$

($ {}^R\vec{V} $ could also have been written directly by examining Figure 3-3.) Take note: this is an equation for the velocity as viewed (or measured) by an observer on the rotating planet-fixed system, but *written in terms of the vehicle-pointing system*!

Another expression for $^R\vec{V}$ can be written by again recalling from dynamics how the derivative of a vector in one frame is related to the derivative in another frame

$$\frac{^1 d\vec{r}}{dt} = \frac{^2 d\vec{r}}{dt} + \vec{\omega}_{2/1} \times \vec{r}$$ **3.30**

where the pre-superscript on the derivative indicates the reference frame for the derivative and the "2/1" subscript indicates the angular rate is for "Frame 2 relative to Frame 1." (This should look familiar since Eq. (3.21) used a special case of this same relationship.) Recognizing $^R\vec{V}$ is the velocity with respect to (i.e., as seen in) the planet-fixed frame ("Frame 1" or, equivalently $OX_1Y_1Z_1$) and making use of Eq. (3.18), Eq. (3.30) can be rewritten:

$$^R\vec{V} = \frac{^2 d\vec{r}}{dt} + \left[\vec{\Omega}\right]_{2/1} \times \vec{r}$$ **3.31**

Substituting for the vectors on the right-hand side, this becomes

$$^R\vec{V} = \dot{r}\,\hat{e}_{x_2} + \begin{bmatrix} \dot{\theta}\sin\phi \\ -\dot{\phi} \\ \dot{\theta}\cos\phi \end{bmatrix}^{\hat{e}_2} \times \begin{bmatrix} r \\ 0 \\ 0 \end{bmatrix}^{\hat{e}_2}$$ **3.32**

when written in $[\hat{e}_2]$ components (which is what the temporary superscript on Eq. (3.32) is intended to emphasize.) The cross-product is straight forward

$$\begin{bmatrix} \dot{\theta}\sin\phi \\ -\dot{\phi} \\ \dot{\theta}\cos\phi \end{bmatrix}^{\hat{e}_2} \times \begin{bmatrix} r \\ 0 \\ 0 \end{bmatrix}^{\hat{e}_2} = \left(r\dot{\theta}\cos\phi\right)\hat{e}_{y_2} + \left(r\dot{\phi}\right)\hat{e}_{z_2}$$ **3.33**

and allows us write an expression for $^R\vec{V}$:

$$^R\vec{V} = \dot{r}\,\hat{e}_{x_2} + \left(r\dot{\theta}\cos\phi\right)\hat{e}_{y_2} + \left(r\dot{\phi}\right)\hat{e}_{z_2} \qquad \textbf{3.34}$$

Equations (3.29) and (3.34) are both valid equations for $^R\vec{V}$ and are both written terms of $\left[\hat{e}_2\right]$ coordinates, so they must be equal, component-by-component:

$$\dot{r} = {}^R V \sin\gamma \qquad \textbf{3.35}$$

$$\dot{\theta} = \frac{{}^R V \cos\gamma\cos\psi}{r\cos\phi} \qquad \textbf{3.36}$$

$$\dot{\phi} = \frac{{}^R V \cos\gamma\sin\psi}{r} \qquad \textbf{3.37}$$

These three differential equations are the kinematic equations. When integrated, they provide the position of the vehicle as seen from the rotating planet.

3.4.2 Force Equations

Equations (3.35) - (3.37) are not a complete set of equations of motion since they only provide three independent quantities (r,θ,ϕ). We now turn our attention to completing the right-hand side of Eq. (3.26):

$$m\frac{{}^R d\left({}^R\vec{V}\right)}{dt} = \vec{F} - 2m\vec{\omega}_\oplus \times {}^R\vec{V} - m\vec{\omega}_\oplus \times (\vec{\omega}_\oplus \times \vec{r}) \qquad \textbf{3.38}$$

The rotation vector of the planet-fixed system with respect to the inertial system has already been found and written in vehicle-pointing coordinates:

$$\vec{\omega}_{\oplus} = \left[\vec{\omega}\right]_{1/0}^{\hat{\mathbf{e}}_2} = \left(\omega_{\oplus} \sin\phi\right)\hat{e}_{x_2} + \left(\omega_{\oplus} \cos\phi\right)\hat{e}_{z_2} \qquad \textbf{3.39}$$

Using Eqs. (3.29) and (3.39) to find the cross product $\vec{\omega}_{\oplus} \times {}^{R}\vec{V}$

$$\vec{\omega}_{\oplus} \times {}^{R}\vec{V} = \begin{vmatrix} \hat{e}_{x_2} & \hat{e}_{y_2} & \hat{e}_{z_2} \\ \omega_{\oplus} \sin\phi & 0 & \omega_{\oplus} \cos\phi \\ {}^{R}V \sin\gamma & {}^{R}V \cos\gamma\cos\psi & {}^{R}V \cos\gamma\sin\psi \end{vmatrix} \qquad \textbf{3.40}$$

and simplifying, we get:

$$\begin{aligned} \vec{\omega}_{\oplus} \times {}^{R}\vec{V} = &-\left({}^{R}V\omega_{\oplus} \cos\varphi\cos\gamma\cos\psi\right)\hat{e}_{x_2} \\ &+ {}^{R}V\omega_{\oplus}\left(\cos\varphi\sin\gamma - \sin\varphi\cos\gamma\sin\psi\right)\hat{e}_{y_2} \\ &+ \left({}^{R}V\omega_{\oplus} \sin\varphi\cos\gamma\cos\psi\right)\hat{e}_{z_2} \end{aligned} \qquad \textbf{3.41}$$

Crossing $\vec{\omega}_{\oplus}$ with Eq. (3.27) gives:

$$\vec{\omega}_{\oplus} \times \vec{r} = \left(r\omega_{\oplus} \cos\phi\right)\hat{e}_{y_2} \qquad \textbf{3.42}$$

This, in turn, yields:

$$\vec{\omega}_{\oplus} \times \left(\vec{\omega}_{\oplus} \times \vec{r}\right) = -\left(r\omega_{\oplus}^2 \cos^2\phi\right)\hat{e}_{x_2} + \left(r\omega_{\oplus}^2 \sin\phi\cos\phi\right)\hat{e}_{z_2} \qquad \textbf{3.43}$$

The aerodynamic force, $\vec{A}$, in Eq. (3.19) can be broken into a drag force $\vec{D}$ acting in a direction opposite of the velocity vector

$$\begin{aligned} \vec{D} = -D\hat{e}_y'' = -D\hat{e}_v \\ = -\left(D\sin\gamma\right)\hat{e}_{x_2} - \left(D\cos\gamma\cos\psi\right)\hat{e}_{y_2} - \left(D\cos\gamma\sin\psi\right)\hat{e}_{z_2} \end{aligned} \qquad \textbf{3.44}$$

43

and a lift force $\vec{L}$ perpendicular to the velocity vector. The bank angle, σ, defines the orientation of $\vec{L}$ relative to the $(\vec{r},\vec{v})$ plane. This is illustrated in Figure 3-6.

In the figure, the unit vector $\hat{e}''_z$ has been shown as well. Realizing $\hat{e}''_z$ lies in a plane perpendicular to the $(\vec{r},\vec{v})$ plane, we can deduce the angle between the lift vector $\vec{L}$ and $\hat{e}''_z$ to be $(90^0 - \sigma)$. Following the process from Section 3.2.4, we can perform a coordinate rotation about the $\vec{v}$ vector (or, equivalently, $\hat{e}''_y = \hat{e}_v$) to align a "new" z-axis $\hat{e}'''_z$ with the lift vector:

$$[\hat{\mathbf{e}}'''] = \mathbf{R}_{y''}(90^\circ - \sigma)\,[\hat{\mathbf{e}}''] \qquad\qquad \textbf{3.45}$$

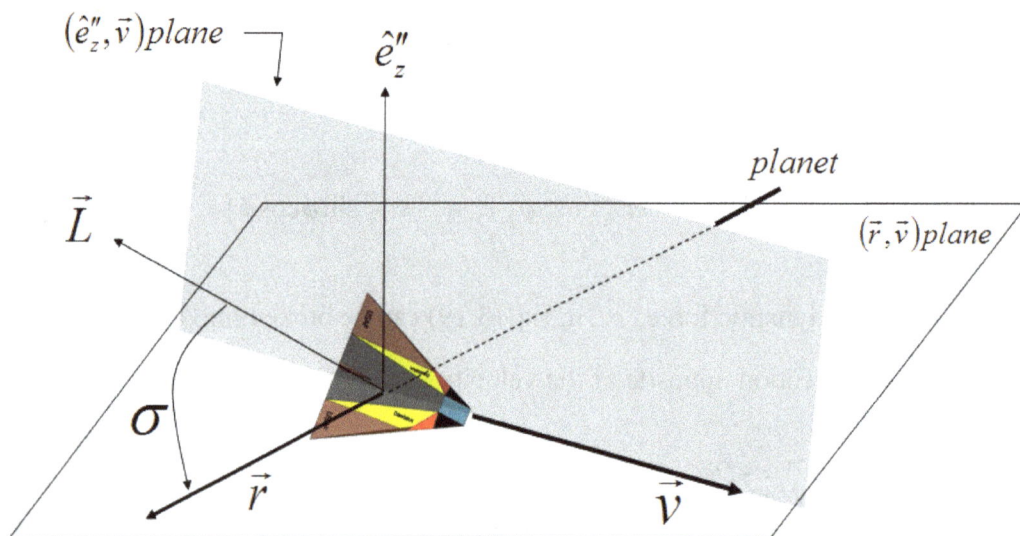

Figure 3-6: Bank Angle Defined

Expanding this out gives:

$$
\begin{bmatrix} \hat{e}_x''' \\ \hat{e}_y''' \\ \hat{e}_z''' \end{bmatrix} = \begin{bmatrix} \hat{e}_x''' \\ \hat{e}_y''' \\ \hat{e}_z''' \end{bmatrix} = \begin{bmatrix} \cos(90° - \sigma) & 0 & -\sin(90° - \sigma) \\ 0 & 1 & 0 \\ \sin(90° - \sigma) & 0 & \cos(90° - \sigma) \end{bmatrix} \begin{bmatrix} \hat{e}_x'' \\ \hat{e}_y'' \\ \hat{e}_z'' \end{bmatrix}
$$

$$
= \begin{bmatrix} \sin\sigma & 0 & -\cos\sigma \\ 0 & 1 & 0 \\ \cos\sigma & 0 & \sin\sigma \end{bmatrix} \begin{bmatrix} \hat{e}_x'' \\ \hat{e}_y'' \\ \hat{e}_z'' \end{bmatrix}
$$

3.46

Equation (3.13) can be substituted into Eq. (3.46) and simplified:

$$
\begin{bmatrix} \hat{e}_x''' \\ \hat{e}_y''' \\ \hat{e}_z''' \end{bmatrix} = \begin{bmatrix} \sin\sigma\cos\gamma & (-\sin\sigma\sin\gamma\cos\psi + \cos\sigma\sin\psi) & (-\sin\sigma\sin\gamma\sin\psi - \cos\sigma\cos\psi) \\ \sin\gamma & \cos\gamma\cos\psi & \cos\gamma\sin\psi \\ \cos\sigma\cos\gamma & (-\cos\sigma\sin\gamma\cos\psi - \sin\sigma\sin\psi) & (-\cos\sigma\sin\gamma\sin\psi + \sin\sigma\cos\psi) \end{bmatrix} \begin{bmatrix} \hat{e}_{x_2} \\ \hat{e}_{y_2} \\ \hat{e}_{z_2} \end{bmatrix}
$$

3.47

Before continuing, it is helpful to examine the $OX_3Y_3Z_3$ frame described by the $[\hat{\mathbf{e}}''']$ unit vectors. $\hat{e}_z'''$ is aligned with the lift (which was the entire reason for the latest coordinate transformation). $\hat{e}_y'''$ is still aligned with the velocity. $\hat{e}_x'''$ completes the system such that $\hat{e}_x''' \times \hat{e}_y''' = \hat{e}_z'''$ and is analogous to the pitch axis of an aircraft as is seen in Figure 3-7. Equation (4.41) below summarizes these "more descriptive" unit vector names:

$$
[\hat{\mathbf{e}}'''] = \begin{bmatrix} \hat{e}_x''' \\ \hat{e}_y''' \\ \hat{e}_z''' \end{bmatrix} = \begin{bmatrix} \hat{e}_p \\ \hat{e}_v \\ \hat{e}_L \end{bmatrix}
$$

3.48

$\hat{e}_p$ has been used to reinforce $\hat{e}_x'''$'s similarity to a pitch axis. We'll call this the $OX_pY_vZ_L$ frame.

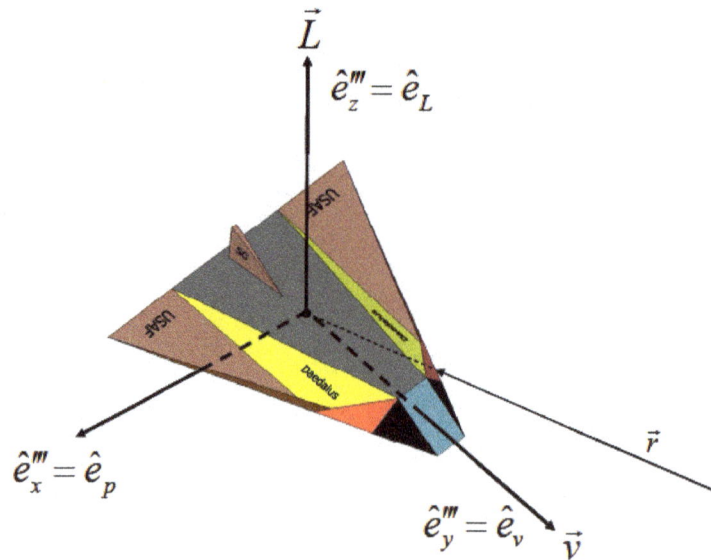

Figure 3-7: Visualizing Unit Vectors of the OX₃Y₃Z₃/ OXₚYᵥZₗ System

Moving forward with these vectors, an expression for lift is easily written as:

$$
\begin{aligned}
\vec{L} &= L\hat{e}_L \\
&= L(\cos\sigma\cos\gamma)\,\hat{e}_{x_2} + L(-\cos\sigma\sin\gamma\cos\psi - \sin\sigma\sin\psi)\,\hat{e}_{y_2} \\
&\quad + L(-\cos\sigma\sin\gamma\sin\psi + \sin\sigma\cos\psi)\,\hat{e}_{z_2}
\end{aligned}
\qquad \textbf{3.49}
$$

The net thrusting force, $\vec{T}$, can be resolved into components in a similar manner. Figure 3-8 shows the general case where the thrust may not be directly aligned with the velocity. By studying the figure (or with another coordinate transformation), the thrust vector can be written as:

$$
\vec{T} = T(\sin\zeta)\hat{e}_p + T(\cos\zeta\cos\varepsilon)\hat{e}_v + T(\cos\zeta\sin\varepsilon)\hat{e}_L
\qquad \textbf{3.50}
$$

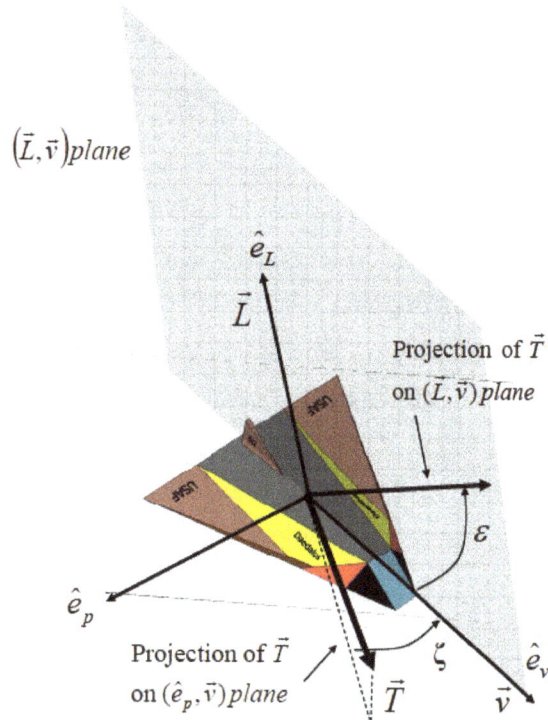

Figure 3-8: Relationship of Thrust Direction to Lift and Velocity Directions

Putting this into vector/matrix form, the thrust vector can be directly written from Eq. (3.50) as:

$$\vec{T} = T\begin{bmatrix}\sin\zeta & \cos\zeta\cos\varepsilon & \cos\zeta\sin\varepsilon\end{bmatrix}\begin{bmatrix}\hat{e}_p \\ \hat{e}_v \\ \hat{e}_L\end{bmatrix} \qquad 3.51$$

Making use of Eq. (3.47), gives the thrust vector in terms of vehicle-pointing coordinates

$$\vec{T} = T\begin{bmatrix}\sin\zeta & \cos\zeta\cos\varepsilon & \cos\zeta\sin\varepsilon\end{bmatrix}$$

$$\bullet \begin{bmatrix}\sin\sigma\cos\gamma & (-\sin\sigma\sin\gamma\cos\psi+\cos\sigma\sin\psi) & (-\sin\sigma\sin\gamma\sin\psi-\cos\sigma\cos\psi) \\ \sin\gamma & \cos\gamma\cos\psi & \cos\gamma\sin\psi \\ \cos\sigma\cos\gamma & (-\cos\sigma\sin\gamma\cos\psi-\sin\sigma\sin\psi) & (-\cos\sigma\sin\gamma\sin\psi+\sin\sigma\cos\psi)\end{bmatrix}\begin{bmatrix}\hat{e}_{x_2} \\ \hat{e}_{y_2} \\ \hat{e}_{z_2}\end{bmatrix}$$

$$3.52$$

which "simplifies" to:

$$
\vec{T} = T \left\{
\begin{array}{l}
(\sin\zeta\sin\sigma\cos\gamma + \cos\zeta\cos\varepsilon\sin\gamma + \cos\zeta\sin\varepsilon\cos\sigma\cos\gamma)\hat{e}_{x_2} \\[2mm]
+ \begin{bmatrix} \sin\zeta(-\sin\sigma\sin\gamma\cos\psi + \cos\sigma\sin\psi) + \cos\zeta\cos\varepsilon\cos\gamma\cos\psi \\ + \cos\zeta\sin\varepsilon(-\cos\sigma\sin\gamma\cos\psi - \sin\sigma\sin\psi) \end{bmatrix}\hat{e}_{y_2} \\[2mm]
+ \begin{bmatrix} \sin\zeta(-\sin\sigma\sin\gamma\sin\psi - \cos\sigma\cos\psi) + \cos\zeta\cos\varepsilon\cos\gamma\sin\psi \\ + \cos\zeta\sin\varepsilon(-\cos\sigma\sin\gamma\sin\psi + \sin\sigma\cos\psi) \end{bmatrix}\hat{e}_{z_2}
\end{array}
\right\}
$$

3.53

The gravity term, $m\vec{g}$ is the only force vector in Eq. (3.19) left to be written in vehicle-pointing coordinates. Since gravity always acts along the radius vector,

$$
m\vec{g} = -mg(r)\hat{e}_{x_2}
$$

3.54

We now have all of elements in Eq. (3.38) written in components of the vehicle-pointing system and can assemble the right-hand side:

$$m\frac{{}^{R}d\left({}^{R}\vec{V}\right)}{dt} = \vec{T} + \vec{L} + \vec{D} + m\vec{g} - 2m\vec{\omega}_{\oplus} \times {}^{R}\vec{V} - m\vec{\omega}_{\oplus} \times (\vec{\omega}_{\oplus} \times \vec{r})$$

$$= \left\{ \begin{array}{l} T\left(\sin\zeta\sin\sigma\cos\gamma + \cos\zeta\cos\varepsilon\sin\gamma + \cos\zeta\sin\varepsilon\cos\sigma\cos\gamma\right) \\ \quad + L(\cos\sigma\cos\gamma) - \left(D\sin\gamma\right) - mg \\ \quad - 2m\left[-\left({}^{R}V\,\omega_{\oplus}\cos\phi\cos\gamma\cos\psi\right)\right] \\ \quad - m\left[-\left(r\omega_{\oplus}^{2}\cos^{2}\phi\right)\right] \end{array} \right\} \hat{e}_{x_2}$$

$$+ \left\{ \begin{array}{l} T\left[\begin{array}{l} \sin\zeta\left(-\sin\sigma\sin\gamma\cos\psi + \cos\sigma\sin\psi\right) + \cos\zeta\cos\varepsilon\cos\gamma\cos\psi \\ \quad + \cos\zeta\sin\varepsilon\left(-\cos\sigma\sin\gamma\cos\psi - \sin\sigma\sin\psi\right) \end{array}\right] \\ + L\left(-\cos\sigma\sin\gamma\cos\psi - \sin\sigma\sin\psi\right) + D\left(-\cos\gamma\cos\psi\right) \\ -2m\,{}^{R}V\,\omega_{\oplus}\left(\cos\phi\sin\gamma - \sin\phi\cos\gamma\sin\psi\right) \end{array} \right\} \hat{e}_{y_2}$$

$$+ \left\{ \begin{array}{l} T\left[\begin{array}{l} \sin\zeta\left(-\sin\sigma\sin\gamma\sin\psi - \cos\sigma\cos\psi\right) + \cos\zeta\cos\varepsilon\cos\gamma\sin\psi \\ \quad + \cos\zeta\sin\varepsilon\left(-\cos\sigma\sin\gamma\sin\psi + \sin\sigma\cos\psi\right) \end{array}\right] \\ + L\left(-\cos\sigma\sin\gamma\sin\psi + \sin\sigma\cos\psi\right) + D\left(-\cos\gamma\sin\psi\right) \\ -2m\left[\left({}^{R}V\,\omega_{\oplus}\sin\phi\cos\gamma\cos\psi\right)\right] - m\left[r\omega_{\oplus}^{2}\sin\phi\cos\phi\right] \end{array} \right\} \hat{e}_{z_2}$$

3.55

In the same manner we wrote a second expression for ${}^{R}\vec{V}$ in Section 3.4 using the relationship between derivatives in different reference frames (i.e., Eq. (3.31)), we can write:

$$\frac{{}^{1}d\left({}^{R}\vec{V}\right)}{dt} = \frac{{}^{2}d\left({}^{R}\vec{V}\right)}{dt} + \left[\vec{\Omega}\right]_{2/1} \times {}^{R}\vec{V}$$

3.56

$\dfrac{^1d\left(^R\vec{V}\right)}{dt}$ in Eq. (3.56) is the same term $\dfrac{^Rd\left(^R\vec{V}\right)}{dt}=\dfrac{^1d\left(^R\vec{V}\right)}{dt}$ in Eq. (3.26).

$\dfrac{^2d\left(^R\vec{V}\right)}{dt}$ can be computed in $\left[\hat{e}_2\right]$ coordinates directly from Eq. (3.29):

$$\frac{^2d\left(^R\vec{V}\right)}{dt}=\left(^R\dot{V}\sin\gamma+{}^RV\dot{\gamma}\cos\gamma\right)\hat{e}_{x_2}$$
$$+\left(^R\dot{V}\cos\gamma\cos\psi-{}^RV\dot{\gamma}\sin\gamma\cos\psi-{}^RV\dot{\psi}\cos\gamma\sin\psi\right)\hat{e}_{y_2}$$
$$+\left(^R\dot{V}\cos\gamma\sin\psi-{}^RV\dot{\gamma}\sin\gamma\sin\psi+{}^RV\dot{\psi}\cos\gamma\cos\psi\right)\hat{e}_{z_2} \qquad \textbf{3.57}$$

The cross-product is:

$$\begin{bmatrix}\dot{\theta}\sin\phi\\-\dot{\phi}\\\dot{\theta}\cos\phi\end{bmatrix}^{\hat{e}_2}\times\begin{bmatrix}^RV\sin\gamma\\^RV\cos\gamma\cos\psi\\^RV\cos\gamma\sin\psi\end{bmatrix}^{\hat{e}_2}=\begin{vmatrix}\hat{e}_{x_2}&\hat{e}_{y_2}&\hat{e}_{z_2}\\\dot{\theta}\sin\phi&-\dot{\phi}&\dot{\theta}\cos\phi\\^RV\sin\gamma&^RV\cos\gamma\cos\psi&^RV\cos\gamma\sin\psi\end{vmatrix}$$

$$={}^RV\left(-\dot{\phi}\cos\gamma\sin\psi-\dot{\theta}\cos\phi\cos\gamma\cos\psi\right)\hat{e}_{x_2}$$
$$+{}^RV\left(\dot{\theta}\cos\phi\sin\gamma-\dot{\theta}\sin\phi\cos\gamma\sin\psi\right)\hat{e}_{y_2}$$
$$+{}^RV\left(\dot{\theta}\sin\phi\cos\gamma\cos\psi+\dot{\phi}\sin\gamma\right)\hat{e}_{z_2} \qquad \textbf{3.58}$$

Before combining Eqs. (3.57) and (3.58), recognize that the kinematic equations give us expressions for $\dot{\theta}$ and $\dot{\phi}$:

$$\dot{\theta}=\frac{^RV\cos\gamma\cos\psi}{r\cos\phi} \qquad \textbf{3.59}$$

$$\dot{\phi}=\frac{^RV\cos\gamma\sin\psi}{r} \qquad \textbf{3.60}$$

Combining Eqs. (3.57) - (3.60):

$$\frac{{}^1 d\left({}^R\vec{V}\right)}{dt} = \left({}^R\dot{V}\sin\gamma + {}^R V\dot{\gamma}\cos\gamma - \frac{{}^R V^2\cos^2\gamma}{r}\right)\hat{e}_{x_2}$$

$$+ \left[\begin{array}{c} {}^R\dot{V}\cos\gamma\cos\psi - {}^R V\dot{\gamma}\sin\gamma\cos\psi - {}^R V\dot{\psi}\cos\gamma\sin\psi \\ + \dfrac{{}^R V^2}{r}\cos\gamma\cos\psi\left(\sin\gamma - \tan\phi\cos\gamma\sin\psi\right) \end{array}\right]\hat{e}_{y_2}$$

$$+ \left[\begin{array}{c} {}^R\dot{V}\cos\gamma\sin\psi - {}^R V\dot{\gamma}\sin\gamma\sin\psi + {}^R V\dot{\psi}\cos\gamma\cos\psi \\ + \dfrac{{}^R V^2}{r}\cos\gamma\left(\cos\gamma\cos^2\psi\tan\phi + \sin\gamma\sin\psi\right) \end{array}\right]\hat{e}_{z_2} \qquad \textbf{3.61}$$

Finally, this can be compared component-by-component to Eq. (3.55) to get three coupled, scalar differential equations:

$$\begin{aligned} {}^R\dot{V}\sin\gamma + {}^R V\dot{\gamma}\cos\gamma - \frac{{}^R V^2\cos^2\gamma}{r} &= \frac{T}{m}\left(\begin{array}{c}\sin\zeta\sin\sigma\cos\gamma + \cos\zeta\cos\varepsilon\sin\gamma \\ + \cos\zeta\sin\varepsilon\cos\sigma\cos\gamma\end{array}\right) \\ &+ \frac{L}{m}\cos\sigma\cos\gamma - \frac{D}{m}\sin\gamma - g \\ &+ 2{}^R V\omega_\oplus\cos\phi\cos\gamma\cos\psi + r\omega_\oplus^2\cos^2\phi \qquad \textbf{3.62}\end{aligned}$$

$${}^R\dot{V}\cos\gamma\cos\psi - {}^R V\dot{\gamma}\sin\gamma\cos\psi - {}^R V\dot{\psi}\cos\gamma\sin\psi$$

$$+ \frac{{}^R V^2}{r}\cos\gamma\cos\psi\left(\sin\gamma - \tan\phi\cos\gamma\sin\psi\right)$$

$$= \frac{T}{m}\left[\begin{array}{c}\sin\zeta\left(-\sin\sigma\sin\gamma\cos\psi + \cos\sigma\sin\psi\right) + \cos\zeta\cos\varepsilon\cos\gamma\cos\psi \\ - \cos\zeta\sin\varepsilon\left(\cos\sigma\sin\gamma\cos\psi + \sin\sigma\sin\psi\right)\end{array}\right]$$

$$- \frac{L}{m}\left(\cos\sigma\sin\gamma\cos\psi + \sin\sigma\sin\psi\right) - \frac{D}{m}\cos\gamma\cos\psi$$

$$- 2{}^R V\omega_\oplus\left(\cos\phi\sin\gamma - \sin\phi\cos\gamma\sin\psi\right) \qquad \textbf{3.63}$$

$$^{R}\dot{V}\cos\gamma\sin\psi - {}^{R}V\dot{\gamma}\sin\gamma\sin\psi + {}^{R}V\dot{\psi}\cos\gamma\cos\psi$$

$$+\frac{^{R}V^{2}}{r}\cos\gamma\left(\cos\gamma\cos^{2}\psi\tan\phi + \sin\gamma\sin\psi\right)$$

$$=\frac{T}{m}\left[\begin{array}{l}-\sin\zeta\left(\sin\sigma\sin\gamma\sin\psi + \cos\sigma\cos\psi\right)+\cos\zeta\cos\varepsilon\cos\gamma\sin\psi\\ +\cos\zeta\sin\varepsilon\left(-\cos\sigma\sin\gamma\sin\psi + \sin\sigma\cos\psi\right)\end{array}\right]$$

$$+\frac{L}{m}\left(-\cos\sigma\sin\gamma\sin\psi + \sin\sigma\cos\psi\right)-\frac{D}{m}\cos\gamma\sin\psi$$

$$-2^{R}V\omega_{\oplus}\sin\phi\cos\gamma\cos\psi - r\omega_{\oplus}^{2}\sin\phi\cos\phi \qquad \textbf{3.64}$$

These can be solved for $^{R}\dot{V}, \dot{\gamma},$ and $\dot{\psi}$

$$^{R}\dot{V} = \frac{T}{m}\left(\cos\zeta\cos\varepsilon\right) - \frac{D}{m} - g\sin\gamma$$

$$+ r\omega_{\oplus}^{2}\cos\phi\left(\cos\phi\sin\gamma - \sin\phi\sin\psi\cos\gamma\right) \qquad \textbf{3.65}$$

$$^{R}V\dot{\gamma} = \frac{T}{m}\left(\sin\zeta\sin\sigma + \cos\zeta\sin\varepsilon\cos\sigma\right)+\frac{L}{m}\cos\sigma - g\cos\gamma$$

$$+\frac{^{R}V^{2}}{r}\cos\gamma + 2^{R}V\omega_{\oplus}\cos\phi\cos\psi$$

$$+ r\omega_{\oplus}^{2}\cos\phi\left(\cos\phi\cos\gamma + \sin\phi\sin\psi\sin\gamma\right) \qquad \textbf{3.66}$$

$$^{R}V\dot{\psi} = \frac{1}{m\cos\gamma}\left[T\left(\cos\zeta\sin\varepsilon\sin\sigma - \sin\zeta\cos\sigma\right)+L\sin\sigma\right]$$

$$-\frac{^{R}V^{2}}{r}\cos\gamma\cos\psi\tan\phi + 2^{R}V\omega_{\oplus}\left(\sin\psi\cos\phi\tan\gamma - \sin\phi\right)$$

$$-\frac{r\omega_{\oplus}^{2}}{\cos\gamma}\sin\phi\cos\phi\cos\psi \qquad \textbf{3.67}$$

Equations (3.65) - (3.67) are the force equations. When solved simultaneously with the kinematic equations ((3.35) - (3.37)), they describe the velocity and its orientation. The three kinematic and three force equations together are the equations of motion and their solutions give the six independent parameters

$(r,\theta,\phi,{}^RV,\gamma,$ and $\psi)$ required to define the motion of a point mass at any time. Using geometric relations (such as the rotation matrices already written), these quantities can transformed to the more familiar $\vec{r}(t)$ and ${}^R\vec{V}(t)$ when necessary.

The equations of motion above cannot be solved in closed form (i.e., analytically) and must be solved by numerical integration subject to the appropriate boundary conditions. While these equations are good for visualizing the problem, in practice, they may not be the most appropriate equations for numerical methods. Instead, it may be better (numerically) to work in Cartesian coordinates directly with the vector equations

$$\frac{{}^Id\vec{r}}{dt} = \vec{V}$$

3.68

$$\frac{{}^Id\vec{V}}{dt} = \vec{f}(\vec{r},\vec{V},t)$$

3.69

where the time derivatives are in the inertial frame and $\vec{f}$ is the sum of the forces on the vehicle (per unit mass). Once solved, the results can then be transformed to whatever coordinate system is desired. Equations (3.35)–(3.37) and (3.65) - (3.67) have the advantage of directly providing the major parameters of interest (once solved), so we will use them for our purposes.

3.5 Summary

Equations (3.35) - (3.37) and (3.65) - (3.67) are not in the forms typically found in other texts for similar three-degree-of-freedom formulations. Most texts will use V instead of ${}^RV$ and assume the reader recognizes it as a velocity *relative* to the rotating atmosphere. Less obvious is our inclusion of thrust force components which may lie outside the lift-drag (or, equivalently, lift-velocity) plane. These components are introduced when $\zeta \neq 0$. To compare the force equations to other texts, set $\zeta = 0$.

Another subtlety that might cause confusion between texts is that we've measured the flight-path angle γ and heading angle ψ with respect to the relative velocity $^{R}\overline{V}$. For complete clarity, we could have written $^{R}\gamma$ and $^{R}\psi$ to distinguish them from values measured relative to an inertial velocity $^{I}\overline{V}$. However, the more cumbersome notation is not necessary since we will rarely mention the inertial quantities. (In Chapter 2, the "orbital" flight-path angle was an inertial value and in Chapter 12 we will see an example where we are only given inertial values for a real-world example.)

The majority of this text will involve working with various simplifications and/or manipulations of these equations. However, before moving on, we should stop and summarize the assumptions made to this point in our derivations. These are listed in Table 3-1.

Finally, we will not address five- or six-degree-of-freedom simulations in this text. (The increased complexity comes from considering such things as roll, pitch, and yaw of the vehicle.) The additional effort does not significantly add to understanding the fundamentals of atmospheric entry. Such simulations are discussed in several of the references (47:369-423; 53:255-261, 414-450; 68:289-361, 367-480).

Table 3-1: Assumptions in Equations of Motion

Assumption	First Introduced
Point mass	Section 3.1
Planet rotation is constant about north pole	Section 3.2.2
Planet-fixed and inertial coordinate systems aligned at t=0	Section 3.2.2 Eq. (3.1)
Constant mass	Section 3.4 Eq. (3.20)
Drag acts in direction opposite the velocity	Section 3.4.2 Eq. (3.44)
Lift force is perpendicular to velocity	Section 3.4.2 Eq. (3.45)
Gravity force is directed along a vector from point mass to the center of the planet	Section 3.4.2 Eq. (3.54)

3.6 Problems

1. Multiply out Eq. (3.6) to find a general expression for the transition matrix between $[\hat{\mathbf{e}}_2]$ and $[\hat{\mathbf{e}}]$.

2. An alien civilization, whose planet has an extremely thick atmosphere, has developed an anti-drag engine for their spacecraft in low orbit. The engine produces a thrust exactly equal to the drag and in the opposite direction; i.e., $\vec{T} = -\vec{D}$. Assuming they design the spacecraft to avoid any lift forces, what are the three "force equations" of motion?

This page intentionally blank.

Chapter 4

Equations for Planar Entry

4.1 Introduction

As a step towards *qualitatively* comparing entry trajectories, we'll look at the relatively simple case of motion confined to a single "entry plane." We'll further simplify the problem by assuming planetary rotation is negligible during the entry phase. These assumptions aren't as restrictive as they might seem. Ballistic entry (e.g., no lift) essentially stays within a plane as does entry where the aerodynamic lift is used only to "pull up" or "pull down" along the radius. Ignoring the rotation of the planet is justified due to the small amount of rotation that occurs during short entry phases. Of course, precise velocity or position requirements would invalidate these assumptions, but they're sufficient for initial study.

4.2 Assumptions for Planar Entry Trajectories

Planar entry is defined as motion confined to the plane of a great circle. Such a plane contains the vehicle's radius and velocity vectors and the planet's center point. Figure 4-1 shows the relevant geometry as viewed perpendicular to the great circle plane and Figure 4-2 shows the three-dimensional view. (The similarity of the entry plane to an orbital plane in the three-dimensional view is not accidental.)

Many books begin with the two-dimensional view in Figure 4-1 and derive the equations of motion all over again for the planar case. That's too simple for us after we spent so much effort finding relatively general equations in Chapter 3! Instead, we will start with Eqs. (3.35) - (3.37) and (3.65) - (3.67) and apply the appropriate assumptions to simplify them.

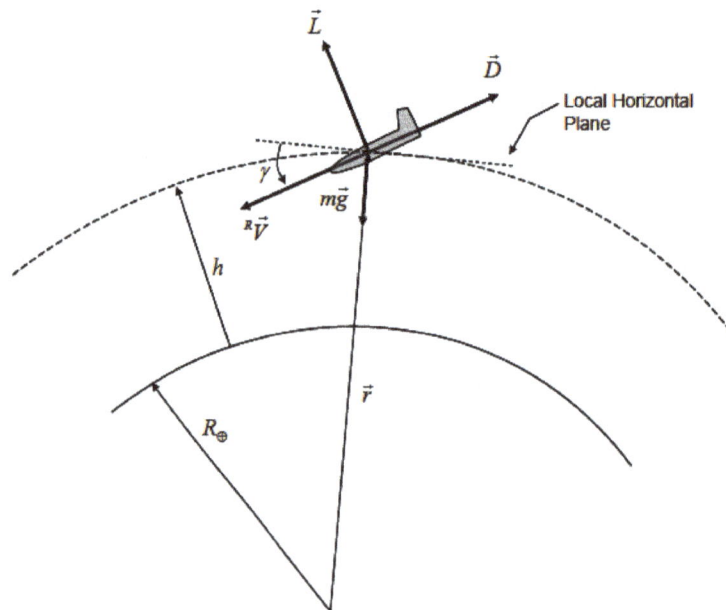

Figure 4-1: Two-Dimensional View of Planar Entry

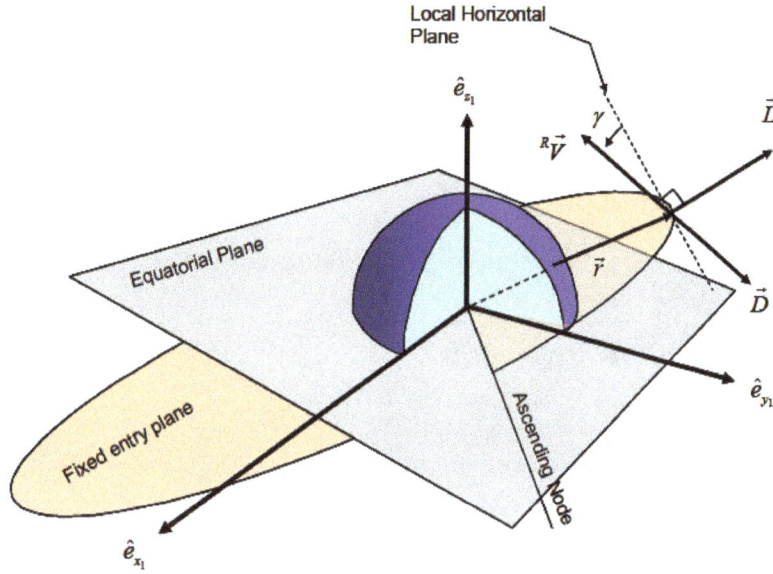

Figure 4-2: Three-Dimensional View of Planar Entry

To maintain planar flight, assume there is no banking of the lift vector (i.e., $\sigma = 0°$) and no thrust force. The equations of motion from Chapter 3 simplify to:

$$\dot{r} = {}^{R}V \sin \gamma \tag{4.1}$$

$$\dot{\theta} = \frac{{}^{R}V \cos \gamma \cos \psi}{r \cos \phi} \tag{4.2}$$

$$\dot{\phi} = \frac{{}^{R}V \cos \gamma \sin \psi}{r} \tag{4.3}$$

$$^{R}\dot{V} = -\frac{D}{m} - g \sin \gamma + r\omega_{\oplus}^{2} \cos \phi (\cos \phi \sin \gamma - \sin \phi \sin \psi \cos \gamma) \tag{4.4}$$

$$^{R}V\dot{\gamma} = \frac{L}{m} - g\cos\gamma + \frac{^{R}V^{2}}{r}\cos\gamma + 2^{R}V\omega_{\oplus}\cos\phi\cos\psi$$
$$+ r\omega_{\oplus}^{2}\cos\phi\left(\cos\phi\cos\gamma + \sin\phi\sin\psi\sin\gamma\right) \qquad \textbf{4.5}$$

$$V\dot{\psi} = -\frac{^{R}V^{2}}{r}\cos\gamma\cos\psi\tan\phi + 2^{R}V\omega_{\oplus}\left(\sin\psi\cos\phi\tan\gamma - \sin\phi\right)$$
$$-\frac{r\omega_{\oplus}^{2}}{\cos\gamma}\sin\phi\cos\phi\cos\psi \qquad \textbf{4.6}$$

To make the problem more analytically palatable, assume the planetary rotation can be ignored (i.e., $\omega_{\oplus} = 0$) and these equations reduce even further:

$$\dot{r} = {}^{R}V\sin\gamma \qquad \textbf{4.7}$$

$$\dot{\theta} = \frac{^{R}V\cos\gamma\cos\psi}{r\cos\phi} \qquad \textbf{4.8}$$

$$\dot{\phi} = \frac{^{R}V\cos\gamma\sin\psi}{r} \qquad \textbf{4.9}$$

$$^{R}\dot{V} = -\frac{D}{m} - g\sin\gamma \qquad \textbf{4.10}$$

$$^{R}V\dot{\gamma} = \frac{L}{m} - g\cos\gamma + \frac{^{R}V^{2}}{r}\cos\gamma \qquad \textbf{4.11}$$

$$^{R}V\dot{\psi} = -\frac{^{R}V^{2}}{r}\cos\gamma\cos\psi\tan\phi \qquad \textbf{4.12}$$

Because we have ignored the rotation of the planet (effectively "stopping it"), the planet relative and inertial velocity are the same. We could (but won't) drop the "R" superscript on the velocity to simplify the appearance of these equations.

It is tempting to make the erroneous assumption the heading angle ψ can be set to zero. (Looking at only the "flat world" view in Figure 4-1 tends to reinforce the belief because the velocity lies in the plane.) Recall, however, that ψ is the angle between the local parallel of latitude and the projection of $^R\vec{V}$ on the local horizontal plane as shown in Figure 4-3. (ψ describes how much the velocity vector contributes to moving "toward and away" from the planetary equator.) ψ is much like the inclination for orbital motion except ψ may not be constant (because the reference – the local parallel of latitude – changes). In fact, at the point where the entry plane and the equatorial plane meet, ψ would measure the same angle as an orbital inclination.

Before leaving the discussion of ψ, it is helpful to take a geometric look at the relationship between ψ, θ, and ϕ. If the entry plane is fixed in space, as it is with our planar entry assumptions (including $\omega_\oplus = 0$), then a vector normal to the plane is fixed in space also (i.e., a constant direction). The fixed entry plane (for planar motion) contains both the radius and the velocity vector. Thus, a unit vector normal to the plane can be defined by

$$\hat{e}_n = \frac{\vec{r} \times {}^R\vec{V}}{\left| \vec{r} \times {}^R\vec{V} \right|} \qquad \textbf{4.13}$$

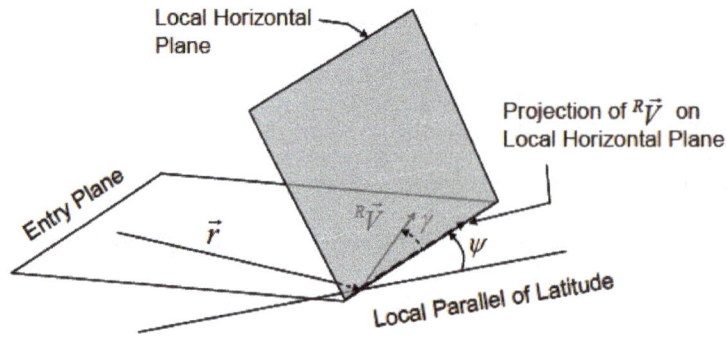

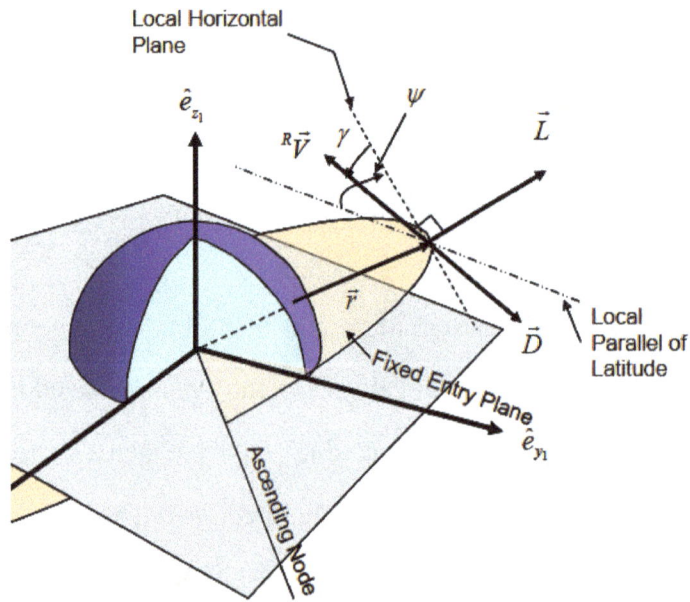

Figure 4-3: Heading Angle during Planar Entry

When written in terms of the planet-fixed $OX_1Y_1Z_1$ coordinates, this is:

$$\hat{e}_n = f_x(\theta,\phi,\psi,\gamma)\,\hat{e}_{x_1} + f_y(\theta,\phi,\psi,\gamma)\,\hat{e}_{y_1} + f_z(\theta,\phi,\psi,\gamma)\,\hat{e}_{z_1} \qquad \textbf{4.14}$$

where f_x, f_y, and f_z are simply scalar equations involving the rotation angles resulting from the coordinate transformations (e.g., Eq. (3.6)). To the accuracy of our assumptions, each component of $\hat{e}_n$ will remain constant when solving Eqs. (4.7) - (4.12). Some would view Eq. (4.14) as a "constraint" equation since it describes the "constraining" of the "no longer independent" variables of ψ, θ, and ϕ. For example, ψ would be "fixed" by the current values of θ and ϕ at any point during the planar entry.

4.3 Basic Planar Equations

Equations (4.7) - (4.12) must be solved simultaneously to obtain the vehicle's complete position and velocity. However, if we aren't concerned with the footprint on the planet (θ and ϕ), then Eqs. (4.7), (4.10), and (4.11) can be solved independent of the other three equations to find the magnitude of velocity and radius (or, equivalently, altitude). This leaves us with the three equations many books start with:

$$\dot{r} = {}^{R}V \sin \gamma \qquad\qquad \textbf{4.15}$$

$$^{R}\dot{V} = -\frac{D}{m} - g \sin \gamma \qquad\qquad \textbf{4.16}$$

$$^{R}V\dot{\gamma} = \frac{L}{m} - g \cos \gamma + \frac{{}^{R}V^{2}}{r} \cos \gamma \qquad\qquad \textbf{4.17}$$

Lift and drag are often replaced with the following

$$L = \frac{\rho C_{L} S}{2} {}^{R}V^{2} \qquad\qquad \textbf{4.18}$$

$$D = \frac{\rho C_{D} S}{2} {}^{R}V^{2} \qquad\qquad \textbf{4.19}$$

where ρ is the atmospheric density. C_L and C_D are the coefficients of lift and drag, respectively and S is a reference area used in calculating these aerodynamic coefficients. In general, C_L and C_D may be complex functions of the angle of attack α, Mach number M, and the Reynolds number Re:

$$C_L = C_L(\alpha, M, \mathrm{Re}) \qquad\qquad \textbf{4.20}$$

$$C_D = C_D(\alpha, M, \mathrm{Re}) \qquad\qquad \textbf{4.21}$$

However, in the case of hypersonic flow (and other special cases), the aerodynamic coefficients are essentially functions of angle of attack only. For our case, we can further assume the angle of attack is held constant, so C_L and C_D are constants.

The gravity term in Eqs. (4.16) and (4.17) is a function of radius and can be written in terms of a reference radius r_0 and the gravity constant at that radius g_0:

$$g = g(r) = g_0 \left(\frac{r_0}{r} \right)^2 \qquad\qquad \textbf{4.22}$$

Finally, with these assumptions, our equations become:

$$\dot{r} = {}^R V \sin \gamma \qquad\qquad \textbf{4.23}$$

$${}^R \dot{V} = -\frac{\rho C_D S\, {}^R V^2}{2m} - g_0 \left(\frac{r_0}{r} \right)^2 \sin \gamma \qquad\qquad \textbf{4.24}$$

$${}^R V \dot{\gamma} = \frac{\rho C_L S\, {}^R V^2}{2m} + \left[\frac{{}^R V^2}{r} - g_0 \left(\frac{r_0}{r} \right)^2 \right] \cos \gamma \qquad\qquad \textbf{4.25}$$

The deceleration expected during an atmospheric entry is an important quantity to know when designing trajectories or entry vehicles. As a vector, the deceleration is defined as:

$$\vec{a}_{decel} = -\vec{a} = -\frac{d}{dt}\left({}^{R}\vec{V}\right)$$

4.26

In terms of the $OX_pY_vZ_L$ coordinate system found in Chapter 3, this can be written:

$$\vec{a}_{decel} = -\frac{d}{dt}\left({}^{R}V\hat{e}_v\right)$$

4.27

The derivative (as seen in the entry plane) is straight-forward

$$\vec{a}_{decel} = -\frac{d}{dt}\left({}^{R}V\hat{e}_v\right)$$

$$= -{}^{R}\dot{V}\hat{e}_v - \dot{\gamma}\,{}^{R}V\hat{e}_L$$

4.28

where our assumption of planar entry ensures there is no component in the $\hat{e}_p$ direction. From this point forward, we will refer to the tangential (along the velocity vector) deceleration as

$$\left(a_{decel}\right)_v = -{}^{R}\dot{V}$$

4.29

and the normal (along the lift vector) deceleration as:

$$\left(a_{decel}\right)_L = -\dot{\gamma}\,{}^{R}V$$

4.30

With the help of Eqs. (4.24) and (4.25), these can also be written as:

$$\left(a_{decel}\right)_v = \frac{\rho C_D S\,^R V^2}{2m} + g_0 \left(\frac{r_0}{r}\right)^2 \sin \gamma$$

4.31

$$\left(a_{decel}\right)_L = \frac{-\rho C_L S\,^R V^2}{2m} - \left[\frac{^R V^2}{r} - g_0\left(\frac{r_0}{r}\right)^2\right]\cos \gamma$$

4.32

4.4 Basic Planar Equations in Non-Dimensional Form

Equations (4.23) – (4.25) cannot be solved analytically in general. Vinh, et al. simplified the equations into a convenient non-dimensional form (58:102-107). Following their procedure, we can obtain a set of differential equations that can be solved analytically for a several realistic entry scenarios.

These three differential equations can be collapsed into two with a change of variables. Let a non-dimensional distance be defined by

$$z = \frac{r_0}{r}$$

4.33

and we can change the independent variable from t to z by dividing Eqs. (4.23) – (4.25) by

$$\frac{dz}{dt} = -\frac{r_0}{r^2}\frac{dr}{dt} = -\frac{^R V \sin \gamma}{r_0} z^2$$

4.34

to get:

$$\frac{dr}{dz} = -\frac{r_0}{z^2}$$

4.35

$$ {}^{R}V \frac{d^{R}V}{dz} = \frac{r_0 \rho C_D S \, {}^{R}V^2}{2mz^2 \sin\gamma} + g_0 r_0 \qquad\qquad \textbf{4.36} $$

$$ -\sin\gamma \frac{d\gamma}{dz} = \frac{r_0 \rho C_L S}{2mz^2} + \left[\frac{1}{z} - \frac{g_0 r_0}{{}^{R}V^2} \right] \cos\gamma \qquad\qquad \textbf{4.37} $$

Examining Eq. (4.36) (and a lot of luck) leads to another substitution. Let the non-dimensional kinetic energy be given by:

$$ T = \frac{1}{2}\left(\frac{{}^{R}V^2}{g_0 r_0} \right) \qquad\qquad \textbf{4.38} $$

Then,

$$ \frac{dT}{dz} = \frac{{}^{R}V}{g_0 r_0} \frac{d^{R}V}{dz} \qquad\qquad \textbf{4.39} $$

Equation (4.36) becomes

$$ \frac{dT}{dz} = \frac{r_0 \rho C_D S T}{mz^2 \sin\gamma} + 1 \qquad\qquad \textbf{4.40} $$

and Eq. (4.37) becomes:

$$ -\sin\gamma \frac{d\gamma}{dz} = \frac{r_0 \rho C_L S}{2mz^2} + \left[\frac{1}{z} - \frac{1}{2T} \right] \cos\gamma \qquad\qquad \textbf{4.41} $$

This last equation can be simplified further with yet another variable substitution, $\xi = \cos\gamma$:

$$ \frac{d\xi}{dz} = \frac{r_0 \rho C_L S}{2mz^2} + \left[\frac{1}{z} - \frac{1}{2T} \right] \xi \qquad\qquad \textbf{4.42} $$

To summarize, our two "reduced" equations for planar entry are:

$$\frac{dT}{dz} = \frac{r_0 \rho C_D S T}{mz^2 \sin \gamma} + 1 \qquad\qquad \textbf{4.43}$$

$$\frac{d\xi}{dz} = \frac{r_0 \rho C_L S}{2mz^2} + \left[\frac{1}{z} - \frac{1}{2T} \right] \xi \qquad\qquad \textbf{4.44}$$

where

$$z = \frac{r_0}{r} \qquad\qquad \textbf{4.45}$$

$$T = \frac{1}{2} \left(\frac{^R V^2}{g_0 r_0} \right) \qquad\qquad \textbf{4.46}$$

$$\xi = \cos \gamma \qquad\qquad \textbf{4.47}$$

The system of Eqs. (4.43) - (4.47) is "exact" *to the accuracy of our approximations* (C_L and C_D constant, no planetary rotation, etc.). Once atmospheric density ρ as a function of z is specified, it can be solved.

A simple, yet reasonable, approximation for atmospheric density is that of an exponential atmosphere:

$$\rho = \rho_s e^{-\beta(r - R_\oplus)} \qquad\qquad \textbf{4.48}$$

In Eq. (4.48), $R_\oplus$ is the planetary radius, ρ_s is the atmospheric density at the surface, and β^{-1} is a scale height selected to best match the atmosphere to the assumed exponential form. We will further assume a *strictly exponential atmosphere* which has a fixed (constant) value of β for any given planet. The height (or altitude) h can be substituted into Eq. (4.48):

$$\rho = \rho_s e^{-\beta h} \tag{4.49}$$

If we follow the customary derivation (or get extremely lucky), we'll see that another change of the independent variable in Eqs. (4.49) will aid in the analytic integration. To do this, we introduce a non-dimensional altitude variable that is proportional to the density:

$$\eta = \frac{\rho S C_D}{2m\beta} \tag{4.50}$$

(Note that our new "altitude" variable η *increases* with *decreasing* altitude.) Taking the logarithmic derivative

$$\frac{d\eta}{\eta} = -\beta \, dr \tag{4.51}$$

and using Eq. (4.35), we arrive at:

$$\frac{d\eta}{dz} = \frac{\beta r_0 \eta}{z^2} \tag{4.52}$$

To change the independent variable in Eqs. (4.43) and (4.44), divide by Eq. (4.52):

$$\frac{dT}{d\eta} = \frac{2T}{\sin\gamma} + \frac{z^2}{\beta r_0 \eta} \tag{4.53}$$

$$\frac{d\xi}{d\eta} = \frac{C_L}{C_D} + \frac{z^2}{\beta r_0 \eta}\left(\frac{1}{z} - \frac{1}{2T}\right)\xi \tag{4.54}$$

Strictly speaking, the variable $z = \dfrac{r_0}{r}$ in Eqs. (4.53) and (4.54) should be replaced with the equivalent value of the new independent variable η. When required, Eq. (4.50) can be solved

$$z = \left\{ \frac{R_\oplus}{r_0} - \frac{1}{\beta r_0}\left[\ln \eta - \ln\left(\frac{\rho_s SC_D}{2m\beta}\right)\right]\right\}^{-1}$$ **4.55**

and z replaced in Eqs. (4.53) - (4.54). However, it is not always necessary to make this somewhat painful substitution. If r_0 is taken to be the initial altitude of entry (or any altitude within the atmosphere), then for planets with atmospheres similar to that of Earth:

$$z = \frac{r_0}{r} \cong 1$$ **4.56**

Mathematically, this can be justified by noting most of the atmospheric effects take place within about 100 km of the Earth's surface. Setting $z = 1$ only introduces a few percent of error – small in comparison to the other assumptions made to this point. (To put this in perspective, if Earth were scaled to the size of a peach, the atmosphere would be about as thick as the fuzz – negligible when measuring the radius of the peach!)

With this latest simplification, we have the basic equations of motion for planar atmospheric entry:

$$\frac{dT}{d\eta} = \frac{2T}{\sin \gamma} + \frac{1}{\beta r_0 \eta}$$ **4.57**

$$\frac{d\xi}{d\eta} = \frac{C_L}{C_D} + \frac{1}{\beta r_0 \eta}\left(1 - \frac{1}{2T}\right)\xi$$ **4.58**

The components of deceleration in Eqs. (4.31) and (4.32) can be similarly written

$$\left(a_{decel}\right)_v = 2\beta r_0 g_0 \eta T \pm g_0 \sqrt{1-\xi^2} \qquad \textbf{4.59}$$

$$\left(a_{decel}\right)_L = -2\beta r_0 g_0 \eta T \left(\frac{C_L}{C_D}\right) + g_0 \left(1-2T\right)\xi \qquad \textbf{4.60}$$

where the choice of sign in Eq. (4.59) coincides with the sign of the flight-path angle. The derivation is left as an exercise.

The basic equations we have derived in this chapter are sufficient for preliminary study of atmospheric entry. They are in a form which will provide analytic insight. In the next chapter, they will be solved using traditional first-order solutions for special cases. Later chapters will develop a second-order theory and investigate numerical solutions to the full equations of motion (Eqs. (3.35) - (3.37) and (3.65) - (3.67)).

4.5 Problems

1. Equations (4.23) - (4.25) can be used to describe an orbit by applying the appropriate simplifications. Apply those simplifications and find the resulting three equations of motion. For the case where the flight-path angle is constant, show that the velocity given by Eq. (4.25) is that typically given in orbital mechanics books for circular orbit speed: $V_c = \sqrt{\mu / r}$, where μ is a gravitational constant. (In our case, $\mu = g_0 r_0^2$.)

2. If the entry plane is fixed in space, then a vector normal to the plane is fixed in space also (i.e., a constant direction). The fixed entry plane (for planar motion) contains both the radius and the velocity vector. Thus, a unit vector normal to the plane can be defined by

$$\hat{e}_n = \frac{\vec{r} \times {}^I\vec{V}}{\left| \vec{r} \times {}^I\vec{V} \right|}$$

where ${}^I\vec{V}$ is the inertial velocity. When written in terms of the inertial OXYZ coordinates, this is

$$\hat{e}_n = f_x(\theta,\phi,\psi,\gamma,\omega\Delta t)\, \hat{e}_x + f_y(\theta,\phi,\psi,\gamma,\omega\Delta t)\, \hat{e}_y + f_z(\theta,\phi,\psi,\gamma,\omega\Delta t)\, \hat{e}_z$$

where f_x, f_y, and f_z are simply scalar equations involving the rotation angles resulting from the coordinate transformations (e.g., Eq. (3.6)). If $\hat{e}_n$ is fixed in space then

$$f_x(\theta,\phi,\psi,\gamma,\omega\Delta t) = C_x$$
$$f_y(\theta,\phi,\psi,\gamma,\omega\Delta t) = C_y$$
$$f_z(\theta,\phi,\psi,\gamma,\omega\Delta t) = C_z$$

(where each C_i is a constant) describe the "constraints" on θ, ϕ, and ψ required to remain planar. Find these three constraints. You do not need to (and probably can't) solve for the variables – just write the expressions for f_x, f_y, f_z. **Simpler option:** Assume the planet does not rotate ($^I\vec{V} = {}^R\vec{V}$) and the OXYZ frame is aligned with the $OX_1Y_1Z_1$.

3. Prove the components of deceleration in Eqs. (4.31) and (4.32) can be approximated as

$$\left(a_{decel}\right)_v = 2\beta r_0 g_0 \eta T \pm g_0 \sqrt{1-\xi^2}$$

$$\left(a_{decel}\right)_L = -2\beta r_0 g_0 \eta T \left(\frac{C_L}{C_D}\right) + g_0 \left(1-2T\right)\xi$$

when $\dfrac{r}{r_0} \approx 1$.

4. Starting with Eq. (4.50), derive Eq. (4.55).

This page intentionally blank.

Chapter 5

Classic Closed-Form Solutions

5.1 Introduction

In this chapter, we'll derive several first-order solutions and a specialized second-order solution to the basic equations for planar entry we found in the last chapter. Each solution will be examined to study how altitude, speed, deceleration, and/or other parameters interact during atmospheric entry. With today's desktop computers, the equations of motion could be solved without the (somewhat painful) derivation of closed-form analytic and semianalytic approximations. However, these solutions (and insights gained in their derivations) are very helpful in understanding the motion *at the conceptual level.* And, while each is a "specialized case," they have the added benefit of being helpful for making preliminary planning decisions.

The differential equations we found in Chapter 4 by following Vinh, et al.'s work are ideal for our analysis. For convenience, they are repeated below:

$$\frac{dT}{d\eta} = \frac{2T}{\sin\gamma} + \frac{1}{\beta r_0 \eta}$$

5.1

$$\frac{d\xi}{d\eta} = \frac{C_L}{C_D} + \frac{1}{\beta r_0 \eta}\left(1 - \frac{1}{2T}\right)\xi$$

5.2

where:

$$\eta = \frac{\rho_s S C_D}{2m\beta} e^{-\beta h}$$

5.3

$$T = \frac{1}{2}\left(\frac{{}^R V^2}{g_0 r_0}\right)$$

5.4

$$\xi = \cos\gamma$$

5.5

In working with these, it will be helpful to understand the physical origin of the individual terms in Eqs. (5.1) and (5.2). The first term on the right side of Eq. (5.1) is from the drag force as can be seen by referring back to Eq. (4.24). The second is the force of gravity along the velocity vector (also seen from Eq. (4.24)). In Eq. (5.2), the first term is, obviously, the lift force. The second can be seen to be the centrifugal force if you refer back to Eq. (4.25). Finally, the third term is the gravity term (also seen by referring back to Eq. (4.25)). These insights are summarized in Figure 5-1.

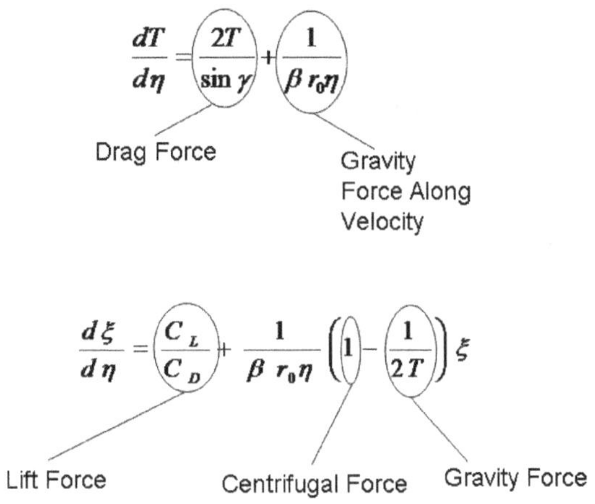

$$\frac{dT}{d\eta} = \underbrace{\left(\frac{2T}{\sin\gamma}\right)}_{\text{Drag Force}} + \underbrace{\left(\frac{1}{\beta\, r_0 \eta}\right)}_{\substack{\text{Gravity} \\ \text{Force Along} \\ \text{Velocity}}}$$

$$\frac{d\xi}{d\eta} = \underbrace{\left(\frac{C_L}{C_D}\right)}_{\text{Lift Force}} + \frac{1}{\beta\, r_0 \eta}\left(\underbrace{(1)}_{\text{Centrifugal Force}} - \underbrace{\left(\frac{1}{2T}\right)}_{\text{Gravity Force}}\right)\xi$$

Figure 5-1: Physical Origin of Terms in the Planar Motion Equations (58:109)

5.2 Shallow Gliding Entry

This first-order analysis assumes the vehicle produces enough lift to maintain a lengthy hypersonic glide at a small flight-path angle. Clearly, this type of entry is an idealization, since in reality it is not practical to maintain a small entry angle at hypersonic speeds all the way to the planet surface! However, it can be used to study a large portion of the entry profile for a gliding entry. As a practical example, the Space Shuttle uses such an entry for the initial phase of reentry -- from entry interface to about 24 km in altitude.

For shallow entry, the main assumption is that the flight-path angle is small and the usual small angle assumptions

$$\sin\gamma \approx \gamma \qquad\qquad \textbf{5.6}$$

$$\cos\gamma \approx 1 \qquad\qquad \textbf{5.7}$$

can be made. Further, if changes in $\xi = \cos\gamma$ are small, we have a quasi-equilibrium glide condition first formulated by Sänger and Bredt (48:82-127). Equation (5.2) becomes simply an algebraic equation:

$$\frac{C_L}{C_D} + \frac{1}{\beta\, r_0 \eta}\left(1 - \frac{1}{2T}\right) = 0 \qquad\qquad \textbf{5.8}$$

This can be solved to relate the kinetic energy to altitude

$$T = \frac{1}{2\left[1 + \beta r_0 \eta\left(\dfrac{C_L}{C_D}\right)\right]} \qquad\qquad \textbf{5.9}$$

or, equivalently, velocity to altitude:

$$\frac{{}^R V^2}{g_0 r_0} = \frac{1}{1 + \beta r_0 \eta\left(\dfrac{C_L}{C_D}\right)} \qquad\qquad \textbf{5.10}$$

A circular orbit at the reference radius has a velocity of $V_0 = \sqrt{g_0 r_0}$. If it is assumed this "reference velocity" is the maximum speed at which entry begins, then Figure 5-2 shows the relationship in Eq. (5.10) over the span of $0 \le {}^R V \le \sqrt{g_0 r_0}$ for various values of the lift-to-drag ratio C_L/C_D. (It is possible to study cases where ${}^R V > \sqrt{g_0 r_0}$; however, for simplicity, we will restrict ourselves to ${}^R V \le \sqrt{g_0 r_0}$.) The resulting range of $\beta r_0 \eta$ values shown in the figure has not been limited to those which can be physically obtained. Recall as altitude approaches zero,

$$\eta \approx \eta_s = \frac{\rho_s S C_D}{2m\beta} \qquad\qquad \textbf{5.11}$$

78

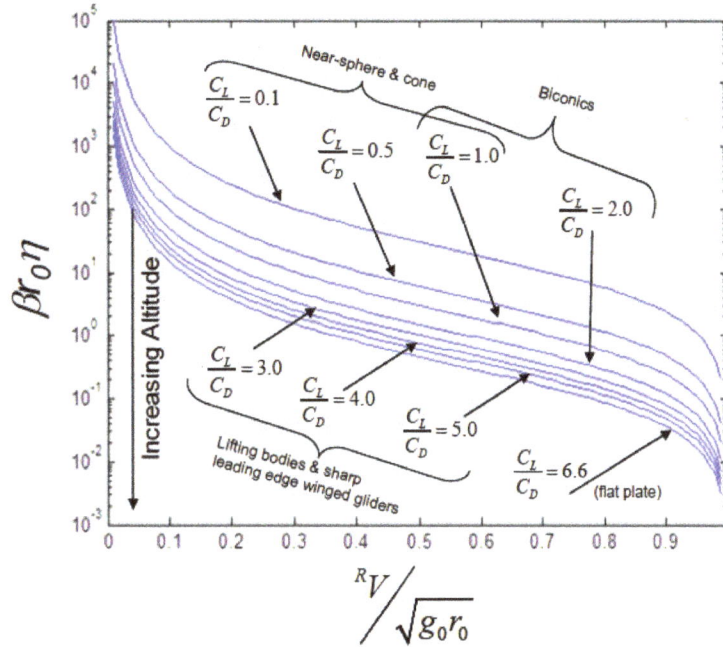

Figure 5-2: Analytic Altitude/Velocity Relationship for Shallow, Gliding Entry

which places a finite upper limit on $\beta r_0 \eta$. For practical values of $\beta r_0 \eta$, and because we assumed hypersonic speeds, we can restrict ourselves to the lower right side of this curve. Figure 5-3 expands the area of interest.

Equation (5.1) also relates the flight-path angle to the kinetic energy. If gravity force along the velocity vector is considered small relative to the drag force, then we can write:

$$\frac{dT}{d\eta} = \frac{2T}{\sin \gamma} \approx \frac{2T}{\gamma} \qquad \textbf{5.12}$$

(This may seem inconsistent with the "practical" values for $\beta r_0 \eta$ in Figure 5-3; however, the choice to ignore $\dfrac{1}{\beta r_0 \eta}$ in Eq. (5.1) was based on its magnitude

79

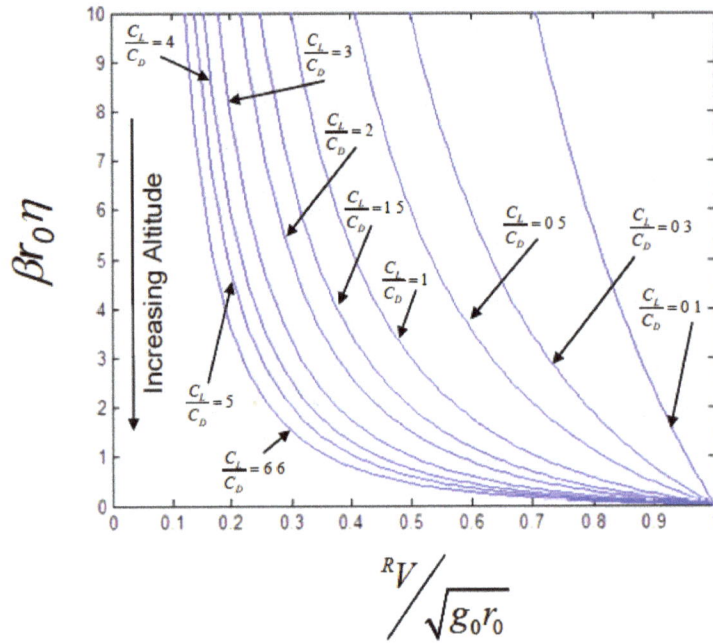

Figure 5-3: Altitude/Velocity Relationship for Shallow, Gliding Entry

relative to the drag force and *not* on it being close to zero!) When differentiated, Eq. (5.9) gives another expression for $\frac{dT}{d\eta}$:

$$\frac{dT}{d\eta} = -2\left(\frac{C_L}{C_D}\right)\beta r_0 T^2$$

5.13

Combining Eqs. (5.12) and (5.13) lets us solve for the flight-path angle:

$$\sin\gamma = \frac{-1}{\left(\dfrac{C_L}{C_D}\right)\beta r_0 T}$$

5.14

Or, using the small angle approximation, we can also write:

$$\gamma = \frac{-1}{\left(\dfrac{C_L}{C_D}\right)\beta r_0 T} \qquad\qquad \textbf{5.15}$$

Note that, while we assumed the flight-path angle was small and its cosine almost constant, we did not (in this case) assume γ itself was constant! Equations (5.14) and (5.15) are consistent with those assumptions. Next, substitute the definition of our kinetic energy variable T into the latter of these expressions:

$$\gamma = \frac{-2g_0}{\left(\dfrac{C_L}{C_D}\right)\beta\,^R V^2} \qquad\qquad \textbf{5.16}$$

From this equation, we can see that larger lift values allow a more shallow entry (smaller magnitude of γ). Rearranging this slightly to introduce the same non-dimensional velocity term we used in Figure 5-2 and Figure 5-3

$$\gamma = \frac{-2}{\left(\dfrac{C_L}{C_D}\right)\beta r_0}\left(\frac{\sqrt{g_0 r_0}}{^R V}\right)^2 \qquad\qquad \textbf{5.17}$$

we can see the flight-path angle tends to become steeper over the course of the glide (if C_L/C_D remains unchanged). Figure 5-4 shows the relationship in Eq. (5.17) for various vehicles entering an Earth-like atmosphere. When considering Eq. (5.17) and Figure 5-4, bear in mind that the assumption of a small flight-path angle begins to breakdown as the angle increases.

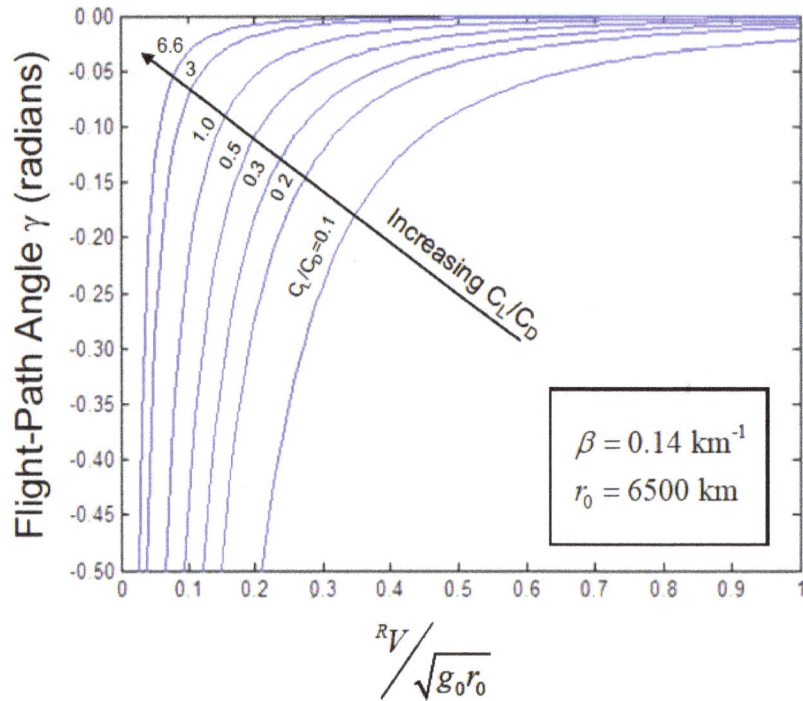

***Figure 5-4: Flight-Path Angle/Velocity Relationship for Shallow, Gliding
Entry***

We can also develop expressions for the distance covered during the glide.
To do this, we start with the expression for the velocity along the trajectory given
by Eq. (4.24). Ignoring the gravity force along the flight path (relative to the drag
force), this becomes:

$$\frac{d\,^R V}{dt} = -\frac{\rho C_D S\,^R V^2}{2m} \tag{5.18}$$

Or, using our dimensionless altitude variable η:

$$\frac{d\,^R V}{dt} = -\beta \eta\,^R V^2 \tag{5.19}$$

If s is the arc length traveled since the initial time, then we can also write:

$$\frac{ds}{dt} = {}^{R}V \qquad\qquad \textbf{5.20}$$

Hence, after dividing Eq. (5.20) by Eq. (5.19), we can relate velocity to arc-length:

$$\frac{ds}{d\,{}^{R}V} = -\frac{{}^{R}V}{\beta\eta\,{}^{R}V^2} \qquad\qquad \textbf{5.21}$$

We can eliminate the product $\beta\eta$ from this relationship by substituting in our solution from Eq. (5.10) and simplifying:

$$\frac{ds}{d\,{}^{R}V} = r_0\left(\frac{C_L}{C_D}\right)\frac{{}^{R}V}{\left({}^{R}V^2 - g_0 r_0\right)} \qquad\qquad \textbf{5.22}$$

For reasons which will be explained shortly, we will designate the entry speed as ${}^{R}V_e$ and integrate Eq. (5.22) between ${}^{R}V_e$ and to ${}^{R}V$ to get the total arc-length during the glide:

$$s = \frac{r_0}{2}\left(\frac{C_L}{C_D}\right)\ln\left(\frac{{}^{R}V^2 - g_0 r_0}{{}^{R}V_e^2 - g_0 r_0}\right) \qquad\qquad \textbf{5.23}$$

Or, grouping terms into non-dimensional elements, this can be written as:

$$\frac{s}{r_0} = \frac{1}{2}\left(\frac{C_L}{C_D}\right)\ln\left(\frac{1 - \left(\dfrac{{}^{R}V}{\sqrt{g_0 r_0}}\right)^2}{1 - \left(\dfrac{{}^{R}V_e}{\sqrt{g_0 r_0}}\right)^2}\right) \qquad\qquad \textbf{5.24}$$

For this statement to be valid, $^{R}V_e \neq \sqrt{g_0 r_0}$, which is the reason we specifically called out the entry speed as $^{R}V_e$ instead of simply $\sqrt{g_0 r_0}$. Once again, for simplicity, we will restrict ourselves to the situations where $^{R}V_e < \sqrt{g_0 r_0}$. Further, for Eq. (5.24) to give non-negative values for the arc-length, $^{R}V \leq ^{R}V_e$, which means the velocity decreases from entry speed to some final velocity.

Equation (5.24) confirms what you might have already expected – the range can be maximized by using the largest lift-to-drag ratio available. A boundary on that maximum can be found by *artificially assuming* the velocity can decrease until it reaches zero. This "limiting" distance is given by

$$\frac{s_{\text{limit}}}{r_0} = \frac{1}{2}\left(\frac{C_L}{C_D}\right) \ln\left(\frac{1}{1 - \left(\dfrac{^{R}V_e}{\sqrt{g_0 r_0}}\right)^2}\right) \qquad \textbf{5.25}$$

and shown in Figure 5-5 for various lift-to-drag ratios and entry velocities.

In a similar manner, Eq. (5.10) can be used in Eq. (5.19) to get it into a form that can be integrated to give flight time:

$$dt = -r_0\left(\frac{C_L}{C_D}\right)\frac{d\,^{R}V}{g_0 r_0 - ^{R}V^2} \qquad \textbf{5.26}$$

After consulting an integral table (or a math graduate student), this can be integrated from the entry to the point of interest:

$$t = \frac{r_0}{2\sqrt{g_0 r_0}}\left(\frac{C_L}{C_D}\right)\ln\left(\frac{\sqrt{g_0 r_0} - V}{\sqrt{g_0 r_0} + V}\right)\Bigg|_{V=^{R}V_e}^{V=^{R}V} \qquad \text{for } V < \sqrt{g_0 r_0} \qquad \textbf{5.27}$$

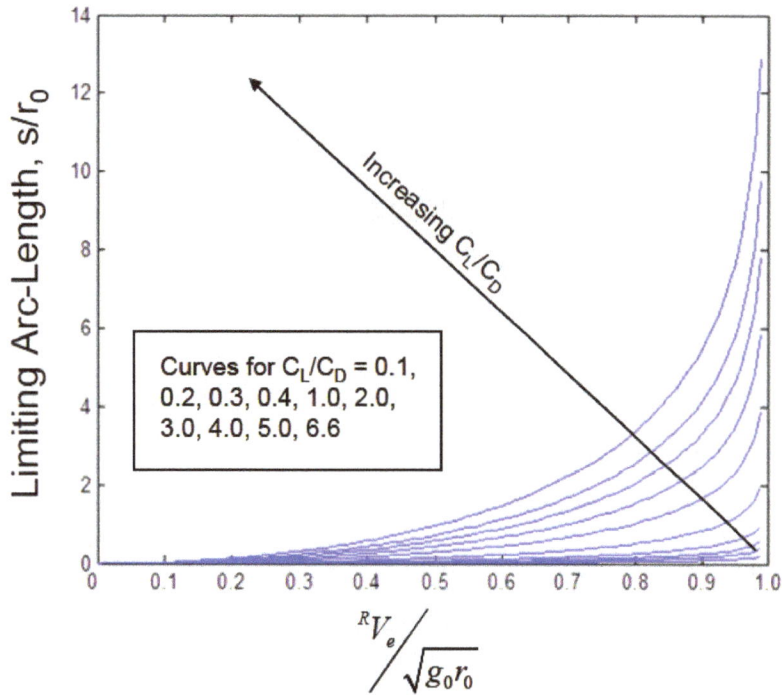

Figure 5-5: Limiting Arc-Length for Shallow, Gliding Entry

Note that the condition on the integration, $V < \sqrt{g_0 r_0}$, is consistent with that in the arc-length calculations. Substituting in the limits on the integration and simplifying the expression:

$$\sqrt{\frac{g_0}{r_0}}t = \frac{1}{2}\left(\frac{C_L}{C_D}\right)\ln\left[\left(\frac{1-\dfrac{^R V}{\sqrt{g_0 r_0}}}{1+\dfrac{^R V}{\sqrt{g_0 r_0}}}\right)\left(\frac{1+\dfrac{^R V_e}{\sqrt{g_0 r_0}}}{1-\dfrac{^R V_e}{\sqrt{g_0 r_0}}}\right)\right] \qquad \textbf{5.28}$$

Taking a final velocity of $^RV = 0$ yields a maximum (limiting) flight time to go along with the limiting arc-length in Eq. (5.25):

$$\sqrt{\frac{g_0}{r_0}}t_{\text{limit}} = \frac{1}{2}\left(\frac{C_L}{C_D}\right)\ln\left[\left(\frac{1+\dfrac{^RV_e}{\sqrt{g_0 r_0}}}{1-\dfrac{^RV_e}{\sqrt{g_0 r_0}}}\right)\right]$$

5.29

Figure 5-6 shows the (non-dimensional) limiting flight time for various lift-to-drag ratios across a range of entry velocities.

A physical parameter of interest to designers is the maximum deceleration the vehicle experiences. The deceleration for this type of entry (a near spiral with $|\dot{\gamma}| \ll 1$) is predominantly along the tangential direction, so we can approximate

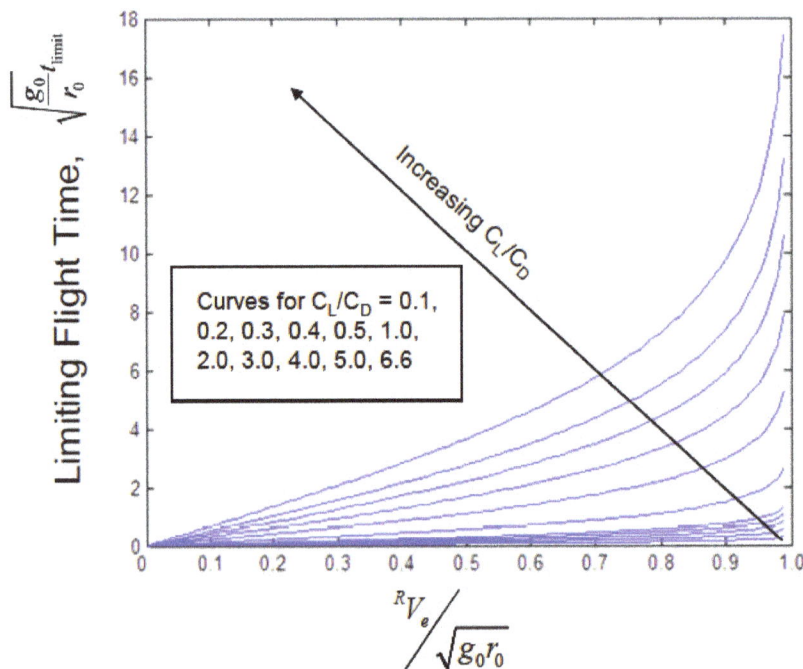

Figure 5-6: Limiting Flight Time for Shallow, Gliding Entry

the total deceleration as simply $a_{decel} \approx \left(a_{decel}\right)_v = -\dfrac{d\,^R V}{dt}$. Using Eq. (5.19), we

quickly get a non-dimensional expression in terms of altitude and velocity:

$$\frac{a_{decel}}{g_0} = \frac{-\dfrac{d\,^R V}{dt}}{g_0} = \frac{\beta\eta\,^R V^2}{g_0} \qquad\qquad \textbf{5.30}$$

This could also have been obtained directly from Eq. (4.59) after noting $\cos\gamma \approx 1$.
Rewriting the right-hand-side slightly:

$$\frac{a_{decel}}{g_0} = \left(\beta r_0 \eta\right)\frac{^R V^2}{g_0 r_0} \qquad\qquad \textbf{5.31}$$

We can get an expression for $\beta r_0 \eta$ from Eq. (5.10)

$$\beta r_0 \eta = \frac{\dfrac{g_0 r_0}{^R V^2} - 1}{\left(\dfrac{C_L}{C_D}\right)} \qquad\qquad \textbf{5.32}$$

and use it in Eq. (5.31) to express the deceleration in terms of velocity only

$$\frac{a_{decel}}{g_0} = \frac{1 - \dfrac{^R V^2}{g_0 r_0}}{\left(\dfrac{C_L}{C_D}\right)} \qquad\qquad \textbf{5.33}$$

or kinetic energy only:

$$\frac{a_{decel}}{g_0} = \frac{1 - 2T}{\dfrac{C_L}{C_D}} \qquad\qquad \textbf{5.34}$$

Both equations show the surprising result that deceleration becomes *larger* as the vehicle *slows* and it continues to become larger throughout the trajectory. The equation also shows lift reduces the maximum deceleration. Figure 5-7 graphically illustrates these observations by plotting Eq. (5.33).

5.3 Medium and Steep Gliding Entry at Near Circular Speed

Section 5.2 looked at the quasi-equilibrium case where lift was sufficient to hold the flight-path angle small and $\cos \gamma \approx$ constant . Lees, et al. examined the case for the situation where neither of these assumptions is valid; specifically, they examined a first-order solution for gliding entry at medium and large flight-path angles (36:633-641).

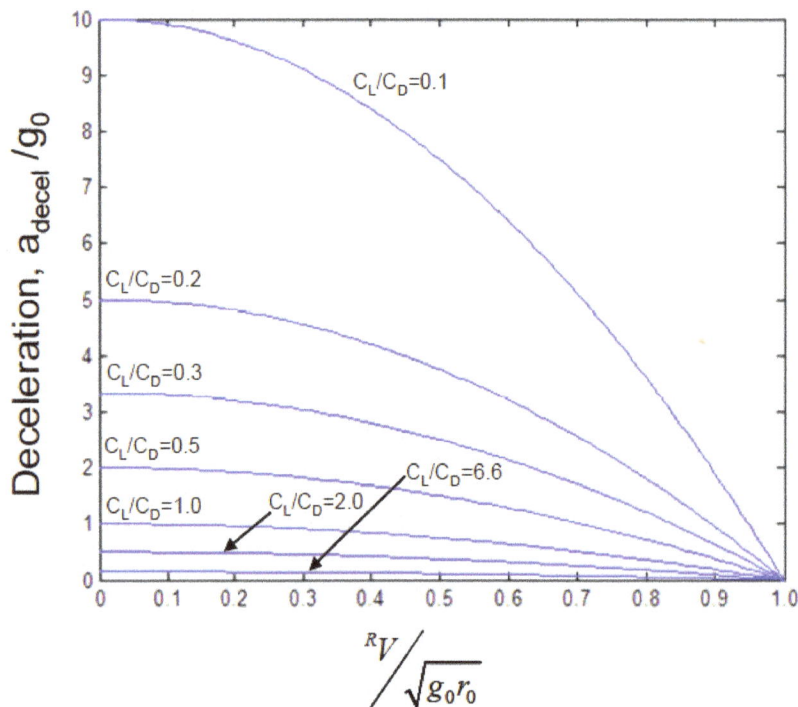

Figure 5-7: Deceleration for Shallow, Gliding Entry

In this case, we can assume the lift force is the dominant term in Eq. (5.2):

$$\frac{d\xi}{d\eta} \approx \frac{C_L}{C_D} \qquad\qquad \textbf{5.35}$$

This is equivalent to saying the difference between the gravity force and the centrifugal force is small compared to the lift. More precisely,

$$\left| \frac{1}{\beta r_0 \eta} \left(1 - \frac{1}{2T}\right) \cos\gamma \right| < \left| \frac{1}{\beta r_0 \eta} \left(1 - \frac{1}{2T}\right) \right| \ll \frac{C_L}{C_D} \qquad\qquad \textbf{5.36}$$

which is the same as saying the velocity is near circular orbital speed ($T \approx \frac{1}{2}$) *and/or* the term $\dfrac{1}{\beta r_0 \eta}$ is small. When $\dfrac{1}{\beta r_0 \eta}$ is small, Eq. (5.35) remains valid even if the velocity is not "as near" to the equivalent circular orbital speed.

Separating the variables and integrating Eq. (5.35) from the entry conditions (denoted by an "*e*" subscript) forward, this becomes:

$$\cos\gamma - \cos\gamma_e = \frac{C_L}{C_D}\left(\eta - \eta_e\right) \qquad\qquad \textbf{5.37}$$

One way of plotting this equation is shown in Figure 5-8. Equation (5.37) contains a few subtleties that should be mentioned. First, since η increases as the vehicle enters the atmosphere, the right-hand side is non-negative for positive lift-to-drag ratios. This means the flight-path angle becomes less steep (less negative) as the vehicle moves along this trajectory! Second, because of the realistic range for flight-path angles on left-hand side, $0 \leq \dfrac{C_L}{C_D}\left(\eta - \eta_e\right) \leq 1$ at all points during the entry. Figure 5-9 illustrates the relationship in terms of physical (dimensional) parameters for two vehicles with the same entry conditions but different lift-to-drag ratios.

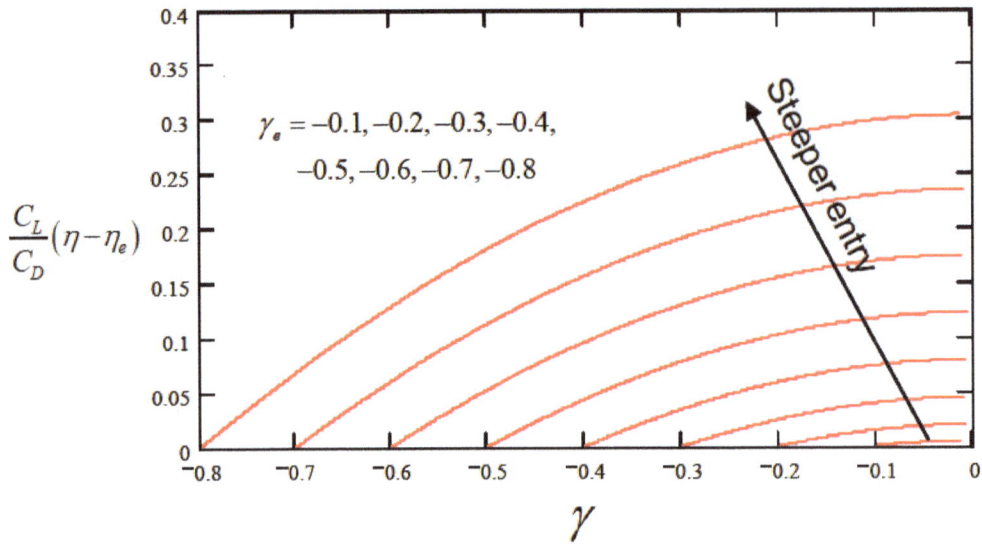

Figure 5-8: Flight-Path Angle/Altitude Relationship for Medium/Steep Gliding Entry

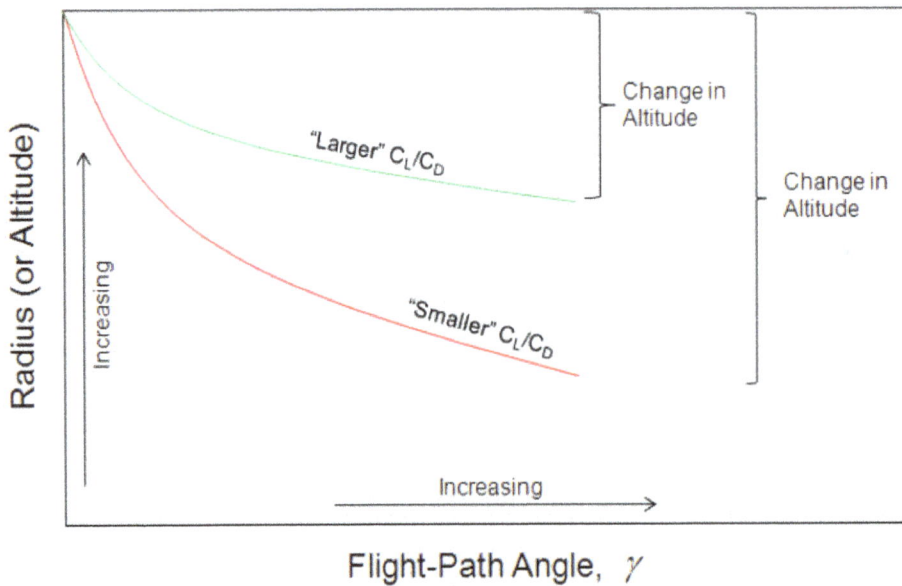

Figure 5-9: Physical Interpretation of Flight-Path Angle/Altitude Relationship for Medium/Steep Gliding Entry

If drag is assumed to be the dominant term in Eq. (5.1), we can write:

$$\frac{dT}{d\eta} \approx \frac{2T}{\sin\gamma}$$

5.38

Equation (5.35) becomes

$$\frac{d\xi}{d\eta} = \frac{d(\cos\gamma)}{d\eta} = -\sin\gamma\frac{d\gamma}{d\eta} = \frac{C_L}{C_D}$$

5.39

which, when combined with Eq. (5.38), eliminates the dimensionless altitude η:

$$\frac{dT}{T} = -\frac{2}{\left(\dfrac{C_L}{C_D}\right)}d\gamma$$

5.40

Again integrating forward from entry conditions, we find the relationship between kinetic energy and flight-path angle:

$$\frac{T}{T_e} = \exp\left[-\frac{2(\gamma-\gamma_e)}{\left(\dfrac{C_L}{C_D}\right)}\right]$$

5.41

Figure 5-10 illustrates this relationship. Equations (5.37) and (5.41) are the first-order solution of Lees et al. for gliding entry at medium positive lift-to-drag ratio and medium flight-path angle (36; 58:113-114).

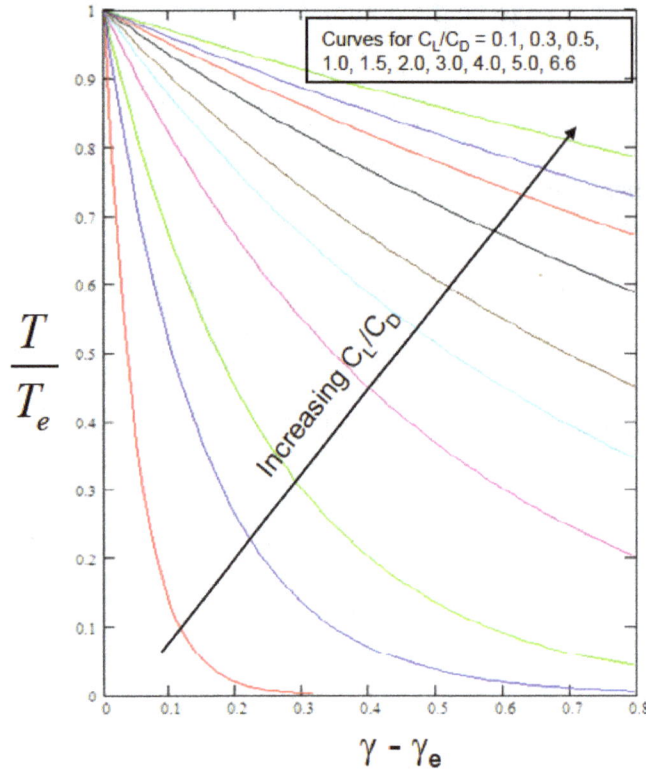

Figure 5-10: Kinetic Energy/Flight-Path Angle Relationship for Medium/Steep Gliding Entry

To examine the deceleration for this type of entry, we begin with Eqs. (4.59) and (4.60):

$$\left(a_{decel}\right)_v = 2\beta r_0 g_0 \eta T \pm g_0 \sqrt{1-\xi^2} \qquad \textbf{5.42}$$

$$\left(a_{decel}\right)_L = -2\beta r_0 g_0 \eta T\left(\frac{C_L}{C_D}\right) + g_0\left(1-2T\right)\xi \qquad \textbf{5.43}$$

For most planets of interest, $\beta r_0 \gg 1$. Combined with our assumption that $T \approx 1/2$, these deceleration components are approximated by

$$\left(a_{decel}\right)_v = 2\beta r_0 g_0 \eta T \qquad \textbf{5.44}$$

and:

$$\left(a_{decel}\right)_L = -2\beta r_0 g_0 \eta T \left(\frac{C_L}{C_D}\right) \qquad \textbf{5.45}$$

It is often assumed $\left|\left(a_{decel}\right)_v\right| \gg \left|\left(a_{decel}\right)_L\right|$ for these types of entries (58:113). However, when comparing the components

$$\left|\frac{\left(a_{decel}\right)_v}{\left(a_{decel}\right)_L}\right| = \frac{C_D}{C_L} \qquad \textbf{5.46}$$

the assumption doesn't always seem justified. It's not much more difficult to avoid making such an assumption by working with the total deceleration magnitude:

$$a_{decel} = \sqrt{\left(a_{decel}\right)_v^2 + \left(a_{decel}\right)_L^2} = 2\beta r_0 g_0 \eta T \sqrt{1 + \left(\frac{C_L}{C_D}\right)^2} \qquad \textbf{5.47}$$

Notice that both components and the total deceleration are proportional, so they experience their respective maxima occur at the same point in the trajectory. Substituting our solutions for η and T from Eqs. (5.37) and (5.41) into Eq. (4.40), we can write an expression for the total deceleration:

$$\left(\frac{a_{decel}}{g_0}\right) = 2\beta r_0 T_e \sqrt{1 + \left(\frac{C_L}{C_D}\right)^2} \left[\frac{\cos\gamma - \cos\gamma_e}{\left(\frac{C_L}{C_D}\right)} + \eta_e\right] \exp\left[-\frac{2(\gamma - \gamma_e)}{\left(\frac{C_L}{C_D}\right)}\right] \qquad \textbf{5.48}$$

Recall $\eta \to 0$ as $h \to \infty$, so it is reasonable to set $\eta_e \approx 0$ for high-altitude entry and rewrite this slightly simpler as:

$$\left(\frac{a_{decel}}{g_0}\right) = \frac{2\beta r_0 T_e \sqrt{1+\left(\frac{C_L}{C_D}\right)^2}\left(\cos\gamma - \cos\gamma_e\right)}{\left(\frac{C_L}{C_D}\right)} \exp\left[-\frac{2\left(\gamma-\gamma_e\right)}{\left(\frac{C_L}{C_D}\right)}\right] \qquad 5.49$$

The point of maximum deceleration occurs when

$$\frac{d\left(\frac{a_{decel}}{g_0}\right)}{d\gamma} = 0 \qquad 5.50$$

Or, after substituting Eq. (5.49) and simplifying

$$\frac{1}{2}\left(\frac{C_L}{C_D}\right)\sin\gamma_* = -\left(\cos\gamma_* - \cos\gamma_e\right) \qquad 5.51$$

where the γ_* signifies the flight-path angle at which the maximum deceleration occurs. (In general, we will denote the "critical" conditions – flight-path angle, altitude, velocity, etc. – corresponding to a maximum or minimum with an asterisk subscript.) Equation (5.51) can be solved analytically for $\sin\gamma_*$.

Start by writing Eq. (5.51) as

$$k\sin\gamma_* - \cos\gamma_e = -\cos\gamma_* \qquad 5.52$$

where

$$k = \frac{1}{2}\left(\frac{C_L}{C_D}\right) \qquad 5.53$$

has been used to simplify the expression we are about to develop. Squaring both sides and simplifying:

$$\left(k^2+1\right)\sin^2\gamma_*+\left(-2k\cos\gamma_e\right)\sin\gamma_*+\left(-\sin^2\gamma_e\right)=0 \qquad \textbf{5.54}$$

This is a quadratic equation in $\sin\gamma_*$. Equation (5.54) has two roots, given by:

$$\sin\gamma_*=\frac{k\cos\gamma_e\pm\sqrt{k^2+\sin^2\gamma_e}}{k^2+1} \qquad \textbf{5.55}$$

To decide which root (or both or neither) is physically realistic, consider that

$$\sin\gamma_*<0 \qquad \textbf{5.56}$$

$$\cos\gamma_e\geq0 \qquad \textbf{5.57}$$

must be true for atmospheric entry. Given the condition in Eq. (5.56), then:

$$k\cos\gamma_e\pm\sqrt{k^2+\sin^2\gamma_e}<0 \qquad \textbf{5.58}$$

For positive lift, $k\cos\gamma_e\geq0$ and the only *possible* way for Eq. (5.58) to be valid is if:

$$k\cos\gamma_e-\sqrt{k^2+\sin^2\gamma_e}<0 \qquad \textbf{5.59}$$

In-other-words, the "+" root cannot yield a result for $\sin\gamma_*<0$ when $\frac{C_L}{C_D}>0$.

Equation (5.59) deserves a little more study. Are there physically realistic values of γ_e for which it would not be satisfied? Rewrite the inequality in Eq. (5.59) as:

$$k\cos\gamma_e<\sqrt{k^2+\sin^2\gamma_e} \qquad \textbf{5.60}$$

Squaring both sides and simplifying, we see

$$-\frac{1}{4}\left(\frac{C_L}{C_D}\right)^2 < 1 \qquad \textbf{5.61}$$

must be true for a valid solution to Eq. (5.55). Since this inequality is always true, we do not have to be concerned that any particular positive lift-to-drag ratio will cause an erroneous solution to Eq. (5.54). (This broad statement assumes the earlier simplifications leading to Eq. (5.54) are valid, of course.) Thus, we find the flight-path angle at the point of maximum deceleration is:

$$\sin \gamma_* = \frac{\dfrac{1}{2}\left(\dfrac{C_L}{C_D}\right)\cos \gamma_e - \sqrt{\dfrac{1}{4}\left(\dfrac{C_L}{C_D}\right)^2 + \sin^2 \gamma_e}}{\dfrac{1}{4}\left(\dfrac{C_L}{C_D}\right)^2 + 1} \qquad \textbf{5.62}$$

Figure 5-11 shows the nearly linear relationship between γ_* and γ_e for a given lift-to-drag ratio.

The altitude at which the maximum deceleration occurs can be found by first evaluating Eq. (5.37) at the corresponding flight-path angle

$$\cos \gamma_* - \cos \gamma_e = \frac{C_L}{C_D}\left(\eta_* - \eta_e\right) \qquad \textbf{5.63}$$

and then using Eq. (5.51) to eliminate the cosine terms:

$$\eta_* - \eta_e = -\frac{\sin \gamma_*}{2} \qquad \textbf{5.64}$$

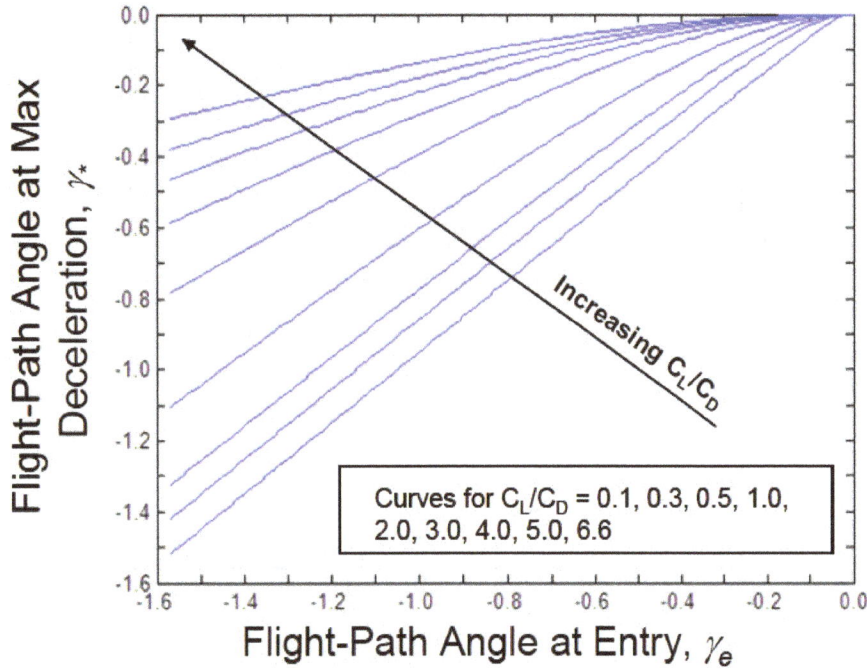

Figure 5-11: Relationship Between Entry and Critical Flight-Path Angles

Or, using Eq. (5.62), this can be written in terms of the entry flight-path angle and the lift-to-drag ratio:

$$\eta_* - \eta_e = \frac{-\dfrac{1}{2}\left(\dfrac{C_L}{C_D}\right)\cos\gamma_e + \sqrt{\dfrac{1}{4}\left(\dfrac{C_L}{C_D}\right)^2 + \sin^2\gamma_e}}{2\left[\dfrac{1}{4}\left(\dfrac{C_L}{C_D}\right)^2 + 1\right]} \qquad \textbf{5.65}$$

In Figure 5-12 we can see vehicles with a larger lift-to-drag ratio experience their maximum deceleration at a higher altitude than those with a lower lift-to-drag ratio. Bear in mind, however, this says *nothing* about the relative magnitudes of

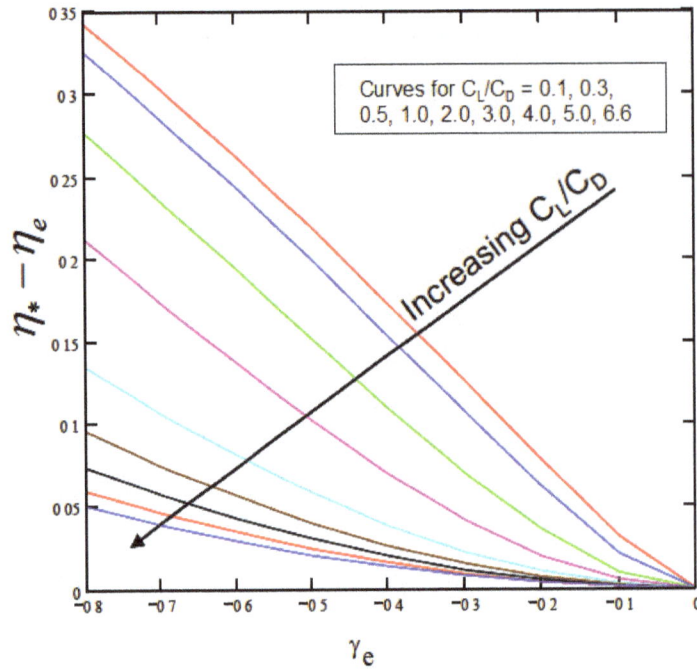

Figure 5-12: Altitude Change Between Entry and Maximum Deceleration

their corresponding maximum decelerations! Although we may suspect the deceleration to be less at higher altitudes, we haven't found equations proving it. The speed at which maximum deceleration occurs can be found from Eq. (5.41) once γ_* is known (from Eq. (5.62)):

$$\frac{^{R}V_*}{^{R}V_e} = \exp\left[-\frac{(\gamma_* - \gamma_e)}{\left(\dfrac{C_L}{C_D}\right)}\right] \qquad \textbf{5.66}$$

The ratio $\dfrac{^{R}V_*}{^{R}V_e}$ is shown as a function of γ_e in Figure 5-13. From the equation (and the figure), we can see that more lift results in less reduction in

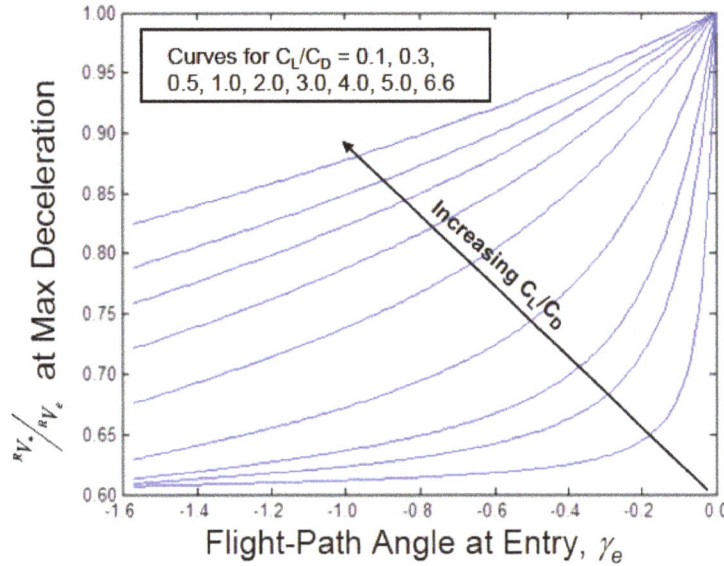

Figure 5-13: Entry Flight-Path Angle/Velocity Relationship at Maximum Deceleration

speed from entry to the point of maximum deceleration. Additionally, the figure demonstrates the dramatic difference between how the equation behaves at shallow entry and steep entry angles for low-lift/high-drag vehicles.

The maximum deceleration can also be found by substituting the appropriate terms from Eqs. (5.41) and (5.64) into Eq. (5.48):

$$
\left(\frac{a_{decel}}{g_0}\right)_{max} = \left(\frac{a_{decel}}{g_0}\right)_* = \frac{2\beta r_0 g_0 \eta_* T_*}{g_0}\sqrt{1+\left(\frac{C_L}{C_D}\right)^2}
$$

$$
= 2\beta r_0 T_e \sqrt{1+\left(\frac{C_L}{C_D}\right)^2}\left(-\frac{\sin\gamma_*}{2}+\eta_e\right)\exp\left[-\frac{2(\gamma_*-\gamma_e)}{\left(\frac{C_L}{C_D}\right)}\right] \qquad \textbf{5.67}
$$

Since the components experience their respective maxima at the same time, they can be evaluated using Eqs. (5.44) and (5.45)

$$\left[\frac{(a_{decel})_v}{g_0}\right]_* = 2\beta r_0 T_e \left(-\frac{\sin \gamma_*}{2} + \eta_e\right) \exp\left[-\frac{2(\gamma_* - \gamma_e)}{\left(\frac{C_L}{C_D}\right)}\right] \qquad \textbf{5.68}$$

$$\left[\frac{(a_{decel})_L}{g_0}\right]_* = -2\beta r_0 T_e \left(\frac{C_L}{C_D}\right)\left(-\frac{\sin \gamma_*}{2} + \eta_e\right) \exp\left[-\frac{2(\gamma_* - \gamma_e)}{\left(\frac{C_L}{C_D}\right)}\right] \qquad \textbf{5.69}$$

where the value for γ_* in Eqs. (5.67) – (5.69) is the same numerical value and is found by solving Eq. (5.62).

For the case where the initial altitude is sufficiently high ($\eta_e \approx 0$), Eq. (5.67) can be simplified as

$$\left(\frac{a_{decel}}{g_0}\right)_* \approx -\beta r_0 T_e \sin \gamma_* \sqrt{1 + \left(\frac{C_L}{C_D}\right)^2} \exp\left[-\frac{2(\gamma_* - \gamma_e)}{\left(\frac{C_L}{C_D}\right)}\right] \qquad \textbf{5.70}$$

and plotted in Figure 5-14 for entry into Earth-like atmospheres. Note how large the total deceleration can be at steep entry angles. For completeness, the components are shown in Figure 5-15 and Figure 5-16. Between the three plots, it is obvious why the normal component of deceleration cannot be ignored in all situations.

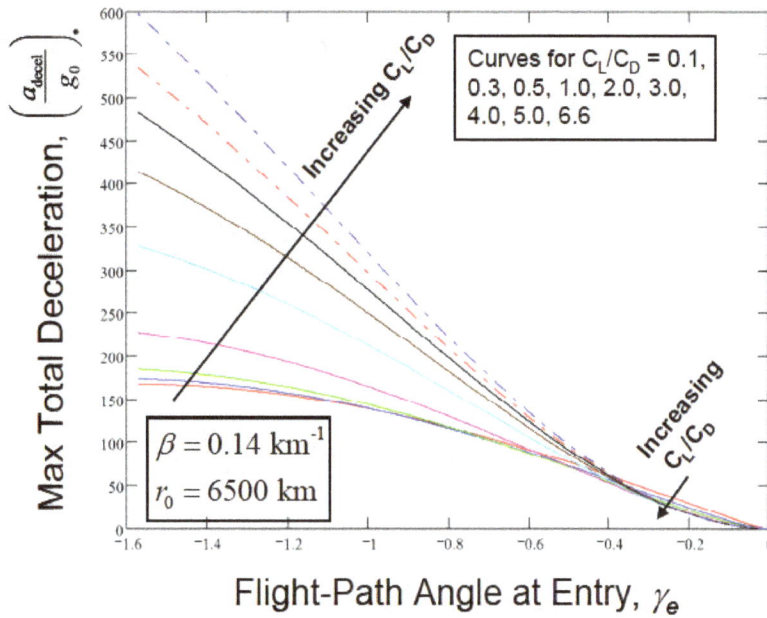

Figure 5-14: Relationship Between Entry Flight-Path Angle and Maximum Total Deceleration, $T_e=1/2$

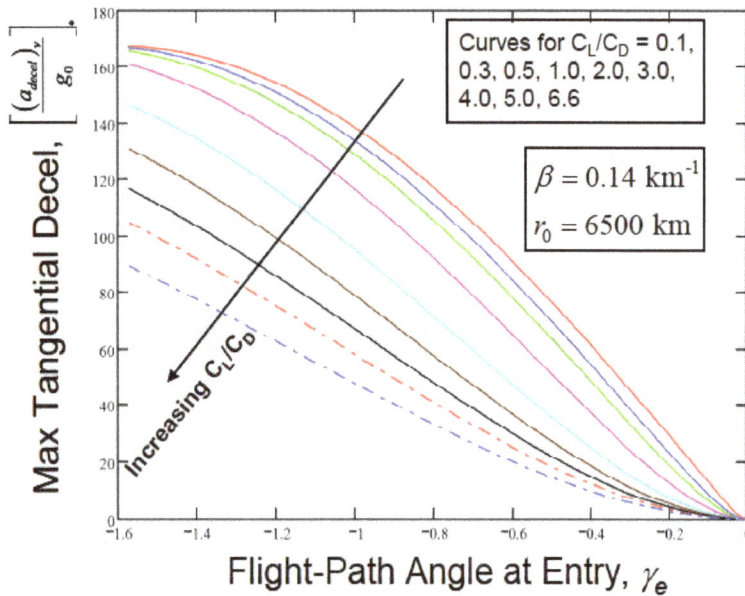

Figure 5-15: Relationship Between Entry Flight-Path Angle and Maximum Tangential Deceleration, $T_e=1/2$

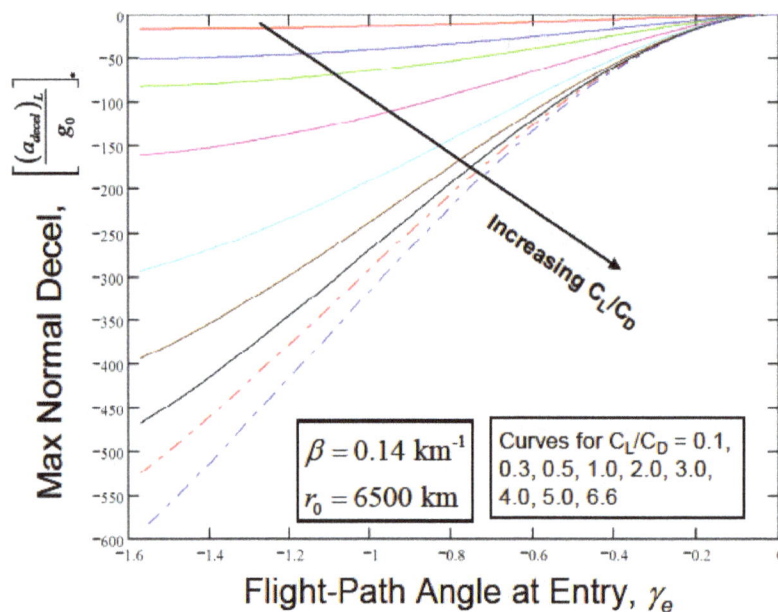

Figure 5-16: Relationship Between Entry Flight-Path Angle and Maximum Normal Deceleration, $T_e=1/2$

To be consistent with the assumption of near circular speeds, T_e should be replaced with $T_e \approx 1/2$ throughout this section; however, we avoided the replacement because we will "reuse" these equations in a later section where $T_e \neq 1/2$. Finally, it should be noted that when Lees, et al. presented their results, they examined entry angles to around 15° (0.26 radians). However, nothing in the equations derived specifically limits the range on γ (as long as the other assumptions are valid). The relations found in this section can be applied to medium and large entry angles (where "medium" and "large" are somewhat subjective).

5.4 Medium Gliding Entry at Supercircular Speed

The equations presented in Section 5.3 were valid when the entry speed was near the equivalent circular speed at the entry radius. (This limitation was due, in large part, to the assumptions called out in Eq. (5.36).) Wang and Ting presented an "approximate" analytic solution for steep gliding entry without this restriction on the entry speed (64:565-566). However, in their derivations, they introduced the restriction that the flight-path angle is "small enough." In this section, we will investigate their approach to an approximate analytic solution.

During the initial portion of the entry trajectory, there is a period where the velocity does not change significantly, particularly for at medium flight-path angles (for example, in Figure 5-13). Wang and Ting used this observation to replace the (non-constant) bracketed term in Eq. (5.2), with a constant value evaluated at entry conditions

$$\left(1 - \frac{1}{2T}\right)\xi \approx \left(1 - \frac{1}{2T_e}\right) \qquad \textbf{5.71}$$

where the flight-path angle has been assumed to be small enough to simultaneously replace $\xi = \cos\gamma$ with unity. Thus, Eq. (5.2) becomes

$$d\xi = \left[\frac{C_L}{C_D} + \frac{1}{\beta r_0 \eta}\left(1 - \frac{1}{2T_e}\right)\right]d\eta \qquad \textbf{5.72}$$

and can be easily integrated from the entry point forward:

$$\xi - \xi_e = \cos\gamma - \cos\gamma_e = \frac{C_L}{C_D}(\eta - \eta_e) + \frac{1}{\beta r_0}\left(1 - \frac{1}{2T_e}\right)\ln\left(\frac{\eta}{\eta_e}\right) \qquad \textbf{5.73}$$

If the cosine terms are replaced by their series approximations (and truncated to the quadratic terms), the following relationship is found:

$$\gamma^2 = \gamma_e^2 - 2\left(\frac{C_L}{C_D}\right)(\eta - \eta_e) - \frac{2}{\beta r_0}\left(1 - \frac{1}{2T_e}\right)\ln\left(\frac{\eta}{\eta_e}\right) \qquad \textbf{5.74}$$

Retaining the expansions through the quadratic terms is one way of allowing the range of γ to be somewhat larger while still assuming $\xi \approx 1$. (Also, the assumption on ξ was "bundled" with the "approximately constant velocity" assumption.) For entry trajectories, we want the negative root of Eq. (5.74):

$$\gamma = -\left[\gamma_e^2 - 2\left(\frac{C_L}{C_D}\right)(\eta - \eta_e) - \frac{2}{\beta r_0}\left(1 - \frac{1}{2T_e}\right)\ln\left(\frac{\eta}{\eta_e}\right)\right]^{\frac{1}{2}} \qquad \textbf{5.75}$$

Taking Eq. (5.1) and, as in the other planar approaches, assuming the drag term is dominant, we have:

$$\frac{dT}{d\eta} = \frac{2T}{\sin\gamma} \qquad \textbf{5.76}$$

Replacing the sine term with a series expansion (through the quadratic term), lets us simplify this as:

$$\frac{dT}{d\eta} = \frac{2T}{\gamma} \qquad \textbf{5.77}$$

After substituting our relationship for γ from Eq. (5.75), we get a differential equation which can be separated for integration:

$$\frac{dT}{T} = \frac{d\eta}{-\frac{1}{2}\left[\gamma_e^2 - 2\left(\frac{C_L}{C_D}\right)(\eta - \eta_e) - \frac{2}{\beta r_0}\left(1 - \frac{1}{2T_e}\right)\ln\left(\frac{\eta}{\eta_e}\right)\right]^{\frac{1}{2}}} \qquad \textbf{5.78}$$

Unfortunately, Eq. (5.78) cannot be solved in closed form. Various approaches can be taken to approximate the solution. Wang and Ting expanded the denominator out in terms of $\left(\dfrac{\eta}{\eta_e}\right)$ and integrated from the entry point to the point of maximum deceleration. Vinh, et al. combined Eqs. (5.72) and (5.76) to get

$$\frac{dT}{T} = -\frac{d\gamma}{\dfrac{1}{2}\left[\dfrac{C_L}{C_D} + \dfrac{1}{\beta r_0 \eta}\left(1 - \dfrac{1}{2T_e}\right)\right]} \qquad \textbf{5.79}$$

and then assumed the denominator was approximately constant for integration purposes (58:115-116). They found:

$$\ln\left(\frac{T}{T_e}\right) = \frac{\gamma_e - \gamma}{\dfrac{1}{2}\left[\dfrac{C_L}{C_D} + \dfrac{1}{\beta r_0 \eta}\left(1 - \dfrac{1}{2T_e}\right)\right]} \qquad \textbf{5.80}$$

The relative accuracy of this assumption is left as an exercise. Equations (5.75) and (5.80) are Wang and Ting's first-order solution for entry at supercircular speeds (58:116).

For the moment, we can avoid choosing between the two approaches for integrating $\dfrac{dT}{T}$ and still find an expression describing when the maximum deceleration occurs. Once again, we can ignore the gravity terms in Eqs. (5.42) and (5.43) and write an expression for the total deceleration as:

$$\left(\frac{a_{decel}}{g_0}\right) = 2\beta r_0 \eta T \sqrt{1 + \left(\frac{C_L}{C_D}\right)^2} \qquad \textbf{5.81}$$

This is a maximum at the altitude satisfied by:

$$\frac{d\left(\dfrac{a_{decel}}{g_0}\right)}{d\eta} = 2\beta r_0 \sqrt{1+\left(\frac{C_L}{C_D}\right)^2}\left(T+\eta\frac{dT}{d\eta}\right) = 0 \qquad \textbf{5.82}$$

Replacing $\dfrac{dT}{d\eta}$ with the aid of Eq. (5.76) and simplifying:

$$T\left(1+\frac{2\eta}{\sin\gamma}\right) = 0 \qquad \textbf{5.83}$$

Since the kinetic energy T is not zero during the period of interest, we can divide it out, eliminating the unknown value/expression for T! Again denoting the critical point with an asterisk subscript, our condition for maximum deceleration then becomes:

$$\sin\gamma_* + 2\eta_* = 0 \qquad \textbf{5.84}$$

Recall, during the derivation of Eq. (5.75), series expansions out to quadratic terms were used. Consistent with that,

$$\sin\gamma_* \approx \gamma_* \qquad \textbf{5.85}$$

to the same order since the sine expansion does not have a quadratic term. Thus, we can simplify Eq. (5.84):

$$\gamma_* + 2\eta_* = 0 \qquad \textbf{5.86}$$

Evaluating Eq. (5.75) at the altitude η_* corresponding to γ_* and substituting it into this equation, we get a new equation of only one unknown, η_*:

$$-\left[\gamma_e^2 - 2\left(\frac{C_L}{C_D}\right)(\eta_* - \eta_e) - \frac{2}{\beta r_0}\left(1-\frac{1}{2T_e}\right)\ln\left(\frac{\eta_*}{\eta_e}\right)\right]^{\frac{1}{2}} + 2\eta_* = 0 \qquad \textbf{5.87}$$

This, as written, must also be solved numerically, but it is a simple algebraic relationship. We also cannot solve for the actual maximum deceleration yet because we do not have a relationship for T* without integrating Eq. (5.78). At this point, we can fall back on the approximation by Vinh, et al. to find the value of the maximum deceleration

$$\left(\frac{a_{decel}}{g_0}\right)_* = 2\beta r_0 \eta_* T_e \sqrt{1+\left(\frac{C_L}{C_D}\right)^2} \exp\left[\frac{\gamma_e - \gamma_*}{\frac{1}{2}\left[\frac{C_L}{C_D} + \frac{1}{\beta r_0 \eta_*}\left(1-\frac{1}{2T_e}\right)\right]}\right]$$

5.88

where the value for η_* is found from Eq. (5.87) and the corresponding γ_* from Eq. (5.75). Because the total and component decelerations differ only by a "scaling" constant, their maxima are, again, at the same point in the trajectory and can be easily written with the aid of Eqs. (5.42) and (5.43):

$$\left[\frac{(a_{decel})_v}{g_0}\right]_* = 2\beta r_0 \eta_* T_e \exp\left[\frac{\gamma_e - \gamma_*}{\frac{1}{2}\left[\frac{C_L}{C_D} + \frac{1}{\beta r_0 \eta_*}\left(1-\frac{1}{2T_e}\right)\right]}\right]$$

5.89

$$\left[\frac{(a_{decel})_L}{g_0}\right]_* = -2\beta r_0 \eta_* T_e \left(\frac{C_L}{C_D}\right) \exp\left[\frac{\gamma_e - \gamma_*}{\frac{1}{2}\left[\frac{C_L}{C_D} + \frac{1}{\beta r_0 \eta_*}\left(1-\frac{1}{2T_e}\right)\right]}\right]$$

5.90

5.5 Planar Skip Entry

A special case of lifting reentry is one in which the vehicle intentionally uses lift to pull back out of the atmosphere – a "skip." During a skip maneuver, a lifting vehicle enters the atmosphere (at hypervelocity), generates lift in the "up direction," and flies back out of the atmosphere as shown in Figure 5-17. (During the solution to some of the computational homework problems, you may already have encountered motion similar to this.) While this type of maneuver could be used to change the orbital plane, we will only consider a constant lift-to-drag ratio and motion confined to the entry plane. Skip maneuvers take place over a short range, so centrifugal motion can be ignored. Further, aerodynamic forces dominate the gravitational forces. With these assumptions, the equations of motion can be simplified from Eqs. (5.1) and (5.2) as:

$$\frac{dT}{d\eta} = \frac{2T}{\sin\gamma} \qquad\qquad \textbf{5.91}$$

$$\frac{d\xi}{d\eta} = \frac{C_L}{C_D} \qquad\qquad \textbf{5.92}$$

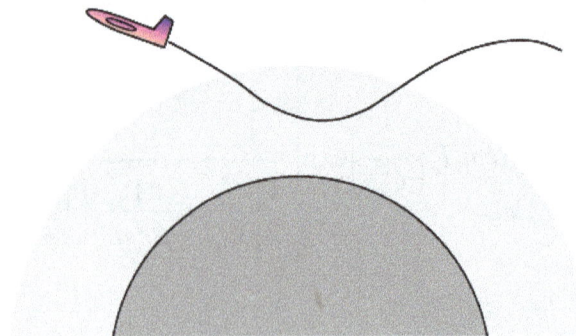

Figure 5-17: Planar Skip Entry

These equations are the same as we found in Section 5.3 for medium/steep gliding entry at near circular speeds, but the assumptions are slightly different. Perhaps most importantly, we did not restrict ourselves to near circular speeds in deriving Eq. (5.92).

Dividing Eq. (5.92) by Eq. (5.91) gives

$$\frac{d\xi}{dT} = \left(\frac{C_L}{C_D}\right)\left(\frac{\sin\gamma}{2T}\right)$$

5.93

which can be separated for integration after replacing $\xi = \cos\gamma$:

$$\frac{dT}{T} = \frac{-2}{\left(\dfrac{C_L}{C_D}\right)}d\gamma$$

5.94

Integrating from the entry point forward to some later point gives:

$$\frac{T}{T_e} = \exp\left[\frac{-2(\gamma - \gamma_e)}{\left(\dfrac{C_L}{C_D}\right)}\right]$$

5.95

In terms of velocity this can be expressed as:

$$\frac{{}^R V}{{}^R V_e} = \exp\left[\frac{-(\gamma - \gamma_e)}{\left(\dfrac{C_L}{C_D}\right)}\right]$$

5.96

Equation (5.92) can be integrated directly to find the flight-path angle as a function of altitude:

$$\cos\gamma - \cos\gamma_e = \frac{C_L}{C_D}(\eta - \eta_e)$$

5.97

Or, solving for the altitude:

$$\eta = \eta_e + \frac{\cos\gamma - \cos\gamma_e}{\left(\dfrac{C_L}{C_D}\right)}$$

5.98

For a "complete" skip (i.e., one that doesn't impact the planet), the final altitude is equal to the entry altitude:

$$\eta_f = \eta_e$$

5.99

By comparing Eqs. (5.98) and (5.99), the final flight-path angle can be easily found

$$\gamma_f = -\gamma_e$$

5.100

as can the final velocity:

$$\frac{{}^RV_f}{{}^RV_e} = \exp\left[\frac{2\gamma_e}{\left(\dfrac{C_L}{C_D}\right)}\right]$$

5.101

We can also develop an equation for the horizontal ("downrange") distance traveled during the skip. To do this, we start with the expression for the velocity along the trajectory given by Eq. (4.24) and again ignore the gravity force along the flight path (relative to the drag force):

$$\frac{d\,{}^{R}V}{dt} = -\frac{\rho C_D S\,{}^{R}V^2}{2m} \qquad \textbf{5.102}$$

(Note, ignoring this component of gravity is also what you would have with a "flat" planet.) Using our dimensionless altitude variable η, Eq. (5.102) becomes:

$$\frac{d\,{}^{R}V}{dt} = -\beta\eta\,{}^{R}V^2 \qquad \textbf{5.103}$$

If X is the horizontal distance traveled since the entry, then we can also write:

$$\frac{dX}{dt} = {}^{R}V\cos\gamma \qquad \textbf{5.104}$$

Hence, after dividing Eq. (5.104) by Eq. (5.103), we can relate velocity to horizontal distance by:

$$\frac{dX}{d\,{}^{R}V} = -\frac{\cos\gamma}{\beta\eta\,{}^{R}V} \qquad \textbf{5.105}$$

Equation (5.94) and the definition $T = \dfrac{{}^{R}V^2}{2g_0 r_0}$, can be used to change the independent variable to γ :

$$\frac{dX}{d\gamma} = \frac{\cos\gamma}{\beta\eta\left(\dfrac{C_L}{C_D}\right)} \qquad \textbf{5.106}$$

(Note, for this type of trajectory, γ is monotonically increasing from γ_e to γ_f, so it is a perfectly acceptable independent variable.) $\eta\left(\dfrac{C_L}{C_D}\right)$ can be replaced in terms of the flight-path angle using Eq. (5.98) to give an equation which can be separated for integration:

$$\beta \frac{dX}{d\gamma} = \frac{\cos\gamma}{\cos\gamma - \left[\cos\gamma_e - \left(\frac{C_L}{C_D}\right)\eta_e\right]} \qquad \textbf{5.107}$$

Separating and integrating from entry forward:

$$\beta X = \int_{\gamma_e}^{\gamma} \frac{\cos\gamma}{\cos\gamma - \left[\cos\gamma_e - \left(\frac{C_L}{C_D}\right)\eta_e\right]} d\gamma$$

$$= \int_{\gamma_e}^{\gamma} d\gamma + \int_{\gamma_e}^{\gamma} \frac{\cos\gamma_e - \left(\frac{C_L}{C_D}\right)\eta_e}{\cos\gamma - \left[\cos\gamma_e - \left(\frac{C_L}{C_D}\right)\eta_e\right]} d\gamma$$

$$= (\gamma + \gamma_e) + \left[\cos\gamma_e - \left(\frac{C_L}{C_D}\right)\eta_e\right] \int_{\gamma_e}^{\gamma} \frac{d\gamma}{\cos\gamma - \left[\cos\gamma_e - \left(\frac{C_L}{C_D}\right)\eta_e\right]} \qquad \textbf{5.108}$$

The last integral can take on several forms, depending on the sign and relative magnitude of the term $\left[\cos\gamma_e - \left(\frac{C_L}{C_D}\right)\eta_e\right]$. For entry from a high-altitude (and/or for very small lift-to-drag ratios), the term is dominated by $\cos\gamma_e$. Therefore, for all but the steepest of entries (which are impractical for skip anyway):

$$1 \geq \left[\cos\gamma_e - \left(\frac{C_L}{C_D}\right)\eta_e\right] \geq 0 \qquad \textbf{5.109}$$

In these situations, the appropriate form of the integral is

$$
\int_{\gamma_e}^{\gamma} \frac{d\gamma}{a + b\cos\gamma} = \frac{1}{\sqrt{b^2 - a^2}} \ln \left[\frac{\sqrt{b^2 - a^2}\,\tan\left(\dfrac{\gamma}{2}\right) + a + b}{\sqrt{b^2 - a^2}\,\tan\left(\dfrac{\gamma}{2}\right) - a - b} \right]_{\gamma_e}^{\gamma}
\qquad \textbf{5.110}
$$

where $a = -\left[\cos\gamma_e - \left(\dfrac{C_L}{C_D}\right)\eta_e\right]$ and $b = 1$. Making the appropriate substitutions,

Eq. (5.108) becomes

$$
\beta X = (\gamma + \gamma_e) + \frac{k}{\sqrt{1-k^2}} \ln \left\{ \left[\frac{\sqrt{1-k^2}\,\tan\left(\dfrac{\gamma}{2}\right) + 1 - k}{\sqrt{1-k^2}\,\tan\left(\dfrac{\gamma}{2}\right) - 1 + k} \right] \left[\frac{\sqrt{1-k^2}\,\tan\left(\dfrac{\gamma_e}{2}\right) - 1 + k}{\sqrt{1-k^2}\,\tan\left(\dfrac{\gamma_e}{2}\right) + 1 - k} \right] \right\}
$$

$$
\textbf{5.111}
$$

where

$$
k = \left[\cos\gamma_e - \left(\frac{C_L}{C_D}\right)\eta_e \right]
\qquad \textbf{5.112}
$$

has been used as shorthand notation. In a slightly different non-dimensional form, the horizontal range is:

$$
\frac{X}{r_0} = \frac{(\gamma + \gamma_e)}{\beta r_0} + \frac{k}{\beta r_0 \sqrt{1-k^2}} \ln \left\{ \left[\frac{\sqrt{1-k^2}\,\tan\left(\dfrac{\gamma}{2}\right) + 1 - k}{\sqrt{1-k^2}\,\tan\left(\dfrac{\gamma}{2}\right) - 1 + k} \right] \left[\frac{\sqrt{1-k^2}\,\tan\left(\dfrac{\gamma_e}{2}\right) - 1 + k}{\sqrt{1-k^2}\,\tan\left(\dfrac{\gamma_e}{2}\right) + 1 - k} \right] \right\}
$$

$$
\textbf{5.113}
$$

To find the horizontal distance covered from atmospheric entry to atmospheric exit, Eq. (5.113) can be evaluated at $\gamma_f = -\gamma_e$:

$$\frac{X_{total}}{r_0} = \frac{2k}{\beta r_0 \sqrt{1-k^2}} \ln \left[\frac{\sqrt{1-k^2} \tan\left(\frac{\gamma_e}{2}\right) - 1 + k}{\sqrt{1-k^2} \tan\left(\frac{\gamma_e}{2}\right) + 1 - k} \right] \qquad \textbf{5.114}$$

It's reasonable to check to see if the skip really is complete by comparing the lowest altitude η_{max} in the trajectory to the radius of the planet. At the surface of the planet,

$$\eta_s = \frac{\rho_s S C_D}{2m\beta} \qquad \textbf{5.115}$$

so we'll want

$$\eta_{max} < \frac{\rho_s S C_D}{2m\beta} \qquad \textbf{5.116}$$

to avoid an undesired deceleration at the planet's surface. (Recall, η increases with decreasing altitude!) To find this maximum, it is sufficient to maximize Eq. (5.98) with respect to the flight-path angle and then use the corresponding "critical" angle γ_* to calculate the maximum. Doing so yields

$$\eta_{max} = \eta_e + \frac{1 - \cos\gamma_e}{\left(\dfrac{C_L}{C_D}\right)} \qquad \textbf{5.117}$$

where $\gamma_* = 0$ is the corresponding angle. Thus, to ensure a successful skip trajectory, a design trade can be made between vehicle's aerodynamic coefficients and the entry flight-path angle:

$$\eta_e + \frac{1-\cos\gamma_e}{\left(\dfrac{C_L}{C_D}\right)} < \frac{\rho_s S C_D}{2m\beta} \qquad \textbf{5.118}$$

The equations of motion (Eqs. (5.95) and (5.97)) are identical (albeit for different assumptions) to those derived for steep gliding entry at near circular speed in Section 5.3. This allows us to "steal" an expression for the total deceleration

$$\left(\frac{a_{decel}}{g_0}\right) = \frac{2\beta r_0 T_e \sqrt{1+\left(\dfrac{C_L}{C_D}\right)^2}\,(\cos\gamma - \cos\gamma_e)}{\left(\dfrac{C_L}{C_D}\right)} \exp\left[-\frac{2(\gamma-\gamma_e)}{\left(\dfrac{C_L}{C_D}\right)}\right] \qquad \textbf{5.119}$$

without deriving it again in this section. Similarly, we can use the previous discussion to give the point at which the maximum deceleration occurs

$$\gamma_* = \sin^{-1}\left[\frac{\dfrac{1}{2}\left(\dfrac{C_L}{C_D}\right)\cos\gamma_e - \sqrt{\dfrac{1}{4}\left(\dfrac{C_L}{C_D}\right)^2 + \sin^2\gamma_e}}{\dfrac{1}{4}\left(\dfrac{C_L}{C_D}\right)^2 + 1}\right] \qquad \textbf{5.120}$$

and the expression for the maximum deceleration itself:

$$\left(\frac{a_{decel}}{g_0}\right)_* = 2\beta r_0 T_e \sqrt{1+\left(\frac{C_L}{C_D}\right)^2}\left(-\frac{\sin\gamma_*}{2}+\eta_e\right)\exp\left[-\frac{2(\gamma_*-\gamma_e)}{\left(\frac{C_L}{C_D}\right)}\right]$$

5.121

Notice we have included both components of deceleration in the total deceleration. Lift is significant, and the normal deceleration may well dominate during portions of the entry. The maximum deceleration as a function of entry flight-path angle is shown in Figure 5-18 for several configurations. The maximum goes up with both steeper and faster entry. At shallow entry angles, lift decreases the maximum, but at steeper entry angles, lift increases the maximum.

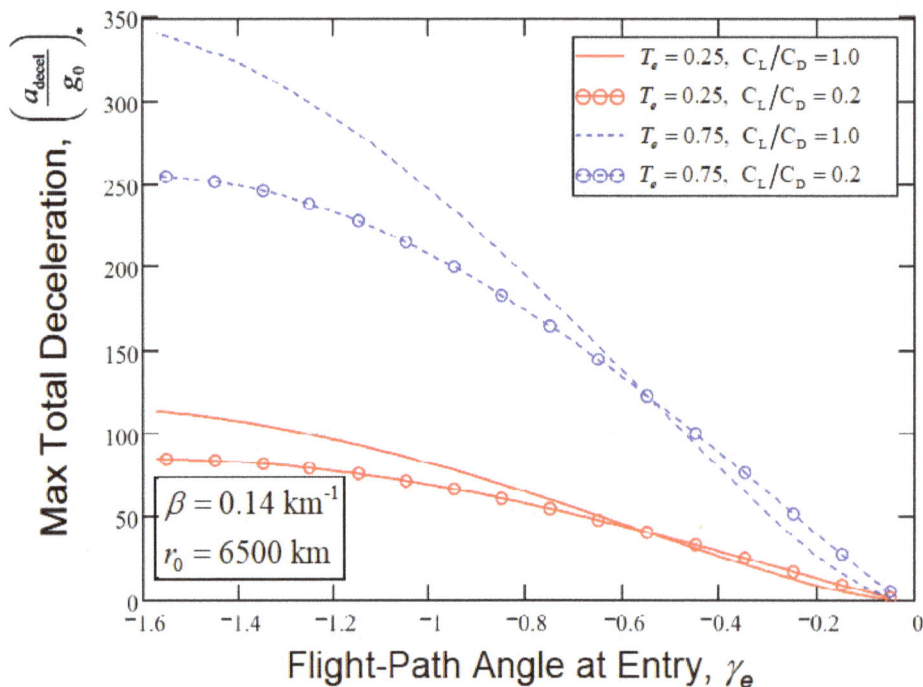

Figure 5-18: Relationship Between Entry Flight-Path Angle and Maximum Total Deceleration

All of the discussion on skip entry to this point has been for one skip (a single entry and exit from the atmosphere). If the vehicle leaves the atmosphere with a speed less than escape velocity, it will return to the atmosphere again. If another skip is performed, the process is repeated (Figure 5-19). Eventually, after multiple skips, the vehicle loses so much kinetic energy it cannot (or need not) pull back out of the atmosphere. At that point, the entry ceases to be skip entry and becomes gliding entry, ballistic entry, or something in between.

5.6 Steep Ballistic Entry

The first ballistic problem to be studied is that of steep entry. This is applicable to such things as ballistic missile reentry, where the trajectory is designed to pass through the atmosphere as quickly and as straight as possible. By coming in fast and relatively straight down, the vehicle minimizes the time spent in the atmosphere and, as a result, minimizes the uncertainty in the trajectory. We will be looking at two solutions for this problem, one which includes gravity and one which does not.

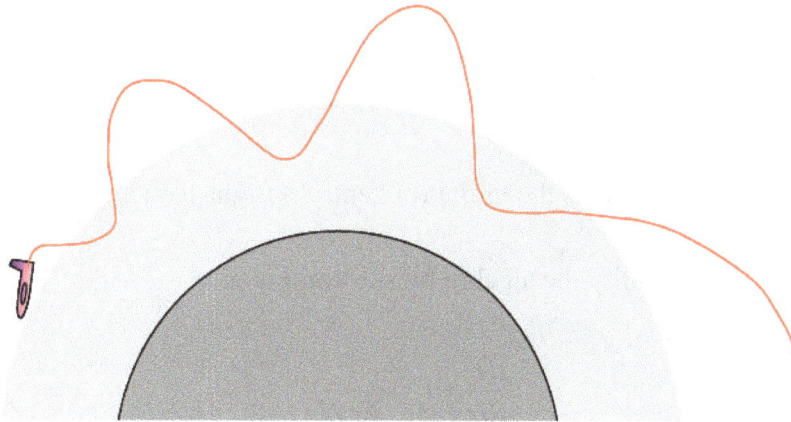

Figure 5-19: Planar Multi-Skip Entry

5.6.1 Analysis Including Gravity

In the case of ballistic entry, lift forces are zero. For steep entry angles, the centrifugal force can also be ignored when compared to the gravity force, so Eq. (5.2) becomes:

$$\frac{d\xi}{d\eta} = \frac{-\xi}{2\beta r_0 \eta T}$$

5.122

If the entry angle is steep enough, $\xi = \cos\gamma$ is small. Further, as the vehicle penetrates the atmosphere, the dimensionless altitude η (which is proportional to atmospheric density) quickly becomes greater than one. Thus, the right-side of Eq. (5.122) quickly becomes small. As a first-order approximation, assume

$$\frac{d\xi}{d\eta} \approx 0$$

5.123

which can be immediately integrated to show that the cosine of the flight-path angle remains constant. While it is possible at times to consider the cosine of an angle to remain approximately constant longer than the angle itself, in this case the flight-path angle itself can also be considered constant. Substituting this into our other basic planar equation gives us a linear, constant-coefficient differential equation for our kinetic energy parameter T

$$\frac{dT}{d\eta} - \frac{2T}{\sin\gamma_e} = \frac{1}{\beta r_0 \eta}$$

5.124

where γ_e has been used as the initial (or "entry") flight-path angle. Before solving, we can simplify the algebra by substituting $\alpha = \frac{-2\eta}{\sin\gamma_e}$:

$$\frac{dT}{d\alpha} + T = \frac{1}{\beta r_0 \alpha}$$

5.125

The homogeneous solution to Eq. (5.125) is given by integrating

$$\int \frac{dT}{T} = -\int d\alpha \qquad\qquad \textbf{5.126}$$

to get

$$T_H(\alpha) = ke^{-\alpha} \qquad\qquad \textbf{5.127}$$

where k is a constant of integration. Numerous authors have shown that the particular solution to Eq. (5.125) is given by

$$T_P(\alpha) = e^{-\alpha}\left[c + \frac{1}{\beta\, r_0} Ei(\alpha)\right] \qquad\qquad \textbf{5.128}$$

with c being a constant of integration. The exponential integral function is defined as:

$$Ei(\alpha) = \int_{-\infty}^{\alpha} \frac{e^x}{x}\, dx \qquad\qquad \textbf{5.129}$$

This solution can be verified by direct substitution of Eq. (5.128) back into the left side of Eq. (5.125) and verifying it produces the right side:

$$\frac{dT_P}{d\alpha} + T_P = \left\{ -e^{-\alpha}\left[c + \frac{1}{\beta r_0} Ei(\alpha)\right] + \frac{e^{-\alpha}}{\beta r_0}\frac{d\left(Ei(\alpha)\right)}{d\alpha}\right\}$$

$$+ \left\{e^{-\alpha}\left[c + \frac{1}{\beta r_0} Ei(\alpha)\right]\right\}$$

$$= \frac{e^{-\alpha}}{\beta r_0}\frac{e^{\alpha}}{\alpha}$$

$$= \frac{1}{\beta r_0 \alpha} \qquad\qquad \textbf{5.130}$$

The total solution is the sum of the homogeneous and particular solutions:

$$T(\alpha) = T_H + T_P = e^{-\alpha}\left[(c+k) + \frac{1}{\beta r_0} Ei(\alpha)\right]$$

5.131

Or, introducing A as a new constant:

$$T(\alpha) = \frac{e^{-\alpha}}{\beta r_0}\left[A + Ei(\alpha)\right]$$

5.132

The value for A can be found by evaluating the initial conditions on the velocity. (This is left as an exercise.) Alternatively, A can be expressed in terms of the entry (initial) kinetic energy and α-value:

$$A = T_e \beta r_0 e^{\alpha_e} - Ei(\alpha_e)$$

5.133

The exponential integral function could easily be a button on a calculator like sine and cosine. However, since we'd be hard-pressed to find a calculator with an $Ei(x)$ button, a Taylor series is often used as a numerical technique to find the value:

$$Ei(x) = \gamma_E + \ln(x) + \sum_{n=1}^{\infty} \frac{x^n}{nn!}$$

5.134

$\gamma_E \approx 0.5772156$ is the Euler constant. Values for $Ei(x)$ are tabulated in reference books. Before declaring victory and moving on to solve another "special case," we can examine Eq. (5.132) in more detail to learn more about the dynamics of steep entry angles.

Equation (5.132) describes the velocity as a function of altitude once variable substitutions are reversed (and the constant is evaluated). When the truncated expansion $Ei(\alpha) \approx \gamma_E + \ln(\alpha)$ is used, this expression is:

$$^{R}V^2 \approx \frac{2g_0 \exp\left(\dfrac{\rho_s SC_D}{m\beta \sin \gamma_e} e^{-\beta h}\right)}{\beta} \left[(A + \gamma_E) + \ln\left(\dfrac{-\rho_s SC_D}{m\beta \sin \gamma_e} e^{-\beta h}\right) \right] \qquad \textbf{5.135}$$

Significantly, the velocity is dependent on an exponential of an exponential. The velocity is *extremely* sensitive to the altitude and slightly less sensitive to the drag factor, mass, and other parameters in the exponential.

With a constant flight-path angle, $\dot{\gamma} = 0$ so there isn't any normal component of deceleration. Thus, we can use Eq. (4.59) to write:

$$a_{decel} = (a_{decel})_v = 2\beta r_0 g_0 \eta T \pm g_0 \sqrt{1 - \xi^2} \qquad \textbf{5.136}$$

For entry into most planets

$$2\beta r_0 g_0 \eta T \gg g_0 \sqrt{1 - \xi^2} \qquad \textbf{5.137}$$

in the region of interest, so we can drop the second term on the right. With the variable change $\alpha = \dfrac{-2\eta}{\sin \gamma_e}$, our non-dimensional deceleration becomes

$$\left(\frac{a_{decel}}{g_0}\right) = -\left(\beta r_0 \alpha \sin \gamma_e\right) T \qquad \textbf{5.138}$$

where $T = T(\alpha)$. This is maximized (with respect to α, a measure of the altitude) when:

$$\frac{d\left(\dfrac{a_{decel}}{g_0}\right)}{d\alpha} = 0 \qquad \textbf{5.139}$$

Substituting Eq. (5.138) into Eq. (5.139) and simplifying, we can get an expression for α_* corresponding to the altitude of maximum deceleration:

$$\left[A + Ei\left(\alpha_*\right)\right]\left(1 - \alpha_*\right) + e^{\alpha_*} = 0 \qquad \textbf{5.140}$$

Once the constant A is determined from initial conditions, α_* can be found numerically. The value of the maximum deceleration is then

$$\left(\frac{a_{decel}}{g_0}\right)_{max} = \left(\frac{a_{decel}}{g_0}\right)_* = \frac{-\alpha_* \sin \gamma_e}{\alpha_* - 1} \qquad \textbf{5.141}$$

where $\alpha_* \neq 1$ is found by solving Eq. (5.140).

It is possible for Eq. (5.140) to yield values of α_* corresponding to altitudes below the planet surface, so care must be taken in solving the equation. Specifically,

$$\alpha_* < \frac{-\rho_s S C_D}{m\beta \sin \gamma_e} \qquad \textbf{5.142}$$

allows the maximum value to be reached *before* impacting the planet surface. If the physical parameters of the problem are such that the inequality in Eq. (5.142) is *not* satisfied, the deceleration increases from the entry point until impact and never reaches the *analytic* maximum given by Eq. (5.141). (Mathematically, this is true. However, other assumptions we've used, such as hypersonic flight, begin to fail before the planet surface is actually reached. None-the-less, it is traditional to find conditions analogous to Eq. (5.142) for this problem.)

In addition to finding the maximum deceleration, we can also find the "terminal" or "limiting" velocity of a high-drag vehicle. During the last phases of entry, the vehicle may approach a point where drag and gravity are essentially balanced and the change in kinetic energy is essentially zero. Thus, Eq. (5.1) becomes:

$$\frac{2T}{\sin\gamma_e} + \frac{1}{\beta r_0 \eta} \approx 0 \qquad \textbf{5.143}$$

Solving this for the velocity

$$^R V_{\text{limit}} = \left(\frac{-2mg_0 \sin\gamma_e}{\rho_s S C_D} \right)^{1/2} e^{\beta h/2} \qquad \textbf{5.144}$$

Note that the limiting velocity is altitude dependent. Skydivers do not notice this altitude dependence since the altitude range they are concerned with is typically such that $0 < \beta h/2 < 0.25$, or a maximum change of about 28%.

5.6.2   Analysis Ignoring Gravity

Equations (5.132) and (5.141) give the solutions for kinetic energy and maximum deceleration, respectively, during steep ballistic entry. Neither equation lends itself to "generalized" plots such as those in Figure 5-2 and Figure 5-14. Equation (5.141) for the maximum deceleration even requires a numerical solution to a transcendental function! They are difficult (to say the least) to use for qualitative comparisons of trajectories with different initial velocities and/or different entry flight-path angles. To simplify the analysis, we need to do something very non-intuitive – we'll ignore gravity! (But, by example, we'll show the solution without gravity closely follows the more complicated solution with gravity.)

When gravity is ignored, Eq. (5.1) becomes

$$\frac{dT}{d\eta} = \frac{2T}{\sin\gamma} \qquad \textbf{5.145}$$

and Eq. (5.2) remains the same as we found in Eq. (5.123):

$$\frac{d\xi}{d\eta} = 0 \qquad\qquad 5.146$$

Equation (5.146) integrates immediately to give:

$$\cos\gamma = \cos\gamma_e \qquad\qquad 5.147$$

Like before, with this type of entry we can assume the flight-path angle remains constant $(\gamma = \gamma_e)$. Substituting this into Eq. (5.145) and solving yields a closed-form solution for the kinetic energy:

$$\frac{T}{T_e} = \exp\left[\frac{2(\eta - \eta_e)}{\sin\gamma_e}\right] \qquad\qquad 5.148$$

Equations (5.147) and (5.148) are the first-order solution for steep ballistic entry presented by Allen and Eggers (Ref. 1), Chapman (Ref. 16), and Gazley (Ref. 26). This solution is shown in Figure 5-20 for entry starting at high altitudes ($\eta_e \approx 0$).

The tangential deceleration can be found by substituting this solution for kinetic energy into Eq. (5.138) (and "unreplacing" $\alpha = \dfrac{-2\eta}{\sin\gamma_e}$):

$$\left(\frac{a_{decel}}{g_0}\right) = (2\beta r_0 \eta T_e)\exp\left[\frac{2(\eta - \eta_e)}{\sin\gamma_e}\right] \qquad\qquad 5.149$$

Deceleration for high-altitude entries are shown in Figure 5-21. To find the maximum value, Eq. (5.149) can be differentiated with respect to altitude and set

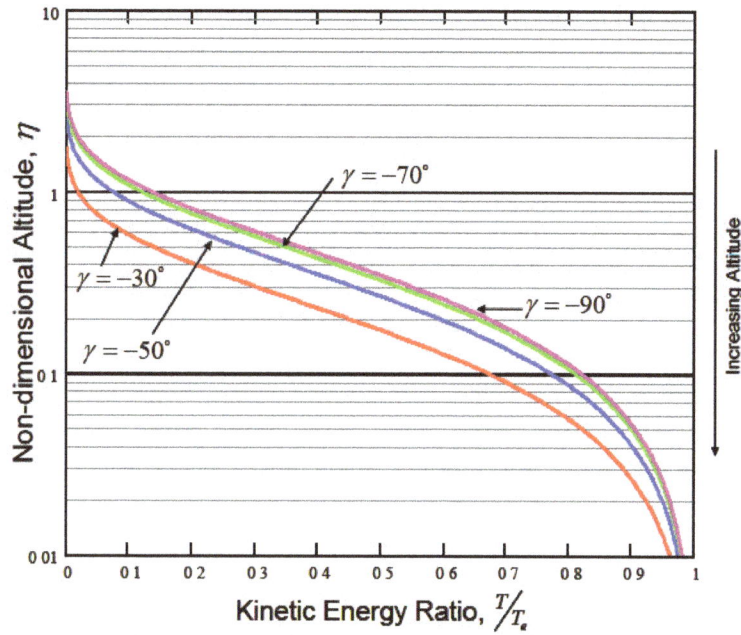

Figure 5-20: Steep Ballistic Entry Solution Ignoring Gravity

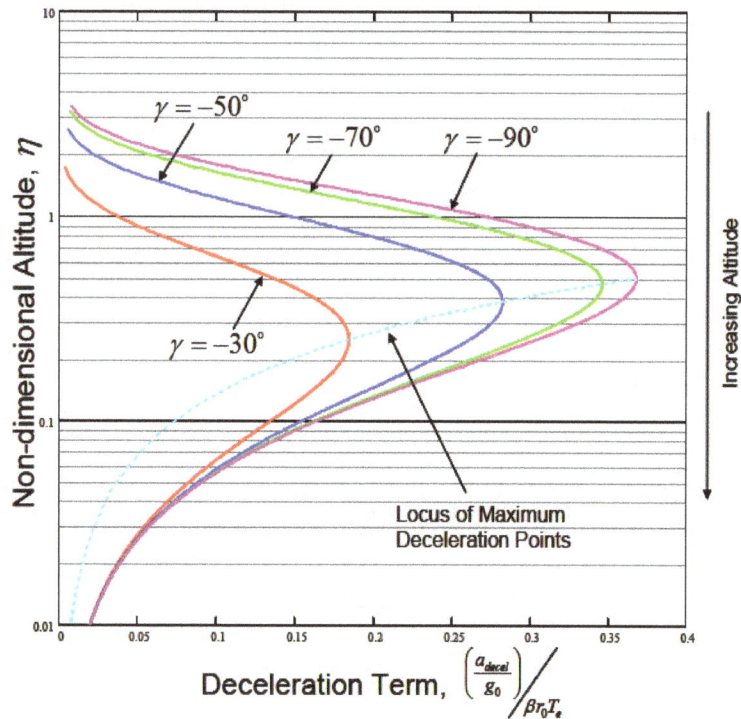

Figure 5-21: Deceleration During Steep Ballistic Entry (Ignoring Gravity)

it equal to zero. Solving gives the altitude of maximum deceleration:

$$\eta_* = -\frac{\sin \gamma_e}{2}$$

5.150

Of course, this altitude is only reached if:

$$\eta_* = -\frac{\sin \gamma_e}{2} \leq \frac{\rho_s SC_D}{2m\beta}$$

5.151

Otherwise, the deceleration will increase (by Eq. (5.149)) until impact with the ground. Assuming the critical altitude is reached before the vehicle hits the surface, the deceleration at η_* is:

$$\left(\frac{a_{decel}}{g_0}\right)_* = (2\beta r_0 \eta_* T_e) \exp\left[\frac{2(\eta_* - \eta_e)}{\sin \gamma_e}\right]$$

$$= (-\beta r_0 T_e \sin \gamma_e) \exp\left[-\left(1 + \frac{2\eta_e}{\sin \gamma_e}\right)\right]$$

5.152

For high-altitude entry, Eq. (5.152) reduces to:

$$\left(\frac{a_{decel}}{g_0}\right)_* = \frac{-\beta r_0 T_e \sin \gamma_e}{e}$$

5.153

This "locus of maximum values" is also shown in Figure 5-21.

5.6.3   Comparison of Analyses

The kinetic energy solutions with and without gravity are given by Eq. (5.132) and Eq. (5.148), respectively. The two are much closer *numerically* than

126

the difference in algebra would seem indicate. Further, it's possible to show Eq. (5.132) can be approximated by Eq. (5.148). Rearranging Eq. (5.132),

$$\frac{T}{T_e} = e^{-\alpha+\alpha_e} + \frac{e^{-\alpha}}{\beta r_0 T_e}\left[-Ei(\alpha_e) + Ei(\alpha)\right]$$

5.154

and substituting the first two terms of a Taylor series expansion for $Ei(\alpha)$ yields:

$$\frac{T}{T_e} \approx e^{-\alpha+\alpha_e} + \frac{e^{-\alpha}}{\beta r_0 T_e}\left\{-\left[\gamma_E + \ln(\alpha_e)\right] + \left[\gamma_E + \ln(\alpha)\right]\right\}$$

$$= \exp\left[\frac{2(\eta-\eta_e)}{\sin\gamma_e}\right] + \exp\left(\frac{2\eta}{\sin\gamma_e}\right)\frac{1}{\beta r_0 T_e}\ln\left(\frac{\eta}{\eta_e}\right)$$

5.155

When $\dfrac{1}{\beta r_0 T_e}$ is sufficiently small (which it is for most planets of interest), this

becomes:

$$\frac{T}{T_e} \approx \exp\left[\frac{2(\eta-\eta_e)}{\sin\gamma_e}\right]$$

5.156

Thus, the two solutions (with and without gravity) produce the same results under many circumstances.

Equations (5.132) and (5.148) can be plotted on the same graph for several cases to see how accurately the "simple" solution without gravity tracks the "more complete" solution with gravity. (In other words, we can use a graph to convince ourselves that $\dfrac{1}{\beta r_0 T_e}$ is sufficiently small.) Two different trajectories (whose specifics are given in Table 5-1) are shown in Figure 5-22 computed with and without gravity. Together, the two cases span a wide range of possible "steep" ballistic entries. The figure shows the solutions represented by Eqs. (5.132) and (5.148) are, for all practical purposes, very close.

Table 5-1: Sample Ballistic Entry Initial Conditions

Case #1	Case #2
$T_e = 0.25$	$T_e = 4.0$
$\gamma_e = -70°$	$\gamma_e = -90°$
$\beta r_0 = 910,\ \eta_e = 0$	

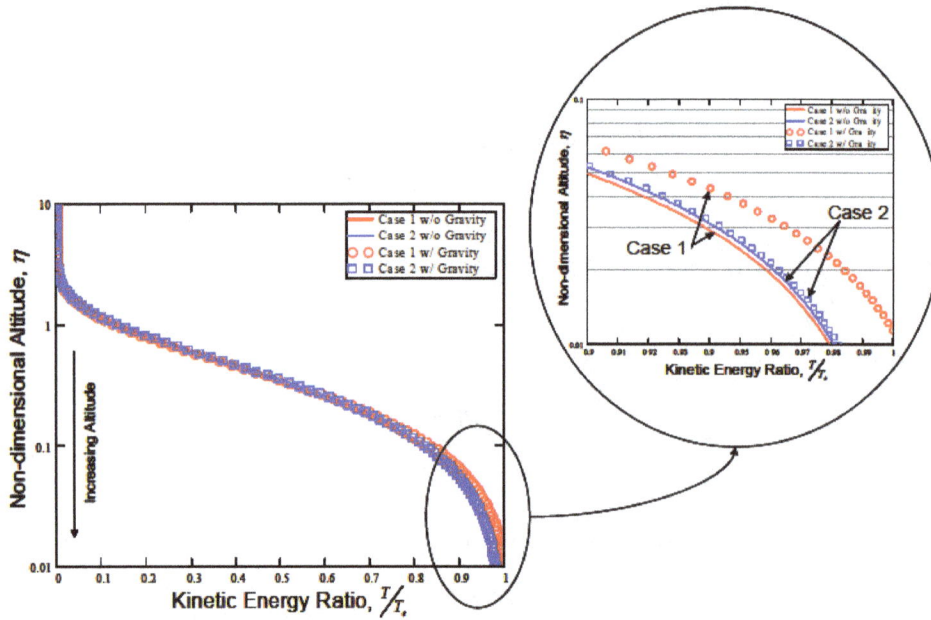

Figure 5-22: Comparison of Steep Ballistic Solutions

Similarly, Figure 5-23 shows how the deceleration varies with altitude for both solution methods. Notice that the qualitative information (the trends) is identical and the quantitative information (the values) is very similar for the two solution methods. In fact, Figure 5-24 shows that the deceleration values differ by less than 5% of the maximum values over the entire solution. (The "steeper"

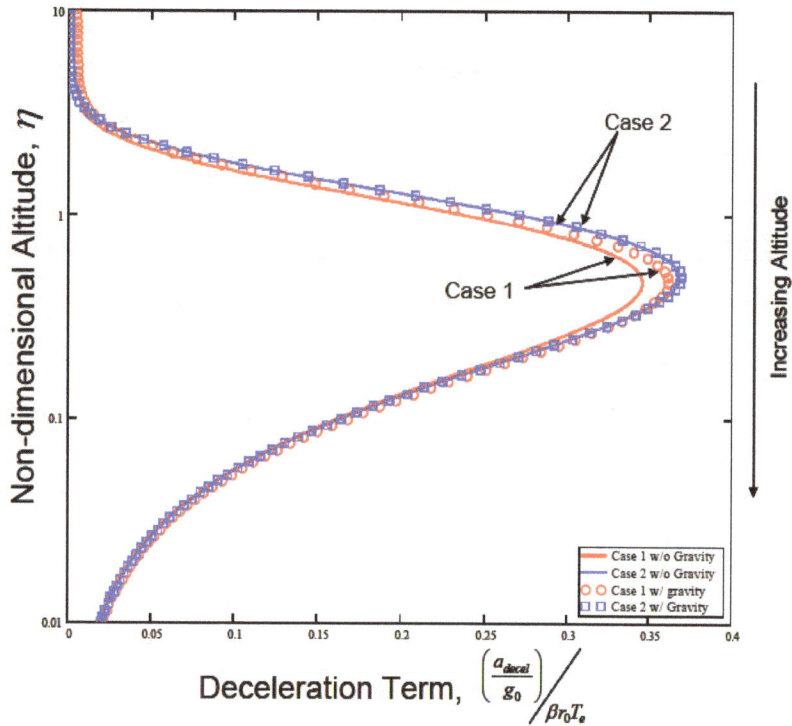

Figure 5-23: Comparison of Deceleration Solutions for Steep Ballistic Entry

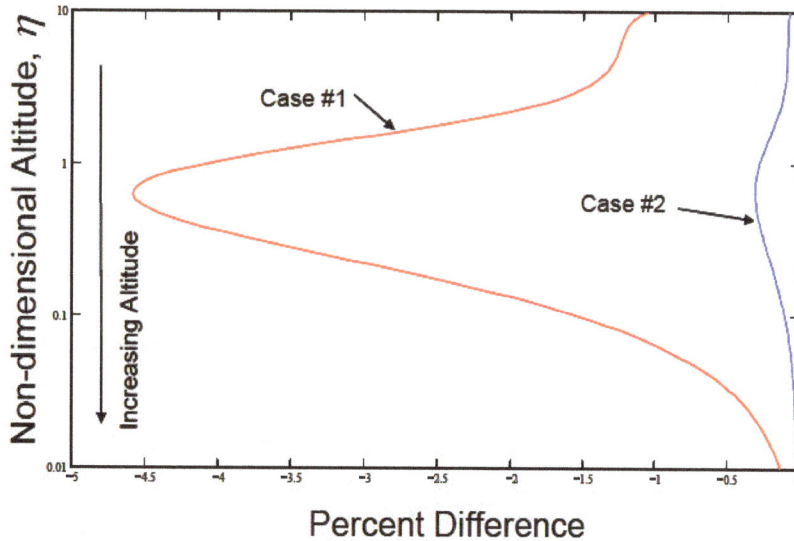

Figure 5-24: Comparison of Deceleration Solutions for Steep Ballistic Entry
(Differences as a Percentage of Maximum Values)

case, Case #2, has less than a 0.5% difference!) For completeness, Figure 5-25 shows that the two methods of computing the altitude of maximum deceleration η_* and maximum deceleration $\left(\dfrac{a_{decel}}{g_0}\right)_*$ are quite close also.

In light of the comparisons made in Figure 5-22 through Figure 5-25, we can make a strong argument that the "simpler" equations derived without gravity are adequate for initial analysis of these types of entries. Indeed, we will use these solutions in a later chapter to help us compare heat transfer between types of reentry trajectories.

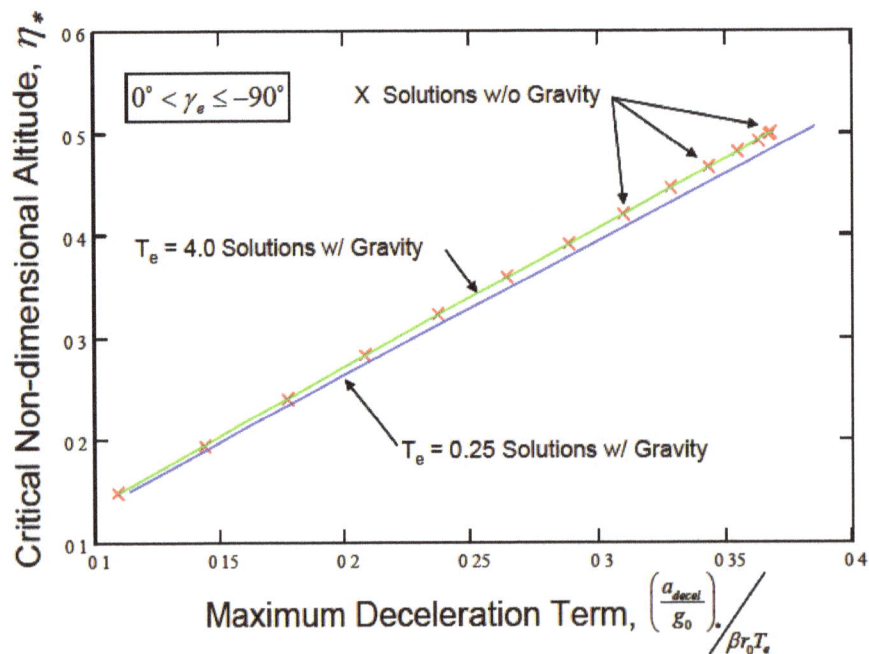

Figure 5-25: Comparison of Critical Deceleration Values for Steep Ballistic Entry

5.7 Shallow Ballistic Entry at Near Circular Entry Speed

In this section, we will cover ballistic entry at shallow angles and nearly orbital speeds. The approach will be one by V. A. Yaroshevskii published in 1964 shortly after Yuri Gagarin's orbital flight. The method is semianalytical rather than strictly analytical and is probably more accurately classified as a second-order solution. Despite this, it fits well with the first-order solutions in this chapter.

Backing up one step and starting with an earlier form of the planar entry equations given by Eqs. (4.23) - (4.25) in Chapter 4, we begin by introducing the small angle approximations $\sin\gamma \approx \gamma$ and $\cos\gamma \approx 1$ along with setting lift to zero:

$$\frac{dr}{dt} = {}^{R}V\gamma \qquad\qquad \textbf{5.157}$$

$$\frac{d\,{}^{R}V}{dt} = -\frac{\rho C_D S\,{}^{R}V^2}{2m} - g_0\left(\frac{r_0}{r}\right)^2\gamma \qquad\qquad \textbf{5.158}$$

$$^{R}V\frac{d\gamma}{dt} = \frac{{}^{R}V^2}{r} - g_0\left(\frac{r_0}{r}\right)^2 \qquad\qquad \textbf{5.159}$$

Unlike for steep ballistic entry, we have retained the centrifugal force term in Eq. (5.159). If we further assume a thin atmosphere and ignore the tangent gravity force in the velocity equation, these equations become:

$$\frac{dr}{dt} = {}^{R}V\gamma \qquad\qquad \textbf{5.160}$$

$$\frac{d\,{}^{R}V}{dt} = -\frac{\rho C_D S\,{}^{R}V^2}{2m} \qquad\qquad \textbf{5.161}$$

$$^{R}V \frac{d\gamma}{dt} = \frac{^{R}V^2}{R_{\oplus}} - g_{\oplus}$$

5.162

The radius change is small, so it has been assumed to be constant on the right side of Eqs. (5.160) - (5.162) and set equal to the sea-level radius $R_{\oplus}$. Similarly, gravity has been assumed to remain constant at the sea-level value $g_{\oplus}$. Using Eq. (5.161) to change the independent variable to ^{R}V:

$$\frac{dr}{d^{R}V} = -\frac{2m\gamma}{\rho C_D S\,^{R}V}$$

5.163

$$^{R}V \frac{d\gamma}{d^{R}V} = \frac{2m}{\rho C_D S\,^{R}V^2}\left(g_{\oplus} - \frac{^{R}V^2}{R_{\oplus}}\right)$$

5.164

Yaroshevskii introduced a new independent variable x and a new dependent variable y defined by

$$x = -\ln \overline{V}$$

5.165

$$y = \frac{\rho C_D S}{2m}\sqrt{\frac{R_{\oplus}}{\beta}}$$

5.166

where $\overline{V}$ is a dimensionless velocity defined relative to the sea-level circular velocity by

$$\overline{V} = \frac{^{R}V}{\sqrt{g_{\oplus} R_{\oplus}}}$$

5.167

and β^{-1} is a scale height selected to best match the atmosphere. For this strictly exponential atmosphere, we have assumed

$$\rho = \rho_s e^{-\beta h}$$

5.168

where $h = r - R_\oplus$ is the altitude, ρ_s is the atmospheric density at the surface, and β is constant. Thus, from Eqs. (5.166) and (5.168), we can write:

$$\frac{dy}{y} = \frac{d\rho}{\rho} = -\beta dr \qquad\qquad \textbf{5.169}$$

Equations (5.165) and (5.167) can be differentiated and combined to give:

$$dx = \frac{-d\bar{V}}{\bar{V}} = \frac{-d^{R}V}{{}^{R}V} \qquad\qquad \textbf{5.170}$$

Substituting Eqs. (5.169) and (5.170) into Eqs. (5.163) and (5.164) to change the dependent and independent variables gives the new equations of motion:

$$\frac{dy}{dx} = -\gamma \sqrt{R_\oplus \beta} \qquad\qquad \textbf{5.171}$$

$$\frac{d\gamma}{dx} = \frac{-1}{y\sqrt{R_\oplus \beta}}\left(\frac{1}{\bar{V}^2} - 1\right) \qquad\qquad \textbf{5.172}$$

Finally, differentiate Eq. (5.171) to get

$$\frac{d^2y}{dx^2} = -\sqrt{R_\oplus \beta}\,\frac{d\gamma}{dx} \qquad\qquad \textbf{5.173}$$

and substitute in Eq. (5.172) for $\dfrac{d\gamma}{dx}$:

$$\frac{d^2y}{dx^2} = \frac{1}{y}\left(\frac{1}{\bar{V}^2} - 1\right) \qquad\qquad \textbf{5.174}$$

If we realize from Eq. (5.165) that $\overline{V} = e^{-x}$, the differential equation of motion becomes surprisingly simple:

$$\frac{d^2 y}{dx^2} = \frac{1}{y}\left(e^{2x} - 1\right)$$

5.175

This second-order, non-linear differential equation is Yaroshevskii's equation for studying entry into planetary atmospheres (specialized to the case of no lift and constant drag coefficient). For the problem at hand, the initial conditions to be used when integrating this equation are:

$$x_i = 0 \quad \text{(entry speed is the circular orbital speed)}$$
$$y_i = 0 \quad \text{(entry begins at high altitude, } \rho \approx 0\text{)}$$
$$\left.\frac{dy}{dx}\right|_i = 0 \quad \text{(circular orbit, } \gamma_i = 0\text{)}$$

5.176

To solve Eq. (5.175), Yaroshevskii turned to a series solution. (Computers were not readily available in the Soviet Union of the early 1960s.) Expanding out the exponential, Eq. (5.175) becomes

$$y'' \approx \frac{2x}{y}$$

5.177

where the primes have been used to denote differentiation with respect to x. This equation is singular near $y \approx 0$ (unless, of course, x is well-behaved and goes to zero at least as fast as y). If we assume a solution of the form $y = Ax^p$ near $y = 0$, then Eq. (5.177) becomes:

$$Ap(p-1)x^{p-2} = \frac{2x}{Ax^p}$$

5.178

Equating the exponent on x on each side gives $p = \frac{3}{2}$. Similarly, equating the coefficients gives $A = \sqrt{\frac{8}{3}}$. Thus, for small x the solution is:

$$y = \sqrt{\frac{8}{3}} x^{3/2}$$

5.179

Equation (5.179) satisfies the initial conditions given by Eqs. (5.176), but is not general enough for studying many of the properties along the trajectory (and it is an approximate solution near $y = 0$). Thus, we generalize Eq. (5.179) to be a series

$$y = \sqrt{\frac{8}{3}} x^{3/2} \left(c_0 + c_1 x + c_2 x^2 + \cdots \right)$$

5.180

where we have yet to determine the unknown coefficients c_i. (However, we should expect to find $c_0 = 1$ so that this solution simplifies to Eq. (5.179) for sufficiently small x near $y = 0$.) Rewriting Eq. (5.175) as

$$yy'' = e^{2x} - 1$$

5.181

and substituting in the assumed series solution, we find

$$yy'' \approx \frac{8}{3} \left[\left(\frac{3}{4} c_0^2 \right) x + \left(\frac{9}{2} c_0 c_1 \right) x^2 + \left(\frac{19}{2} c_0 c_2 + \frac{15}{4} c_1^2 \right) x^3 \right]$$

5.182

when terms smaller than x^4 are ignored. (Recall that x is small.) Expanding the exponential in Eq. (5.181) and combining with Eq. (5.182) gives an algebraic equation in x. Ignoring the terms smaller than x^4, this expression is simply:

$$\frac{8}{3} \left[\left(\frac{3}{4} c_0^2 \right) x + \left(\frac{9}{2} c_0 c_1 \right) x^2 + \left(\frac{19}{2} c_0 c_2 + \frac{15}{4} c_1^2 \right) x^3 \right] = 2x + 2x^2 + \frac{4}{3} x^3$$

5.183

For this equation to be true, the coefficients in front of like powers of x must be equal on the left and right sides. Thus, $c_0 = 1$, $c_1 = \frac{1}{6}$, and $c_2 = \frac{1}{24}$ and the approximate solution to the second-order differential equation becomes:

$$y = \sqrt{\frac{8}{3}} x^{3/2} \left(1 + \frac{1}{6} x + \frac{1}{24} x^2 \right)$$

5.184

This solution is sufficient to analyze the properties of interest in our problem.

For this "near spiral" trajectory, the deceleration is very nearly all along the tangential direction ($\dot{\gamma} \approx 0$). Thus, our usual non-dimensional term can be written using Eq. (5.161)

$$
\left(\frac{a_{decel}}{g_\oplus} \right) = \left[\frac{(a_{decel})_v}{g_\oplus} \right]
$$

$$
= \left[\frac{-\left(\dfrac{d^R V}{dt} \right)}{g_\oplus} \right]
$$

$$
= \frac{\left(\dfrac{\rho C_D S^R V^2}{2m} \right)}{g_\oplus}
$$

5.185

where we have used $g_\oplus$ instead of g_0 since gravity has been assumed constant.

Grouping terms allows us to simplify this with our definitions of y and $\bar{V}$:

$$\left(\frac{a_{decel}}{g_\oplus}\right) = \frac{\left(\dfrac{\rho C_D S}{2m}\sqrt{\dfrac{R_\oplus}{\beta}}\right)\sqrt{\dfrac{\beta}{R_\oplus}}}{g_\oplus}\left(^R V\right)^2 = \frac{y}{g_\oplus}\sqrt{\frac{\beta}{R_\oplus}}\left(\bar{V}\sqrt{g_\oplus R_\oplus}\right)^2$$

$$= y\bar{V}^2\sqrt{\beta R_\oplus} \qquad\qquad \textbf{5.186}$$

We have approximate answers for y and $\bar{V}$, so this simplifies to

$$\left(\frac{a_{decel}}{g_\oplus}\right) = \sqrt{\frac{8\beta R_\oplus}{3}}\,x^{3/2}\left(1+\frac{1}{6}x+\frac{1}{24}x^2\right)e^{-2x} \qquad \textbf{5.187}$$

Taking the derivative with respect to x and setting it equal to zero gives an expression for where the maximum deceleration occurs:

$$\left(\frac{3}{2}x^{1/2}-\frac{19}{12}x^{3/2}-\frac{3}{16}x^{5/2}-\frac{1}{12}x^{7/2}\right)e^{-2x}=0 \qquad \textbf{5.188}$$

For $x\neq 0$ and $x\neq\infty$, this can be simplified:

$$4x_*^3+9x_*^2+76x_*-72=0 \qquad\qquad \textbf{5.189}$$

The solution to this cubic equation is $x_*\approx 0.835$. The corresponding value for the velocity is $\bar{V}_*\approx 0.434$ from Eq. (5.165) and for the dependent variable $y_*\approx 1.46$ from Eq. (5.184). Finally, the maximum deceleration is given by Eq. (5.186):

$$\left(\frac{a_{decel}}{g_\oplus}\right)_* \approx 0.275\sqrt{\beta R_\oplus} \qquad \textbf{5.190}$$

Note that this maximum deceleration and the velocity at which it occurs is *independent of the vehicle drag coefficient!* High- and low-drag vehicles experience the same maximum deceleration at the same speed during reentry. For Earth, these values are

$$\left(a_{decel}\right)_* \approx 8.2 g_\oplus$$

$${}^R V_* \approx 3.4 \text{ km/sec}$$

Only the altitude at which the maximum occurs changes as a result of the vehicle configuration (as seen in Eq. (5.166)).

Decelerating at eight "gees" is about the limit the human body can endure. An eight gee deceleration also imposes significant structural requirements on the vehicle. It is possible to reduce the maximum deceleration by adding lift and staying in the upper atmosphere longer. The first Soviet Vostok and American Mercury capsules were purely ballistic and the equations of this section are applicable. However, the capsules that followed could produce lift to reduce the structural (and human) durability requirements.

Before leaving this section, it is important to note we have only covered a small portion of Yaroshevskii's theory. For a more complete treatment, Vinh, et al. provide a more in-depth (and general) look at the theory (58:157-177). In particular, they present a general expression for the unknown c_i coefficients for series expansions beyond x^3 as well as a discussion on the radius of convergence for the series. They also present Yaroshevskii's theory as it applies to non-circular entry and lifting entry.

5.8 Summary

This chapter found a variety of classic closed-form solutions for specific types of entry trajectories. For each, we derived explicit equations relating flight-path angle, kinetic energy, and altitude that were, for the most part, independent of requirements to specify details about the vehicle (other than C_L/C_D) and the atmosphere. We found similarly general expressions for the deceleration and maximum deceleration in each case. Depending on the type of entry, we also found other "convenient" solutions such as maximum range and time-of-flight.

Each solution in this chapter applies to a specific set of assumptions (e.g., shallow lifting entry or steep ballistic entry). As such, the solutions fail to model reality outside of the region for which they were intended. (For the interested reader, a few examples comparing several of these closed-form solutions to much more complex numerical solutions are presented in Ref. 32.) The next chapter presents a solution which gives up the simplicity of the explicit relationships we found here in favor of a set of transcendental equations valid over a wide range of entry trajectories.

5.9 Problems

Material Understanding:

1. Why is it an "artificial assumption" that the velocity in Eq. (5.24) can decrease to zero to give the limiting value of arc-length in Eq. (5.25)?

2. Show that the following two expressions are equivalent:

$$k \sin \gamma_* - \cos \gamma_e = -\cos \gamma_*$$

$$\left(k^2 + 1\right)\sin^2 \gamma_* + \left(-2k \cos \gamma_e\right)\sin \gamma_* + \left(-\sin^2 \gamma_e\right) = 0$$

3. Prove that the flight-path angle corresponding to the minimum altitude during a skip entry is $\gamma_* = 0$ (assuming, of course, the vehicle doesn't impact the planet surface first).

4. Evaluate the constant A in the steep-entry angle energy equation (Eq. (5.132)) if the initial altitude h_e (or, equivalently, an initial radius r_e) and initial velocity $^{R}V_e$ are known.

5. Prove Eq. (5.142).

6. Yaroshevskii's solution for shallow ballistic entry assumed a solution of the form:

$$y = \sqrt{\frac{8}{3}} x^{3/2} \left(c_0 + c_1 x + c_2 x^2 + c_3 x^3 + c_4 x^4 \cdots \right)$$

In Section 5.7, we discussed solving this for a series truncated after $c_2 x^2$ (i.e., $c_3, c_4, c_5 \cdots = 0$). Specifically, we found values for $c_0, c_1,$ and c_2 to give us a solution $y = \sqrt{\frac{8}{3}} x^{3/2} \left(1 + \frac{1}{6} x + \frac{1}{24} x^2 \right)$. While this was sufficient for circular entries, at least one more term is needed to study elliptical entries. Expand the solution to include c_3 and solve for the constants $c_0, c_1, c_2,$ and c_3.

Computational Insights:

7. Investigate how velocity changes with altitude for a gliding vehicle entering the atmosphere at near circular speed and at a medium flight-path angle. Specifically, for an entry flight-path angle of $\gamma_e = -.20 \text{ radians}$, combine Eq. (5.37)

$$\cos\gamma - \cos\gamma_e = \frac{C_L}{C_D}(\eta - \eta_e)$$

and Eq. (5.41)

$$\frac{T}{T_e} = \exp\left(-\frac{2(\gamma - \gamma_e)}{\left(\frac{C_L}{C_D}\right)}\right)$$

to plot altitude change, $\eta - \eta_e$, versus velocity change, $\frac{{}^R V}{{}^R V_e}$, for a range of lift-to-drag ratios. You do *not* need to analytically combine the equations.

8. Equation (5.78)

$$\frac{dT}{T} = \frac{d\eta}{-\frac{1}{2}\left[\gamma_e^2 - 2\left(\frac{C_L}{C_D}\right)(\eta - \eta_e) - \frac{2}{\beta r_0}\left(1 - \frac{1}{2T_e}\right)\ln\left(\frac{\eta}{\eta_e}\right)\right]^{\frac{1}{2}}}$$

can be numerically integrated to find $T(\eta)$ and compared to the closed-form approximation in Eq. (5.80)

$$\ln\left(\frac{T}{T_e}\right) = \frac{\gamma_e - \gamma}{\frac{1}{2}\left[\frac{C_L}{C_D} + \frac{1}{\beta r_0 \eta}\left(1 - \frac{1}{2T_e}\right)\right]}$$

where γ is given by Eq. (5.75):

$$\gamma = -\left[\gamma_e^2 - 2\left(\frac{C_L}{C_D}\right)(\eta - \eta_e) - \frac{2}{\beta r_0}\left(1 - \frac{1}{2T_e}\right)\ln\left(\frac{\eta}{\eta_e}\right)\right]^{\frac{1}{2}}$$

a. Select two specific examples with supercircular entry speeds ($T_e > 0.5$), "medium" entry angles, and realistic lift-to-drag ratios and graphically compare the results using Eqs.(5.78) and (5.80) for $T(\eta)$. You can assume the entry conditions and atmosphere are such that $\beta r_0 = 910$, and $\eta_e = 1x10^{-4}$.

b. How constant is the "constant"

$$\frac{C_L}{C_D} + \frac{1}{\beta r_0 \eta}\left(1 - \frac{1}{2T_e}\right)$$

for each trajectory?

9. During supercircular gliding entry at medium flight-path angles, Eqs. (5.87) and (5.88) give the altitude and magnitude of the maximum deceleration. Compare the effect of changing the entry angle γ_e for an lunar-return capsule ($^R V_e \approx 1.4\,^R V_0$ and $\frac{C_L}{C_D} \approx 0.2$) on the altitude η_* and magnitude $\left(\frac{a_{decel}}{g_0}\right)_*$ of the maximum deceleration. Use the Earth-like values of $\beta = 0.14$ km^{-1}, $r_0 = 6500$ km, $\rho_s = 1.225$ kg/m^3, and $g_s = 9.81$ m/sec^2 and assume $\eta_e = 1 \times 10^{-4}$. Suggestion: Create plots such as the ones below to capture the answer.

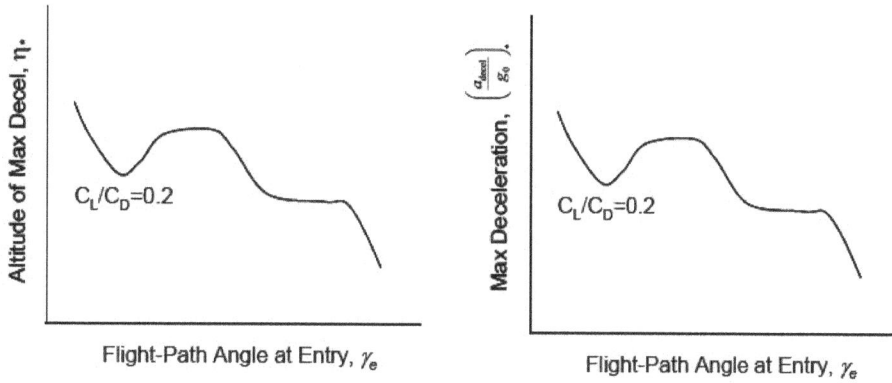

10. For steep ballistic entry (ignoring gravity), plot $\beta r_0 \eta$ as a function of $\dfrac{^R V}{^R V_e}$ for several entry flight-path angles in the range $-\dfrac{\pi}{2} < \gamma_e < 0$. (Note: the equations are not necessarily valid for the entire range – this is for illustration purposes only.) Use the Earth-like values of $\beta = 0.14$ km^{-1}, $r_0 = 6500$ km, $\rho_s = 1.225$ kg/m^3, and $g_s = 9.81$ m/sec^2. If required, assume $\eta_e = 1 \times 10^{-4}$. When $^R V_e = \sqrt{g_0 r_0}$ and γ_e is "small," this plot can be compared to Figure 5-2 for shallow gliding entry with small values of lift. Compare the major similarities/differences. Does it appear the equations would give similar answers near the overlap? What does this say about understanding the type of entry prior to picking the equations to use?

This page intentionally blank.

Chapter 6

Loh's Second-Order Solution

6.1 Introduction

In Chapter 4, we derived dimensionless equations for planar entry (Eqs. (4.53) and (4.54)). Those equations could not be solved analytically so we spent a great deal of effort in Chapter 5 solving for several closed-form approximate solutions (all but one of which were "first-order"). Each of those was for a specific (and somewhat limited) type of entry. Loh, however, derived a more general solution which covers all of the special cases in Chapter 5 (Refs. 39, 58) as well as more general situations. His solution is usually referred to as *Loh's Second-Order Solution*. Loh's theory is empirical, based on the results of extensive numerical integrations of entry trajectories. Even so, it turns out to be quite accurate (limited to the accuracy of the assumptions made in the original derivation of the differential equations, of course) (58:226).

We can begin with the equations of planar entry from Chapter 4, Eqs. (4.53) and (4.54), as are repeated below

$$\frac{dT}{d\eta} = \frac{2T}{\sin\gamma} + \frac{z^2}{\beta r_0 \eta}$$

6.1

$$\frac{d\xi}{d\eta} = \frac{C_L}{C_D} + \frac{z^2}{\beta r_0 \eta}\left(\frac{1}{z} - \frac{1}{2T}\right)\xi$$

6.2

where

$$z = \frac{r_0}{r}$$

6.3

$$T = \frac{1}{2}\left(\frac{{}^R V^2}{g_0 r_0}\right)$$

6.4

$$\xi = \cos\gamma$$

6.5

$$\eta = \frac{\rho\, SC_D}{2m\beta}$$

6.6

Loh's first assumption was to assume a thin atmosphere and set $z = \frac{r_0}{r} = 1$, simplifying the equations to the same ones we used throughout the majority of Chapter 5:

$$\frac{dT}{d\eta} = \frac{2T}{\sin\gamma} + \frac{1}{\beta\, r_0 \eta}$$

6.7

$$\frac{d\xi}{d\eta} = \frac{C_L}{C_D} + \frac{1}{\beta r_0 \eta}\left(1 - \frac{1}{2T}\right)\xi$$

6.8

146

Loh noted through numerical integration of different types of entry trajectories that

$$G = \frac{-1}{\beta r_0 \eta}\left(1 - \frac{1}{2T}\right)\xi \qquad \textbf{6.9}$$

remained nearly constant for each trajectory, even for varying lift-to-drag ratio (39:28; 47:211-212; 58:129). Therefore, although G is actually a function of η, T, and ξ, Loh considered it a constant for integration with respect to η or γ.

6.2 Loh's Unified Solution for Entry Trajectories

Treating G as a constant, our equation of motion for the flight-path angle, Eq. (6.8) becomes:

$$\frac{d\xi}{d\eta} = \frac{C_L}{C_D} - G \qquad \textbf{6.10}$$

For constant lift-to-drag ratio, this can be integrated to give

$$\cos\gamma - \cos\gamma_e = \left(\frac{C_L}{C_D} - G\right)(\eta - \eta_e) \qquad \textbf{6.11}$$

where γ_e and η_e are constants of integration and can be evaluated at the initial point (which may or may not be the same altitude as the reference radius r_0). Replacing G with its definition and rearranging, this becomes:

$$\cos\gamma = \frac{\cos\gamma_e + \dfrac{C_L}{C_D}(\eta - \eta_e)}{1 + \dfrac{1}{\beta r_0 \eta}\left(\dfrac{1}{2T} - 1\right)(\eta - \eta_e)} \qquad \textbf{6.12}$$

To simplify derivations later, this can be rewritten as:

$$\cos\gamma = \frac{\cos\gamma_e + \frac{C_L}{C_D}\eta\left(1 - \frac{\eta_e}{\eta}\right)}{1 + \frac{1}{\beta r_0}\left(\frac{1}{2T} - 1\right)\left(1 - \frac{\eta_e}{\eta}\right)}$$

6.13

To integrate Eq. (6.7), we'll need to change the independent variable to γ. Begin by substituting $\xi = \cos\gamma$ into Eq. (6.10) to relate $d\gamma$ and $d\eta$

$$-\sin\gamma\frac{d\gamma}{d\eta} = \frac{C_L}{C_D} - G$$

6.14

and rearranging:

$$\frac{d\gamma}{d\eta} = \frac{-1}{\sin\gamma}\left(\frac{C_L}{C_D} - G\right)$$

6.15

Dividing Eq. (6.7) by Eq. (6.15) changes the independent variable to γ:

$$\frac{dT}{d\gamma} = \frac{-2T}{\frac{C_L}{C_D} - G} - \frac{\sin\gamma}{\beta r_0\eta\left(\frac{C_L}{C_D} - G\right)}$$

6.16

To complete the variable change, we rearrange Eq. (6.11) as

$$\eta = \eta_e + \frac{\cos\gamma - \cos\gamma_e}{\frac{C_L}{C_D} - G}$$

6.17

and use it to eliminate η in Eq. (6.16):

$$\frac{dT}{d\gamma} + \frac{2T}{\frac{C_L}{C_D} - G} = \frac{-\sin\gamma}{\beta r_0\left[\eta_e\left(\frac{C_L}{C_D} - G\right) + \cos\gamma - \cos\gamma_e\right]}$$

6.18

If the entry starts at high altitude, we can assume $\eta_e \approx 0$ (at least for the purposes of the integration). This reduces the equation for T to:

$$\frac{dT}{d\gamma} + \frac{2T}{\dfrac{C_L}{C_D} - G} = \frac{\sin\gamma}{\beta r_0 (\cos\gamma_e - \cos\gamma)} \qquad \textbf{6.19}$$

This is a non-homogeneous, linear differential equation for T that can be written as

$$\frac{dT}{d\gamma} + KT = f(\gamma) \qquad \textbf{6.20}$$

where

$$K = \frac{2}{\dfrac{C_L}{C_D} - G} \qquad \textbf{6.21}$$

and:

$$f(\gamma) = \frac{\sin\gamma}{\beta r_0 (\cos\gamma_e - \cos\gamma)} \qquad \textbf{6.22}$$

If K is considered a constant, then Eq. (6.20) can be integrated using integrating factors to give:

$$T = Ce^{-K\gamma} + F(\gamma) \qquad \textbf{6.23}$$

where

$$F(\gamma) = e^{-K\gamma} \int e^{K\gamma} f(\gamma) d\gamma \qquad \textbf{6.24}$$

and C is a constant of integration that can be evaluated based on initial conditions. Note that, once the integral in Eq. (6.24) is computed (with K treated as a constant), then

$$K = \frac{2}{\frac{C_L}{C_D} - G} = \frac{2}{\frac{C_L}{C_D} + \frac{1}{\beta r_0 \eta}\left(1 - \frac{1}{2T}\right)\xi} \qquad \textbf{6.25}$$

should be substituted into Eq. (6.23) to make both terms on the right-hand side functions of T, γ, and η. G (and, hence, K) is only considered to be a constant *for the purposes of integration* with respect to γ and η and *not* for evaluating the resulting equations of motion. Evaluating Eq. (6.23) at the entry conditions allows us to solve for the constant of integration

$$C = \left[T_e - F(\gamma_e)\right]e^{K_e \gamma_e} \qquad \textbf{6.26}$$

where:

$$K_e = \frac{2}{\frac{C_L}{C_D} + \frac{1}{\beta r_0 \eta_e}\left(1 - \frac{1}{2T_e}\right)\cos\gamma_e} \qquad \textbf{6.27}$$

Finally, combining these constants gives the solution for kinetic energy:

$$T = T_e e^{(K_e \gamma_e - K\gamma)} + \left[F(\gamma) - F(\gamma_e)e^{(K_e \gamma_e - K\gamma)}\right] \qquad \textbf{6.28}$$

Strictly speaking,

$$K = K(T,\gamma,\eta) = \frac{2}{\frac{C_L}{C_D} + \frac{1}{\beta r_0 \eta}\left(1 - \frac{1}{2T}\right)\xi} \qquad \textbf{6.29}$$

should be used in writing Eq. (6.28), but for simplicity (and to keep the type size big enough to read) the shorthand K has been used. Equations (6.13) and (6.28) constitute Loh's Unified Solution for Entry. Between them, we can solve for any two variables (from T, γ, and η) as a function of the remaining one.

A detail about this solution should be noted. The value of the drag parameter C_D alone is only required in the computation of η_e. If entry is from high altitude, $\eta_e \approx 0$ then the need to know C_D is eliminated (except when recovering the actual altitude from η is required). However, setting $\eta_e = 0$ introduces a singularity in Eq. (6.27); therefore, the constant K should be evaluated at some other point along the trajectory (e.g., a non-zero reference altitude η_0).

6.3 Loh's Second-Order Solution for Entry

In general, Loh's Unified Solution, Eqs. (6.13) and (6.28), are too tedious to use since they both are transcendental in T, γ, and η. Luckily, they can be simplified without destroying their universality. Except in cases where accuracy is paramount (and the previous assumptions in the derivations haven't already negated the accuracy required), we can take advantage of the fact $\dfrac{1}{\beta r_0} \ll 1$ for atmospheres such as Earth, Venus, Mars, and Jupiter (with $\beta r_0 \approx 900$, 500, 350, and 3000, respectively). For $\dfrac{1}{\beta r_0} \ll 1$, Eq. (6.22) becomes

$$f(\gamma) \approx 0 \qquad\qquad \textbf{6.30}$$

and we can completely avoid the numerical integration in Eq. (6.24). Equation (6.20) then becomes simply a homogeneous differential equation:

$$\frac{dT}{d\gamma} + KT = 0 \qquad\qquad \textbf{6.31}$$

This can be separated

$$\frac{dT}{T} = -K d\gamma \qquad\qquad \textbf{6.32}$$

and integrated:

$$\ln\left(\frac{T}{T_e}\right) = -K\left(\gamma - \gamma_e\right) \qquad\qquad \textbf{6.33}$$

Substituting in the definition of K from Eq. (6.25) gives

$$\ln\left(\frac{T}{T_e}\right) = \frac{-2\left(\gamma - \gamma_e\right)}{\dfrac{C_L}{C_D} + \dfrac{1}{\beta r_0 \eta}\left(1 - \dfrac{1}{2T}\right)\cos\gamma} \qquad\qquad \textbf{6.34}$$

For convenience later, this can be rewritten as:

$$\gamma = \gamma_e - \frac{1}{2}\ln\left(\frac{T}{T_e}\right)\left[\frac{C_L}{C_D} + \frac{1}{\beta r_0 \eta}\left(1 - \frac{1}{2T}\right)\cos\gamma\right] \qquad\qquad \textbf{6.35}$$

Equation (6.13) can be left as already found:

$$\cos\gamma = \frac{\cos\gamma_e + \dfrac{C_L}{C_D}\eta\left(1 - \dfrac{\eta_e}{\eta}\right)}{1 + \dfrac{1}{\beta r_0}\left(\dfrac{1}{2T} - 1\right)\left(1 - \dfrac{\eta_e}{\eta}\right)} \qquad\qquad \textbf{6.36}$$

Equations (6.34) (or, equivalently, (6.35)) and (6.36), are Loh's Second-Order Solution for Entry. They are, however, *still* transcendental in T, γ, and η!

The two equations for Loh's solution must be solved simultaneously. A common technique (usually employed when solving by hand) is to "serially" solve one equation then the other and iterate until the solution converges.

However, a word of caution is in order. Equation (6.34) tends to be sensitive to errors when solving for T or T_e, while Eq. (6.36) tends to introduce round-off error when solving for γ or γ_e. The order of the "serial" process can make the difference between converging to a realistic solution and iterating forever. The key thing to realize is that, if one method fails to converge, try rearranging the equations. The same idea also applies to *true* simultaneous solution also; e.g., if Eq. (6.34) causes problems, try rewriting it as Eq. (6.35).

The computational effort to solve the equations can be reduced when η_e corresponds to a high-altitude entry. In this case, $\dfrac{\eta_e}{\eta} \approx 0$ and Eq. (6.36) can be simplified further:

$$\cos\gamma = \frac{\cos\gamma_e + \dfrac{C_L}{C_D}\eta}{1 + \dfrac{1}{\beta r_0}\left(\dfrac{1}{2T} - 1\right)}$$

6.37

Conveniently, even though we have assumed η_e is small, we don't need to worry about the singularity in Eq. (6.27) mentioned in Section 6.2 because the value of K_e never needs to be computed! Solving this (flight-path angle) equation for η:

$$\eta = \frac{\left[1 + \dfrac{1}{\beta r_0}\left(\dfrac{1}{2T} - 1\right)\right]\cos\gamma - \cos\gamma_e}{\dfrac{C_L}{C_D}}$$

6.38

and substituting into the kinetic energy equation Eq. (6.35) gives a relationship for γ in terms of T only (albeit, transcendental). This relation is:

6.39

$$\gamma = \gamma_e - \frac{1}{2}\left(\frac{C_L}{C_D}\right)\ln\left(\frac{T}{T_e}\right)\left\{1 + \frac{1}{\beta\, r_0}\left[\frac{\left(\frac{1}{2T}-1\right)}{1-\frac{\cos\gamma_e}{\cos\gamma}}\right]\right\}^{-1}$$

For specified values of T, this equation can be solved for γ which can then be used in Eq. (6.38) along with T to solve for η. It is possible to restore some accuracy in the solution by rewriting Eq. (6.36) as

$$\eta = \eta_e + \left(\frac{C_L}{C_D}\right)^{-1}\left[\left(\cos\gamma - \cos\gamma_e\right) + \frac{1}{\beta r_0}\left(1-\frac{\eta_e}{\eta}\right)\left(\frac{1}{2T}-1\right)\cos\gamma\right] \qquad \textbf{6.40}$$

and using it to solve for η instead of Eq. (6.38). Even when the assumption

$\dfrac{\eta_e}{\eta} \approx 0$ is used,

$$\eta = \eta_e + \left(\frac{C_L}{C_D}\right)^{-1}\left[\left(\cos\gamma - \cos\gamma_e\right) + \frac{1}{\beta r_0}\left(\frac{1}{2T}-1\right)\cos\gamma\right] \qquad \textbf{6.41}$$

this still retains some accuracy over Eq. (6.38) because the (possibly non-zero) value of η_e is used in the calculation (when known).

Loh has shown that the second-order solution, as derived, is very accurate compared to the numerical integration of Eqs. (6.7) and (6.8). The underlying basis for this solution is that the term G is nearly a constant. Since the assumption is based on observations of extensive numerical integrations of Eqs. (6.7) and (6.8), the accuracy is to be expected. There are physical explanations as to why G remains constant and they will be shown in the upcoming sections.

6.4 Reduction to First-Order Solutions

Chapter 5 presented a series of first-order solutions for specific (and somewhat limited) cases. In each case, the appropriate approximations were made and the equations of motion simplified to the point they could be solved analytically. Each solution is valid only for the specified type of trajectory. Loh's second-order solution, on-the-other-hand, is valid for all types of trajectories. As such, Loh's solution should reduce to the first-order solutions when the corresponding simplifications and approximations are made. Several of these reductions are included in the following sections.

6.4.1 Shallow Gliding Entry

When the flight-path angle is small and the entry altitude high:

$$\sin \gamma \approx \gamma \qquad\qquad\qquad \textbf{6.42}$$

$$\cos \gamma \approx 1 \qquad\qquad\qquad \textbf{6.43}$$

$$\eta_e \approx 0 \qquad\qquad\qquad \textbf{6.44}$$

Thus, Loh's second-order equation Eq. (6.36) becomes:

$$1 = \frac{1 + \dfrac{C_L}{C_D}\eta}{1 + \dfrac{1}{\beta r_0}\left(\dfrac{1}{2T} - 1\right)} \qquad\qquad\qquad \textbf{6.45}$$

Or, when simplified:

$$\frac{1}{\beta r_0}\left(\frac{1}{2T} - 1\right) = \frac{C_L}{C_D}\eta \qquad\qquad\qquad \textbf{6.46}$$

Solving for kinetic energy, this can be written as:

$$T = \frac{1}{2\left[1 + \beta r_0 \eta\left(\frac{C_L}{C_D}\right)\right]} \qquad \text{6.47}$$

This equation is identical to the kinetic energy equation found in the first-order solution for shallow gliding entry (Eq. (5.9)). When $\cos\gamma \approx 1$, the expression for G becomes:

$$G = \frac{-1}{\beta r_0 \eta}\left(1 - \frac{1}{2T}\right) \qquad \text{6.48}$$

Or, slightly rearranging:

$$G\eta = \frac{1}{\beta r_0}\left(\frac{1}{2T} - 1\right) \qquad \text{6.49}$$

Comparing this to Eq. (6.46) shows us:

$$G = \frac{C_L}{C_D} \qquad \text{6.50}$$

In this case, the "constant" G turns out to be equal (or approximately equal) to the constant lift-to-drag ratio (which explains G remains constant along the trajectory).

Turning to the other of Loh's second-order equations, this time in the form of Eq. (6.39) we can write:

$$\frac{\gamma - \gamma_e}{\cos\gamma - \cos\gamma_e} = \frac{-\frac{1}{2}\left(\frac{C_L}{C_D}\right)\ln\left(\frac{T}{T_e}\right)}{(\cos\gamma - \cos\gamma_e) + \frac{1}{\beta r_0}\left(\frac{1}{2T} - 1\right)\cos\gamma} \qquad \text{6.51}$$

When $\cos\gamma \approx \cos\gamma_e \approx 1$, the right-hand side becomes:

$$\frac{-\frac{1}{2}\left(\frac{C_L}{C_D}\right)\ln\left(\frac{T}{T_e}\right)}{(\cos\gamma - \cos\gamma_e) + \frac{1}{\beta r_0}\left(\frac{1}{2T} - 1\right)\cos\gamma} \approx -\frac{\beta r_0}{2}\left(\frac{C_L}{C_D}\right)\ln\left(\frac{T}{T_e}\right)\left(\frac{1}{2T} - 1\right)^{-1} \qquad \textbf{6.52}$$

When the entry speed is near circular speed, $T_e \approx \frac{1}{2}$, so the logarithm term in Eq. (6.52) can be rewritten as:

$$\ln\left(\frac{T}{T_e}\right) = \ln(2T) = -\ln\left(\frac{1}{2T}\right) \qquad \textbf{6.53}$$

At and below circular speed, $\frac{1}{2T} \geq \frac{1}{2}$ so a series expansion for the natural logarithm takes the form:

$$\ln\left(\frac{T}{T_e}\right) = -\ln\left(\frac{1}{2T}\right)$$

$$= -\left(\frac{\frac{1}{2T} - 1}{\frac{1}{2T}} + \cdots\right)$$

$$\approx -2T\left(\frac{1}{2T} - 1\right) \qquad \textbf{6.54}$$

Using this in Eq. (6.51) reduces the right-hand side to:

$$\frac{-\frac{1}{2}\left(\frac{C_L}{C_D}\right)\ln\left(\frac{T}{T_e}\right)}{\left(\cos\gamma - \cos\gamma_e\right) + \frac{1}{\beta r_0}\left(\frac{1}{2T}-1\right)\cos\gamma} \approx \beta r_0\left(\frac{C_L}{C_D}\right)T \qquad \textbf{6.55}$$

The left-hand side requires a little more care, since

$$\frac{\gamma - \gamma_e}{\cos\gamma - \cos\gamma_e} \approx \frac{0}{0} \qquad \textbf{6.56}$$

as $\gamma \to \gamma_e$. Thankfully, l'Hospital's Rule can be used to evaluate the expression in the limit:

$$\lim_{\gamma \to \gamma_e}\left(\frac{\gamma - \gamma_e}{\cos\gamma - \cos\gamma_e}\right) = \frac{-1}{\sin\gamma} \qquad \textbf{6.57}$$

(Technically, the γ on the right-hand-side of Eq. (6.57) should be replaced with γ_e, but since $\gamma \approx \gamma_e$, we will leave it as written.) Finally, replacing the right-hand side of Eq. (6.51) with Eq. (6.55) and the left-hand side with Eq. (6.57) yields (when simplified):

$$\sin\gamma = \frac{-1}{\left(\frac{C_L}{C_D}\right)\beta r_0 T} \qquad \textbf{6.58}$$

This is identical to the corresponding first-order equation (Eq. (5.14)) in Chapter 5 for the flight-path angle.

Equations (6.47) and (6.58) prove Loh's second-order solutions will reduce to the first-order solution for shallow gliding entry. Further, we have seen Loh's "constant" G is essentially equal to the constant lift-to-drag ratio in this situation, giving a physical justification for his empirical observation.

6.4.2 Medium and Steep Gliding Entry at Near Circular Speed

When the flight-path angle is not small, Eq. (6.36), remains the same. However, if we ignore the terms multiplied by $\dfrac{1}{\beta r_0}$, then it reduces to simply:

$$\cos\gamma - \cos\gamma_e = \frac{C_L}{C_D}(\eta - \eta_e)$$

6.59

With the same assumption, Loh's other second-order equation, Eq. (6.34), becomes:

$$\ln\left(\frac{T}{T_e}\right) = \frac{-2(\gamma - \gamma_e)}{\left(\dfrac{C_L}{C_D}\right)}$$

6.60

These equations correspond to Eqs. (5.37) and (5.41) in Chapter 5's first-order solution to medium and steep gliding entry. Note also, these correspond to setting Loh's constant $G \approx 0$, which means the second-order assumption assuming G is constant is reasonable.

6.4.3 Steep Skip Entry

In steep skip entry, two terms in Loh's constant G conspire to make it small: $\cos\gamma \ll 1$ and $\dfrac{1}{\beta r_0} \ll 1$. So, if it is assumed G is small relative to the lift term in Eq. (6.11), we immediately get a solution for flight-path angle:

$$\cos\gamma - \cos\gamma_e = \frac{C_L}{C_D}(\eta - \eta_e)$$

6.61

If the same assumption ($G \ll 1$) is used in Eq. (6.34), we can get a solution for kinetic energy:

$$\ln\left(\frac{T}{T_e}\right) = \frac{-2(\gamma - \gamma_e)}{\left(\dfrac{C_L}{C_D}\right)}$$

6.62

With some simple rearranging, this becomes:

$$\frac{T}{T_e} = \exp\left[\frac{-2(\gamma - \gamma_e)}{\left(\dfrac{C_L}{C_D}\right)}\right]$$

6.63

Equations (6.61) and (6.63) are identical to the first-order solutions we found in Section 5.5 for skip entry. Again in this case, Loh's constant is $G \approx 0$.

It is important to note Loh's equations are not very accurate for the "pull-out" portion of a skip trajectory with a shallow entry angle or large amounts of lift (39:55-59). While Eq. (6.36) appears to give the correct first-order answer of $\cos\gamma_f = \cos\gamma_e$ for $\eta_f = \eta_e$, when compared to the numerically integrated solution, Loh's equations completely miss the oscillatory nature of skip trajectories. However, the "inbound" portion of the skip trajectory is well-matched.

6.4.4   Steep Ballistic Entry

By definition, $C_L = 0$ for ballistic entry. This, combined with ignoring the terms multiplied by $\dfrac{1}{\beta r_0}$, Eq. (6.36) reduces to:

$$\cos\gamma \approx \cos\gamma_e$$

6.64

Like in Eq. (5.123) in Chapter 5, Eq. (6.64) above says the flight-path angle remains essentially constant. However, we can retain a little more accuracy (and eventually get the answer we want) by keeping the terms multiplied by $1/\beta r_0$:

$$\cos\gamma - \cos\gamma_e = -\frac{1}{\beta r_0}\left(\frac{1}{2T}-1\right)\left(1-\frac{\eta_e}{\eta}\right)\cos\gamma$$

$$= \frac{1}{\beta r_0 \eta}\left(1-\frac{1}{2T}\right)(\eta-\eta_e)\cos\gamma \qquad \text{6.65}$$

Turning to Loh's other equation, Eq. (6.35), and setting lift to zero gives us

$$\gamma - \gamma_e = -\frac{1}{2}\ln\left(\frac{T}{T_e}\right)\frac{1}{\beta r_0 \eta}\left(1-\frac{1}{2T}\right)\cos\gamma \qquad \text{6.66}$$

after some minor simplification. When $\cos\gamma \neq 0$, we can divide Eq. (6.66) by Eq. (6.65):

$$\frac{\gamma - \gamma_e}{\cos\gamma - \cos\gamma_e} = \frac{-\frac{1}{2}\ln\left(\frac{T}{T_e}\right)}{\eta - \eta_e} \qquad \text{6.67}$$

We've already seen that the terms on the left-hand side can be rewritten as

$$\frac{\gamma - \gamma_e}{\cos\gamma - \cos\gamma_e} \approx \frac{-1}{\sin\gamma} \qquad \text{6.68}$$

when $\gamma \approx \gamma_e$. Thus, Eq. (6.67) can be solved as:

$$\frac{T}{T_e} = \exp\left[\frac{2(\eta-\eta_e)}{\sin\gamma}\right] = \exp\left[\frac{2(\eta-\eta_e)}{\sin\gamma_e}\right] \qquad \text{6.69}$$

This is not (algebraically, at least) equal to the first-order solution for kinetic energy we found in Section 5.6.1 (Eq. (5.132)). It is, however, the "simplified" solution we found when we ignored gravity in Section 5.6.2 (Eq. (5.148))! In Section 5.6.3 we showed this was a good approximation for the "more complicated" solution found by including gravity. So, have shown Loh's second-order solution reduces to the first-order solution for steep ballistic entry.

We can also examine Loh's constant G in this case. By comparing Eqs. (6.65) and the definition of G

$$G = \frac{-1}{\beta r_0 \eta}\left(1 - \frac{1}{2T}\right)\cos\gamma \qquad \textbf{6.70}$$

we see we can rewrite Eq. (6.65) as:

$$\cos\gamma - \cos\gamma_e = -G(\eta - \eta_e) \qquad \textbf{6.71}$$

Or, if we solve for G:

$$G = \frac{-(\cos\gamma - \cos\gamma_e)}{(\eta - \eta_e)} \qquad \textbf{6.72}$$

When the flight-path angle is nearly constant (as it is in this type of entry), $G \approx 0$. Thus, the constancy of G is again explained with physical justification.

6.5 Second-Order Estimate for Maximum Deceleration

If we, once again, assume the drag force is the dominant term in the tangential direction and lift is dominant in the normal direction, we can write non-dimensional deceleration terms as:

$$\frac{(a_{decel})_v}{g_0} = 2\beta r_0 \eta T \qquad \textbf{6.73}$$

$$\frac{(a_{decel})_L}{g_0} = -2\beta r_0 \eta T \left(\frac{C_L}{C_D}\right) \qquad \textbf{6.74}$$

$$\frac{a_{decel}}{g_0} = 2\beta r_0 \eta T \sqrt{1 + \left(\frac{C_L}{C_D}\right)^2} \qquad \textbf{6.75}$$

Since these are all proportional, the maxima occur at the same point in the trajectory. It will suffice to maximize just one to find that point. In this case, we will solve:

$$\frac{d\left(\dfrac{a_{decel}}{g_0}\right)}{dT} = 0 \qquad \textbf{6.76}$$

Expanding out the derivative, this becomes

$$\eta_* + T_* \left(\frac{d\eta}{dT}\right)_* = 0 \qquad \textbf{6.77}$$

with the "*" subscript denoting the value at the critical point of maximum deceleration. Equation (6.7) can be used to replace $\left(\dfrac{d\eta}{dT}\right)_*$:

$$\eta_* + \frac{\sin \gamma_*}{2} = 0 \qquad \textbf{6.78}$$

(In solving Eq. (6.7) for $\left(\dfrac{d\eta}{dT}\right)$, it was assumed drag forces dominated the equation.) Rearranging, we get the solution

$$\eta_* = -\frac{\sin \gamma_*}{2} \qquad \textbf{6.79}$$

relating the altitude to the flight-path angle at the maximum deceleration point.

We can use Loh's Second-Order Solution to relate the altitude, kinetic energy, and flight-path angle at this critical point. For high-altitude entry, Eqs. (6.35) and (6.37) are appropriate:

$$\gamma_* = \gamma_e - \frac{1}{2}\ln\left(\frac{T_*}{T_e}\right)\left[\frac{C_L}{C_D} + \frac{2}{\beta r_0 \sin\gamma_*}\left(\frac{1}{2T_*}-1\right)\cos\gamma_*\right] \qquad 6.80$$

$$\cos\gamma_* = \frac{\cos\gamma_e - \left(\dfrac{\sin\gamma_*}{2}\right)\left(\dfrac{C_L}{C_D}\right)}{1 + \dfrac{1}{\beta r_0}\left(\dfrac{1}{2T_*}-1\right)} \qquad 6.81$$

Equations (6.79) - (6.81) comprise three equation in three unknowns, and when solved simultaneously (numerically), T_*, η_*, and γ_* can all be found. Thus, with the help of Eq. (6.75), we have the altitude of maximum deceleration η_*, the velocity at which it occurs $^RV_*^2 = 2g_0r_0T_*$, the flight-path angle at that instant γ_*, and the magnitude of the maximum deceleration itself:

$$\left(\frac{a_{decel}}{g_0}\right)_{max} = \left(\frac{a_{decel}}{g_0}\right)_* = 2\beta r_0\eta_* T_* \sqrt{1+\left(\frac{C_L}{C_D}\right)^2} \qquad 6.82$$

It is left as an exercise to compare these values with the corresponding first-order estimates.

6.6 Problems

Computational Insights:

1. For gliding entry from a high altitude into an atmosphere with $\beta r_0 = 910$, graphically compare the first-order, second-order (Loh's), and the "exact" solutions for kinetic energy as a function of altitude and flight-path angle as a function of altitude. Use the parameters $\frac{C_L}{C_D} = 0.2$, $\gamma_e = -3.0°$, $\eta_e = 1x10^{-6}$, and $T_e = 0.5$ to describe the vehicle and entry conditions. Use the solutions found in Section 5.2 for first-order theory and those in Section 6.3 for Loh's second order solution. Numerically integrate Eqs. (6.7) and (6.8) for the "exact" solution.

2. For a high-altitude ballistic entry into an atmosphere with $\beta r_0 = 910$, compare the first-order and second-order estimates of maximum deceleration for the two cases given in the table below.

Case 1	$T_e = 0.5$	$\gamma_e = -1.0°$
Case 2	$T_e = 0.75$	$\gamma_e = -85°$

This page intentionally blank.

Chapter 7

Aerodynamic Heating

7.1 Introduction

The overall focus of this text is to present an understanding of atmospheric entry from a dynamics point-of-view. As such, we've avoided details of the aerodynamics as much as possible. Unfortunately, aerodynamic heating is one of the major tradeoffs in designing (or choosing) a reentry scheme, so we can't avoid it entirely. The goal in this chapter is to provide a basic understanding of the thermal problems encountered during entry without delving into the aerodynamic and heat transfer physics any more than absolutely necessary.

7.2 Fundamentals of Entry Heating

To this point, we've only considered particle dynamics, defining the vehicle completely by C_L/C_D and $C_D S/m$. The thermal dynamics and thermal loads experienced are equally important (even though we'll choose not to study them in as much detail). When a vehicle enters the atmosphere from space, it has a tremendous amount of total energy, due to its kinetic energy as well as its potential energy. As the vehicle enters, a shock wave forms ahead of it, creating a high-temperature region between the shock and body. Further, the relative velocity of the fluid ("air") drops to zero at the vehicle's surface, causing an even greater increase in the static enthalpy of the fluid. (This is called a "zero slip" condition.) Thus, the fluid temperature a short distance from the vehicle surface may be much greater, so thermal energy may be transferred to the vehicle. (The opposite is also true, of course. If the vehicle is hotter than the surrounding fluid, energy is transferred *out of the vehicle* and *to the fluid*.) Two mechanisms move energy between the vehicle and the surrounding environment: convection and radiation.

Convection by energy transport in the boundary layer moves heat between the fluid and the vehicle. Radiation (radiant heating) moves energy from the hot gas to the vehicle and away from the hot surface of the vehicle to cooler areas of the surrounding environment.

Thermal control is a significant design challenge. As Regan notes, the specific kinetic energy dissipated during entry from low-Earth orbit is on the order of $10^7 \ J/kg$ (46:135). This is sufficient to vaporize a heat shield made of pure carbon and equal to half of the initial vehicle mass! A good vehicle design will divert all but a few percent of this energy to the atmosphere rather than the vehicle.

7.3 Thermal Protection Systems

Even if only a small fraction of the initial total energy reaches the vehicle as heat, there can be a significant amount of energy to be dealt with. (A small fraction of a very large number can still be large!) Thermal protection systems can be designed to *absorb* the energy or *reject* it. (Both methods can be used simultaneously.)

With *heat sinks*, a large mass is used to "soak up" heat energy during the entry. The initial suborbital Mercury capsule design used this approach with a beryllium heat shield. The larger energy dissipation requirements for the Mercury orbital flights forced a change to an ablative heat shield. *Ablative techniques* absorb the energy and the dissipate (or reject) it through the vaporizing of an expendable material. Ablative techniques are generally less massive (and more practical) than heat sink techniques. Figure 7-1 shows two examples of ablative systems.

The space shuttles (both US and Soviet) use *radiative techniques* to reject heat. The skin of the vehicle is allowed to absorb heat (through convection) and literally become "red hot." As it heats, energy is lost through radiation. Once an equilibrium is reached, equal amounts of energy are absorbed and rejected with the surface maintaining relatively "safe" temperatures. Note, however, the hot skin must be very well insulated from the rest of the vehicle or else the heat will be transferred to the rest of the vehicle through conduction. The silica tiles on the

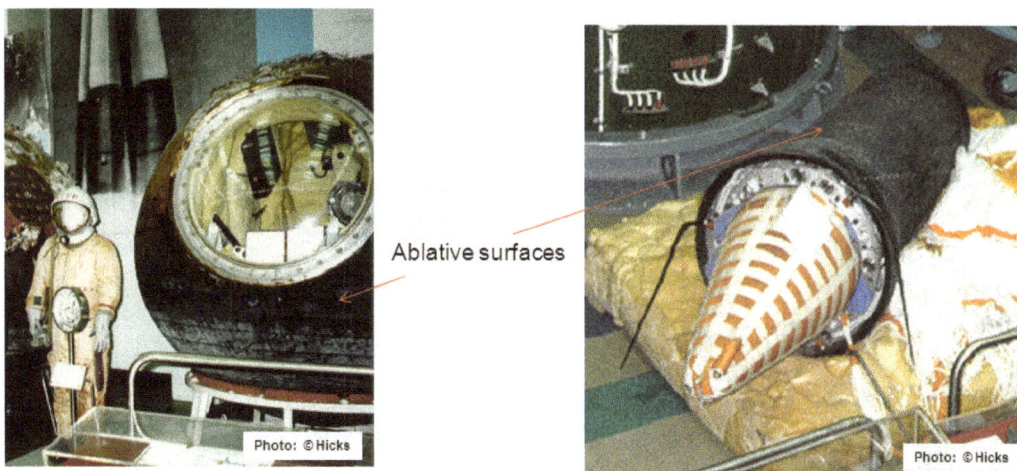

Figure 7-1: Examples of Ablative Systems (Yuri Gagarin's Vostok-1 and a Raduga Ballistic Return Capsule)

space shuttle are designed to be just that type of insulating skin. Figure 7-2 shows the thermal protection tiles on the Soviet shuttle *Buran*.

Heat sinks are best suited to brief, high-drag entry. Radiative cooling techniques are better for long, "gliding" entries where the heating rates are smaller and there is sufficient time for the surface to reach equilibrium. Ablative cooling offers more flexibility in the entry profile than either of these, but sacrifices some reusability (30:300-301).

7.4 Heat Flow into the Vehicle

Two of the most important parameters of the entry trajectory (for thermal analysis) are the total heat input and the maximum rate of heating. The total heat input is important for scaling the cooling system as well as for determining the average temperature rise in the vehicle during the entry. The heating rates are a concern because they impact the maximum instantaneous heat rejection requirements. Tradeoffs between these two are often necessary. For example, long flights at high altitude (e.g., shallow, gliding entry) reduce the heating rates but last longer so the total heating increases.

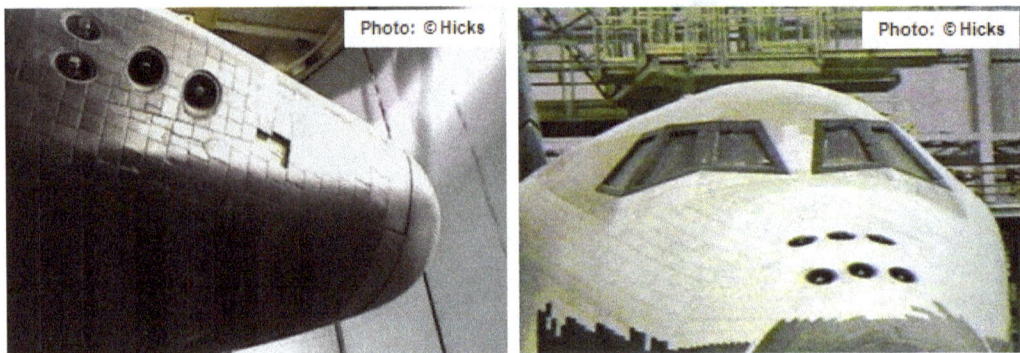

Figure 7-2: Thermal Protection Tiles on Buran Orbiter

Many authors have already looked at these and formulated empirical, theory-based, and/or hybrid approaches to estimating each parameter. While the approaches differ in the numerical factors, range of validity, etc., the basics have remained the same. For simplicity, we'll follow one of the earlier formulations by Allen and Eggers (1; 30:301-305; 58:141-144).

To simplify their analysis, Allen and Eggers made five assumptions:

1.   Convective heat transfer is the dominant form of energy transfer and all radiation transfer can be ignored. This simplification is justified because most vehicle materials will experience similar maximum surface temperatures regardless of the shape, so the radiation away from the vehicle will be approximately the same. Thus, ignoring radiation will not alter the qualitative results even when comparing relative heating between types of entry profiles. (It will, however, impact the quantitative results!)

2.   Real gas effects, particularly dissociation, can be ignored. For speeds below 3 km/sec, this is a good assumption. At higher speeds, this is a conservative assumption resulting in estimates of higher heating rates that are actually experienced (58:141).

3.   Shock-wave boundary-layer interaction can be ignored. This assumption has been shown to hold under speeds of about 6 km/sec (Refs. 37, 38).

4.   Reynolds' analogy is applicable. Reynolds' analogy is one of the most powerful methods for obtaining simple heating estimates (31:81). The analogy is exact only when the Prandtl number is one (Pr=1), but gives useful trends for many other cases (58:142). It also has the distinct advantage of producing a result which contains constants that can be adjusted to improve the accuracy (when compared to experimental results).

5. Prandtl number is one. This allows Reynolds' analogy to be invoked.

Using Allen and Eggers' approach, the primary source of the energy input is assumed to be convective heating from the laminar boundary-lay flow over the vehicle. Thus, we can equate the heat flux (a rate per unit area) at the wall, $\dot{q}_w$, with the change in total enthalpy across the boundary layer:

$$\dot{q}_w = \kappa \frac{\partial \mathcal{T}}{\partial y}\bigg)_w = \left(\frac{\kappa}{C_P}\right)\left(\frac{Nu_L}{L}\right)(H_{oe} - H_w)$$

$$= \left(\frac{\kappa}{C_P}\right)\left(\frac{Nu_L}{L}\right)H_{oe}\left(1 - \frac{H_w}{H_{oe}}\right)$$

$$= \left(\frac{Nu_L}{\text{Pr}}\right)\left(\frac{\mu}{L}\right)H_{oe}\left(1 - \frac{H_w}{H_{oe}}\right) \qquad \textbf{7.1}$$

where:

$$Nu_L = \frac{h^* L}{\kappa} \text{ Nusselt number}$$

$$\text{Pr} = \frac{\mu C_P}{\kappa} = \text{ Prandtl number}$$

κ = coefficient of thermal conductivity

L = reference length

h^* = heat transfer coefficient of fluid

C_P = fluid heat capacity at constant pressure

μ = fluid viscosity

$\mathcal{T}$ = absolute temperature

y = distance from wall

$$H = \frac{V^2}{2} + C_P \mathcal{T} = \text{ total enthalpy}$$

$( )_w$ = conditions at wall of the vehicle

$( )_{oe}$ = conditions at "outer edge" of boundary layer

Total enthalpy is conserved for inviscid flow across the normal shock wave (and one most certainly will exist), so we can relate the total enthalpy at the edge of the boundary layer to conditions just ahead of the vehicle's bow wave:

$$H_{oe} = H_{\infty} = h_{\infty} + \frac{V_{\infty}^2}{2}$$

$$= C_P \mathcal{T}_{\infty} + \frac{V_{\infty}^2}{2} \qquad \textbf{7.2}$$

The "∞" subscript denotes conditions upstream of the vehicle and its effects and $h_{\infty} = C_P \mathcal{T}_{\infty}$ is specific enthalpy upstream. For reasonably fast entry into Earth's atmosphere, $C_P \mathcal{T}_{\infty} \ll \left(\frac{V_{\infty}^2}{2} \right)$ (30:303). Thus, Eq. (7.2) reduces to simply:

$$H_{oe} \approx \frac{V_{\infty}^2}{2} \qquad \textbf{7.3}$$

We can regroup the terms in Eq. (7.1) as

$$\dot{q}_w = (\rho V)_{oe} \left(\frac{Nu_L}{Pr \cdot Re_L} \right) H_{oe} \left(1 - \frac{H_w}{H_{oe}} \right) = (\rho V)_{oe} St \cdot H_{oe} \left(1 - \frac{H_w}{H_{oe}} \right) \qquad \textbf{7.4}$$

where:

$$St = \frac{Nu_L}{Pr \cdot Re_L} = \text{ Stanton number}$$

$$Re_L = \frac{\rho VL}{\mu} = \text{ Reynold's number based on a reference length } L$$

Reynold's analogy for laminar boundary layers lets us approximate the Stanton number in terms of an average skin friction coefficient c_f as $St \approx \frac{c_f}{2}$. Using this and

Eq. (7.3), the relationship for the heating rate at the wall becomes:

$$\dot{q}_w = \frac{c_f}{4}(\rho V)_{oe} V_\infty^2 \left(1 - \frac{H_w}{H_{oe}}\right)$$

7.5

For us, $V_{oe} = V_\infty \approx {}^R V$ and $\frac{H_w}{H_{oe}} \ll 1$, so we arrive at the simple, albeit approximate,

equation for the rate at which heat (per unit area) is transferred to/from the vehicle:

$$\dot{q}_w = \frac{1}{4} c_f \rho\, {}^R V^3$$

7.6

Notice that this energy flux is independent of the vehicle temperature. This flux is an *average* for the vehicle. There will be, in all probability, areas on the vehicle with higher and lower values for $\dot{q}_w$ than those predicted by Eq. (7.6).

It is well known that for any blunt body the bow wave shock (Figure 7-3) is detached and a stagnation region exists on the vehicle at the "nose." Stagnation heating at this point on the body can be approximated by

$$\dot{q}_s = \frac{K}{\sqrt{R}} \rho^n\, {}^R V^m$$

7.7

where R is the radius of curvature for the forward portion of the vehicle, K, n, and m are constants (Refs. 31, 43, 58). K, n, and m can be adjusted to reflect laminar or turbulent flow as well as to match observed data. For our purposes, we will limit ourselves to laminar, incompressible flow behind the shock where the values $n = 1/2$ and $m = 3$ are characteristic. If we slightly redefine the constant in Eq. (7.7), we arrive at the relationship:

$$\dot{q}_s = \frac{k}{\sqrt{R}} \left(\frac{\rho}{\rho_0}\right)^{1/2} \left(\frac{{}^R V}{\sqrt{g_0 r_0}}\right)^3$$

7.8

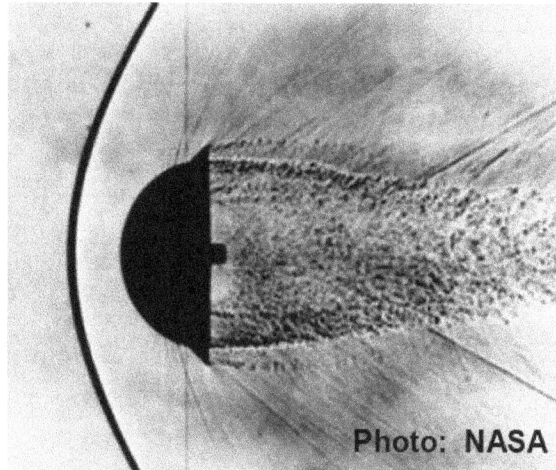

Figure 7-3: Detached Shockwave Forward of Blunt Body

As usual, the "0" subscript denotes our reference point (which, more often than not, is the atmospheric entry point). The stagnation point is a "local" point, so $\dot{q}_s$ can be appropriately termed a "local heat flux" or "local heating rate per unit area." Since the stagnation point tends to be the "hot spot" on the vehicle, $\dot{q}_s$ often represents the maximum heat flux into the vehicle.

Integrating the average heat flux over the vehicle surface gives us the total heating rate (power input) to the body

$$\dot{Q} = \iint\limits_{\substack{surface \\ area}} \dot{q}_w dA = \frac{1}{4} c_f \rho^R V^3 A \qquad \qquad \textbf{7.9}$$

where A is the total surface area of the vehicle. Equation (7.9) can be integrated for the duration of the atmospheric entry to find the total energy transferred into the vehicle.

175

When integrating Eq. (7.9) to find the total energy transfer, it will be easier to change the independent variable to our kinetic energy variable T. Begin with Eq. (4.24):

$$^R\dot{V} = \frac{d}{dt}\left(^RV\right) = -\frac{\rho C_D S \, ^RV^2}{2m} - g_0\left(\frac{r_0}{r}\right)^2 \sin\gamma \qquad \textbf{7.10}$$

During the time when heat transfer is significant, drag is significantly larger than the tangential component of gravity, so the second term can be ignored relative to the first term:

$$\frac{d}{dt}\left(^RV\right) = -\frac{\rho C_D S \, ^RV^2}{2m} \qquad \textbf{7.11}$$

Substituting $^RV = \sqrt{2g_0 r_0}\sqrt{T}$ and simplifying yields:

$$\frac{dt}{dT} = -\frac{1}{2\beta\sqrt{2g_0 r_0}}\frac{1}{\eta T^{3/2}} \qquad \textbf{7.12}$$

This relationship can be used to change the independent variable in Eq. (7.9) to T:

$$\frac{dQ}{dT} = \frac{dQ}{dt}\frac{dt}{dT}$$

$$= \left(\frac{1}{4}c_f \rho \, ^RV^3 A\right)\left(-\frac{1}{2\beta\sqrt{2g_0 r_0}}\frac{1}{\eta T^{3/2}}\right) \qquad \textbf{7.13}$$

This reduces to the simple relationship:

$$\frac{dQ}{dT} = -\left(\frac{c_f A}{2}\right)\left(\frac{m}{C_D S}\right)(g_0 r_0) \qquad \textbf{7.14}$$

This equation is separable and can be immediately integrated to give the total heat transfer to the vehicle:

$$\Delta Q = -\left(\frac{c_f A}{2}\right)\left(\frac{m}{C_D S}\right)(g_0 r_0)\int dT \qquad 7.15$$

Integrating from the entry conditions gives the net heat transfer for the entire trajectory up to the point of interest:

$$\Delta Q = -\left(\frac{c_f A}{2}\right)\left(\frac{m}{C_D S}\right)(g_0 r_0)\int_{T_e}^{T} dT$$

$$= -\left(\frac{c_f A}{2}\right)\left(\frac{m}{C_D S}\right)(g_0 r_0)(T - T_e) \qquad 7.16$$

Equation (7.16) gives us a somewhat expected answer – the heat transfer to the vehicle is directly proportional to the change in kinetic energy during atmospheric entry. For simplicity, ΔQ will be written as simply Q for the remainder of this text.

7.5 Non-Dimensional Heating Equations

Like in the earlier chapters, the equations we found in Section 7.4 can be written in terms of non-dimensional variables. By doing this, we can better focus on the *trends* in the heating relationships without having to sort through the vehicle- or atmosphere-specific characteristics. Since our heating equations are *very* approximate anyway, we are more interested in the qualitative results rather than the quantitative results.

Inserting our definitions for kinetic energy

$$T = \frac{1}{2}\left(\frac{{}^R V^2}{g_0 r_0}\right) \qquad 7.17$$

and altitude η

$$\eta = \frac{\rho SC_D}{2m\beta} \qquad\qquad \textbf{7.18}$$

into Eq. (7.6) and simplifying, we get:

$$\dot{q}_w = \frac{\sqrt{2}m\beta\left(g_0 r_0\right)^{3/2}}{SC_D} c_f \left(\eta T^{3/2}\right) \qquad\qquad \textbf{7.19}$$

The non-dimensional group of terms $\eta T^{3/2}$ is free of characteristics of the vehicle and the atmosphere. In effect, it describes the general nature of the heat transfer per unit time per unit area at the vehicle surface (wall). So, we define

$$\bar{\dot{q}}_w = \eta T^{3/2} \qquad\qquad \textbf{7.20}$$

as the non-dimensional heating rate per unit area (i.e., heat flux) at the wall. Remember, $\bar{\dot{q}}_w$ is an "average" value, not specific to any particular point on the vehicle.) The surface heat flux "profile" can be determined once the appropriate relationships for η and T are substituted into Eq. (7.20). These relationships describe the characteristics of the entry trajectory; therefore, they dictate the atmospheric heating experienced!

Similarly, we can write the stagnation heat flux (Eq. (7.8)) as:

$$\dot{q}_s = \frac{4k}{\sqrt{R}}\left(\frac{m\beta}{SC_D \rho_0}\right)^{1/2}\left(\eta^{1/2} T^{3/2}\right) \qquad\qquad \textbf{7.21}$$

If we define $\bar{\dot{q}}_s$ to be the non-dimensional stagnation heat flux, then

$$\dot{q}_s = \frac{4k}{\sqrt{R}}\left(\frac{m\beta}{SC_D \rho_0}\right)^{1/2}\bar{\dot{q}}_s \qquad\qquad \textbf{7.22}$$

where:

$$\dot{\bar{q}}_s = \eta^{\frac{1}{2}} T^{\frac{3}{2}}$$

7.23

As with the surface heat flux, a stagnation heat flux "profile" can be found once the trajectory is described by providing solutions for η and T.

Finally, we can form an expression for a non-dimensional total heat transfer variable $\bar{Q}$. To do so, we can simply divide Eq. (7.16) by an appropriate constant (with energy dimensions). Choosing that constant isn't quite as obvious, because two logical choices come to mind. These are the initial kinetic energy at entry, $\frac{1}{2}m^R V_e^2$, and the "reference" kinetic energy using the same reference point as we've used in forming our other non-dimensional terms in earlier chapters, $\frac{1}{2}m^R V_0^2 = \frac{1}{2}mg_0 r_0$. We'll choose the latter because it simplifies the algebra later. (And, quite often, the reference conditions *are* the entry conditions!) Thus, we write:

$$\bar{Q} = \frac{\Delta Q}{\frac{1}{2}m^R V_0^2} = -\left(\frac{c_f A}{C_D S}\right)(T - T_e)$$

7.24

At times, we will find it more helpful to leave Eq. (7.24) a little "less simplified" as:

$$\bar{Q} = \frac{\Delta Q}{\frac{1}{2}m^R V_0^2} = -\left(\frac{c_f A}{C_D S}\right)\left(\frac{g_0 r_0}{{}^R V_0^2}\right)(T - T_e)$$

7.25

7.6 Heat Transfer for Various Entry Profiles

In Chapter 5, we found several first-order trajectory profiles which we can use to begin examining the heat transfer during atmospheric entry. In this section, we will examine several of those.

7.6.1 Shallow Gliding Entry

In Section 5.2, we examined atmospheric entry where the vehicle produces enough lift to maintain a hypersonic glide at small flight-path angles for extended periods. For this shallow gliding entry, we derived an expression for kinetic energy as a function of altitude. This relationship was:

$$T = \frac{1}{2\left[1 + \beta r_0 \eta \left(\dfrac{C_L}{C_D}\right)\right]} \qquad \text{7.26}$$

Immediately, we can write the total heat transfer to the vehicle during the entry trajectory as

$$\bar{Q}_f = -\left(\frac{c_f A}{C_D S}\right)\left(T_f - T_e\right) \qquad \text{7.27}$$

where T_f and T_e denote the final and entry kinetic energy, respectively, found by evaluating Eq. (7.26). If entry begins at high altitude ($\eta_e \approx 0$), and nearly circular speed ($T_e \approx \frac{1}{2}$), then this can be written as:

$$\bar{Q}_f = -\left(\frac{c_f A}{C_D S}\right)\left(T_f - \frac{1}{2}\right) \qquad \text{7.28}$$

The limiting case for this type of trajectory is entry at high altitude and flight until $T_f \approx 0$ (or at least until $T_f \ll T_e$). Thus:

$$\bar{Q}_{f_{\text{limit}}} = \frac{1}{2}\left(\frac{c_f A}{C_D S}\right)$$

7.29

$\bar{Q}_{f_{\text{limit}}}$ represents the total energy (heat) transferred to the vehicle during entry. To minimize this total, the skin friction and exposed surface area ("wetted area") can reduced and/or the drag quantity $(C_D S)$ increased. Taken together, small $\left(c_f A\right)$ and large $\left(C_D S\right)$ is equivalent to saying that a blunt vehicle will minimize the total heat absorbed.

The average heating rate per unit area (heat flux) is given by Eq. (7.20). Solving Eq. (7.26) for η

$$\eta = \frac{1 - 2T}{2\beta r_0 \left(\dfrac{C_L}{C_D}\right)T}$$

7.30

and substituting into Eq. (7.20) yields the heat flux as a function of kinetic energy:

$$\dot{\bar{q}}_w = \eta T^{3/2} = \frac{(1-2T)T^{1/2}}{2\beta r_0 \left(\dfrac{C_L}{C_D}\right)}$$

7.31

Alternatively, we could have found the flux as a function of altitude:

$$\dot{\bar{q}}_w = \eta \left\{ \frac{1}{2\left[1 + \beta r_0 \eta \left(\dfrac{C_L}{C_D}\right)\right]} \right\}^{3/2}$$

7.32

Figure 7-4 shows this heat flux as a function of altitude for an Earth-like atmosphere ($\beta r_0 = 910$). Notice that it only takes a small amount of lift can dramatically decrease the peak flux experienced.

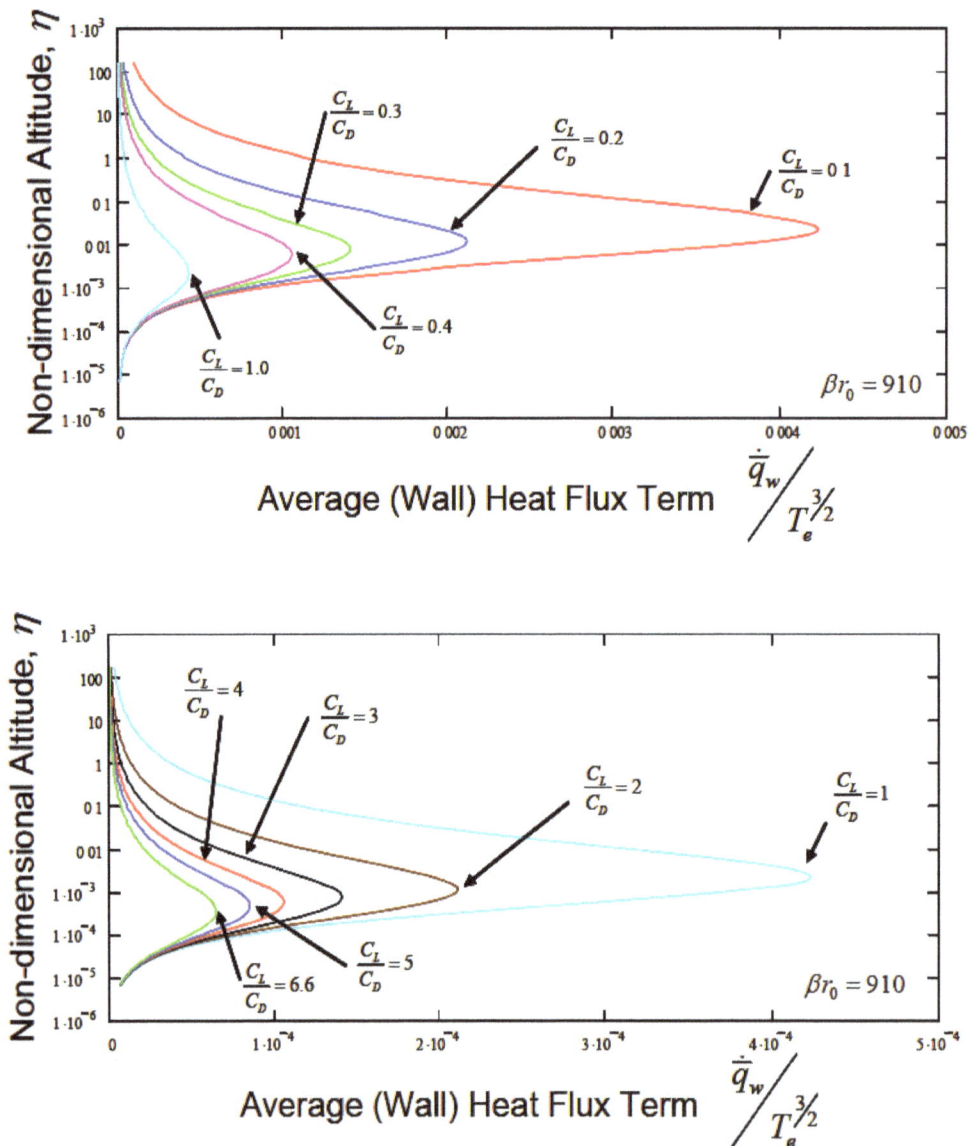

Figure 7-4: *Average (Wall) Heat Flux for Shallow Gliding Entry at Near Circular Speeds*

The maximum value of $\dot{\bar{q}}_w$ is important in analyzing the peak "average" heat load the vehicle thermal system will need to handle. It is easiest to find the point at which this maximum occurs from Eq. (7.31). Solving

$$\frac{d\dot{\bar{q}}_w}{dT} = 0 \qquad\qquad \textbf{7.33}$$

gives the kinetic energy where the maximum occurs:

$$T_* = \frac{1}{6} \qquad\qquad \textbf{7.34}$$

This has the surprising result of being independent of vehicle characteristics! The corresponding maximum is

$$\dot{\bar{q}}_{w_{max}} = \dot{\bar{q}}_{w_*} = \frac{1}{3\sqrt{6}\beta r_0\left(\dfrac{C_L}{C_D}\right)} \qquad\qquad \textbf{7.35}$$

and occurs at an altitude of

$$\eta_* = \frac{2}{\beta r_0\left(\dfrac{C_L}{C_D}\right)} \qquad\qquad \textbf{7.36}$$

and velocity of :

$$\frac{^R V_*}{\sqrt{g_0 r_0}} = \frac{\sqrt{3}}{3} \qquad\qquad \textbf{7.37}$$

Similarly, the stagnation heat flux profile can be found using Eqs. (7.23), (7.26), and (7.30). In terms of kinetic energy the profile is

$$\dot{\bar{q}}_s = \eta^{1/2} T^{3/2} = \left(\frac{1-2T}{2\beta r_0 \left(\dfrac{C_L}{C_D} \right)} \right)^{1/2} T \qquad \textbf{7.38}$$

and in terms of the altitude it is:

$$\dot{\bar{q}}_s = \frac{\eta^{1/2}}{2\sqrt{2}} \left[1 + 2\beta r_0 \eta \left(\frac{C_L}{C_D} \right) \right]^{-3/2} \qquad \textbf{7.39}$$

This is plotted in Figure 7-5.

The maximum stagnation flux $\dot{\bar{q}}_{s*}$ is useful for estimating the peak heat load the vehicle thermal system will need to handle (albeit over a small area). Solving

$$\frac{d\dot{\bar{q}}_s}{dT} = 0 \qquad \textbf{7.40}$$

gives the kinetic energy where the maximum occurs:

$$T_* = \frac{1}{3} \qquad \textbf{7.41}$$

As before, note that this value is completely independent of vehicle characteristics!

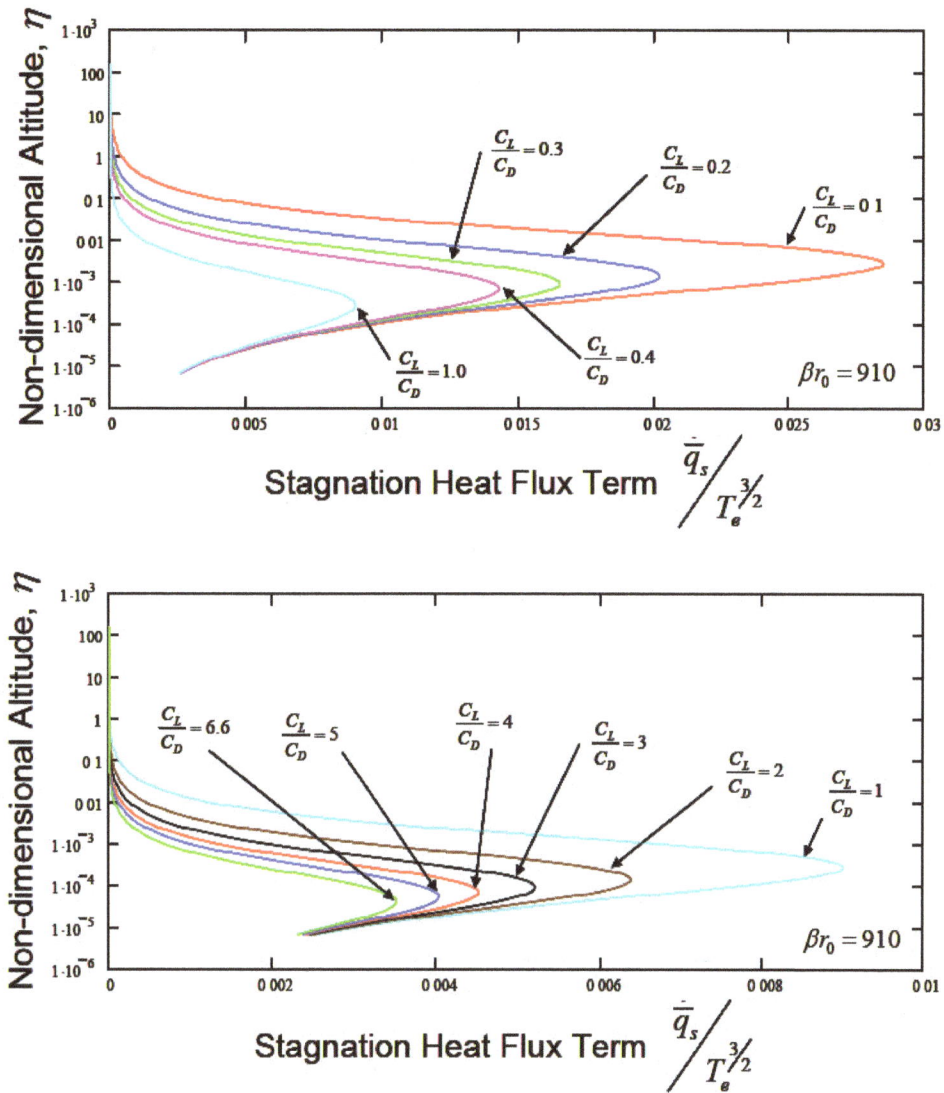

*Figure 7-5: **Stagnation Heat Flux for Shallow Gliding Entry at Near Circular Speeds***

The corresponding maximum is

$$\dot{\bar{q}}_{s_{max}} = \dot{\bar{q}}_{s_*} = \frac{1}{3\sqrt{6\beta r_0 \left(\dfrac{C_L}{C_D}\right)}}$$

7.42

185

at an altitude of

$$\eta_* = \cfrac{1}{2\beta r_0 \left(\cfrac{C_L}{C_D}\right)}$$

7.43

and velocity of :

$$\frac{{}^R V_*}{\sqrt{g_0 r_0}} = \frac{\sqrt{6}}{3}$$

7.44

Comparing Eqs. (7.36) and (7.43) reveals the point of maximum stagnation heat flux occurs earlier (higher) in the entry trajectory than does the point of maximum average heat flux. Similarly, Eqs. (7.35) and (7.42) confirm the stagnation heating is, under realistic conditions, larger than the average heating. Figure 7-6 combines both heat flux terms for a representative entry trajectory.

The maximum heating fluxes and where they occur during shallow gliding entry are summarized in Table 7-1.

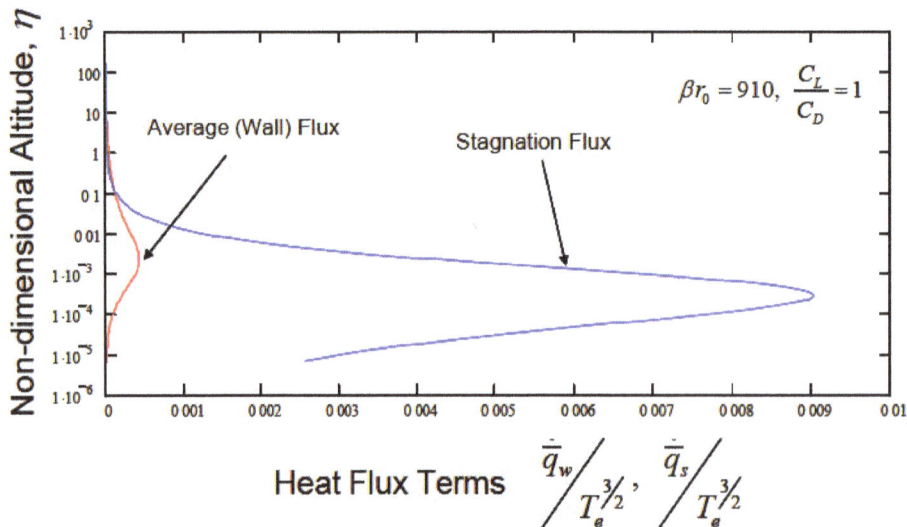

Figure 7-6: Typical Heat Flux for Shallow Gliding Entry

Table 7-1: Points of Maximum Heat Flux for Shallow Gliding Entry

	$\dot{q}_{w_*}$	$\dot{q}_{s_*}$
Maximum Value	$\dfrac{1}{3\sqrt{6}\beta r_0\left(\dfrac{C_L}{C_D}\right)}$	$\dfrac{1}{3\sqrt{6\beta r_0\left(\dfrac{C_L}{C_D}\right)}}$
T_*	$\dfrac{1}{6}$	$\dfrac{1}{3}$
η_*	$\dfrac{2}{\beta r_0\left(\dfrac{C_L}{C_D}\right)}$	$\dfrac{1}{2\beta r_0\left(\dfrac{C_L}{C_D}\right)}$
$\dfrac{{}^R V_*}{\sqrt{g_0 r_0}}$	$\dfrac{\sqrt{3}}{3}$	$\dfrac{\sqrt{6}}{3}$

7.6.2 Medium and Steep Gliding Entry at Near Circular Speed

We examined medium and steep gliding entry at near circular speeds in Section 5.3. In that section, we found the kinetic energy in terms of the flight-path angle and the entry conditions:

$$T = T_e \exp\left[-\frac{2(\gamma - \gamma_e)}{\left(\dfrac{C_L}{C_D}\right)}\right] \qquad \textbf{7.45}$$

The flight-path angle, in turn, could be written in terms of the altitude and entry conditions:

$$\eta = \frac{\cos\gamma - \cos\gamma_e}{\left(\dfrac{C_L}{C_D}\right)} + \eta_e \qquad\qquad 7.46$$

For high-altitude entry, $\eta_e \approx 0$ and this is simplified:

$$\eta \approx \frac{\cos\gamma - \cos\gamma_e}{\left(\dfrac{C_L}{C_D}\right)} \qquad\qquad 7.47$$

With the aid of Eq. (7.24), we can formulate the total heat transfer to the vehicle as

$$\bar{Q}_f = \left(\frac{c_f A}{C_D S}\right) T_e \left\{ 1 - \exp\left[-\frac{2(\gamma_f - \gamma_e)}{\left(\dfrac{C_L}{C_D}\right)} \right] \right\} \qquad\qquad 7.48$$

where the final flight-path angle γ_f is found by solving Eq. (7.47) at the final altitude:

$$\gamma_f = \cos^{-1}\left[\left(\frac{C_L}{C_D}\right)\eta_f + \cos\gamma_e \right] \qquad\qquad 7.49$$

Care should be taken when solving this inverse cosine to insure the angle found satisfies the physical requirement:

$$\gamma_f \leq 0 \qquad\qquad 7.50$$

Unlike for the case of shallow gliding entry, we *can't* assume $T_f \approx 0$ in order to find a limiting value for $\bar{Q}_f$. With steep entry angles, it is entirely possible the vehicle will impact the ground before the kinetic energy is significantly decreased from its entry value. We can, however, see from Eq. (7.48) that blunt bodies *appear* to reduce the total heat absorbed. It is left as an exercise to prove if that is actually the case.

Turning to examining the heating rates, we can start with the average wall flux, given by Eq. (7.20):

$$\dot{\bar{q}}_w = \eta T^{3/2} = \left[\frac{T_e^{3/2}\left(\cos\gamma - \cos\gamma_e\right)}{\left(\frac{C_L}{C_D}\right)} \right] \exp\left[-\frac{3\left(\gamma - \gamma_e\right)}{\left(\frac{C_L}{C_D}\right)} \right] \qquad \textbf{7.51}$$

Figure 7-7 plots Eqs. (7.47) and (7.51) simultaneously for an Earth-like atmosphere. While the plots are for $\gamma_e = -90°$, the basic shape and trends are the same for less steep entry. (In the figure, the curves end abruptly for larger values

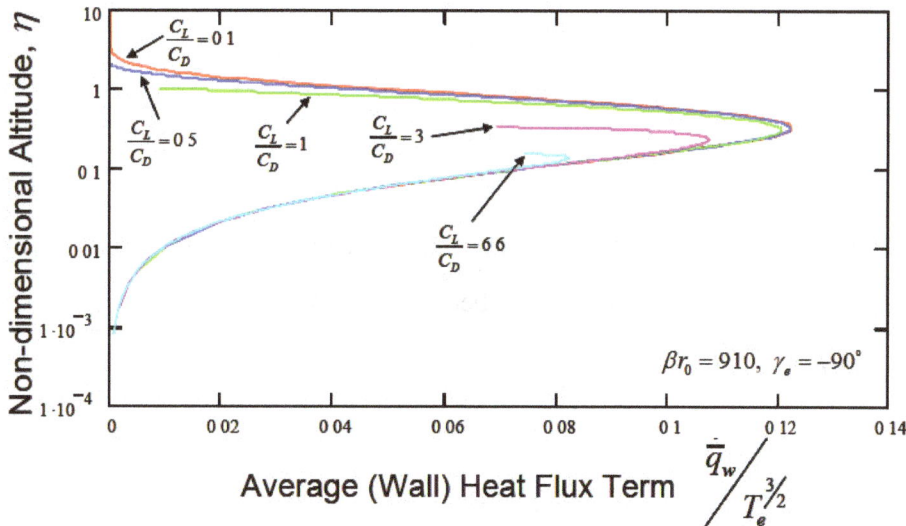

Figure 7-7: Average (Wall) Heat Flux for Steep Gliding Entry at Near Circular Speeds (High-Altitude Entry)

of C_L/C_D because there is sufficient lift to increase the flight-path angle from γ_e to zero. When the flight-path angle reaches zero, the curve stops because the vehicle begins to climb.)

The point where $\ddot{q}_w$ is a maximum is found in terms of the flight-path angle by solving:

$$\frac{d\dot{q}_w}{d\gamma} = 0 \qquad\qquad \textbf{7.52}$$

Taking this derivative and equating to zero doesn't give us as simple of a solution as in the previous section. Instead, we need to solve

$$\sin\gamma_* + \frac{3}{\left(\dfrac{C_L}{C_D}\right)}\left(\cos\gamma_* - \cos\gamma_e\right) = 0 \qquad\qquad \textbf{7.53}$$

to find γ_*. Equation (7.53) can be solved numerically, or with a little work, a "simplified" closed-form solution can be found:

$$\gamma_* = 2\tan^{-1}\left[\frac{\left(\dfrac{C_L}{C_D}\right) - \sqrt{\left(\dfrac{C_L}{C_D}\right)^2 + 9\sin^2\gamma_e}}{3\left(1+\cos\gamma_e\right)}\right] \qquad\qquad \textbf{7.54}$$

The numerator in the arctangent is negative, so the principle value of the will have the proper sign ($\gamma_* < 0$). Once γ_* has been found, the corresponding kinetic energy is found with Eq. (7.45):

$$T_* = T_e \exp\left[-\frac{2\left(\gamma_* - \gamma_e\right)}{\left(\dfrac{C_L}{C_D}\right)}\right] \qquad\qquad \textbf{7.55}$$

Using the definition of our kinetic energy term in this, we can write the relative change in velocity from entry to this point as

$$\frac{^R V_*}{^R V_e} = \exp\left[-\frac{(\gamma_* - \gamma_e)}{\left(\frac{C_L}{C_D}\right)}\right]$$ 7.56

Equations (7.47) and (7.53) combine to give the corresponding altitude:

$$\eta_* = -\frac{\sin\gamma_*}{3}$$ 7.57

Finally, the maximum value for the average heat flux at the vehicle surface ("wall") can be computed:

$$\dot{\bar{q}}_{w_{max}} = \dot{\bar{q}}_{w_*} = -\frac{T_e^{3/2}\sin\gamma_*}{3}\exp\left[-\frac{3(\gamma_* - \gamma_e)}{\left(\frac{C_L}{C_D}\right)}\right]$$ 7.58

Following the same process, the stagnation heat flux is written as

$$\dot{\bar{q}}_s = \eta^{1/2}T^{3/2}$$

$$= T_e^{3/2}\left[\frac{(\cos\gamma - \cos\gamma_e)}{\left(\frac{C_L}{C_D}\right)}\right]^{1/2}\exp\left[-\frac{3(\gamma - \gamma_e)}{\left(\frac{C_L}{C_D}\right)}\right]$$ 7.59

in terms of the flight-path angle. Figure 7-8 shows the altitude/stagnation heat flux relationship found by evaluating Eqs. (7.47) and (7.59) simultaneously (with $\gamma_e = -90°$). The maximum stagnation flux $\overline{\dot{q}}_{s_*}$ is again found by solving

$$\frac{d\overline{\dot{q}}_s}{d\gamma} = 0 \qquad\qquad \textbf{7.60}$$

for the critical flight-path angle γ_*. In this case, the equation to be solved for γ_* is:

$$\sin \gamma_* + \frac{6}{\left(\dfrac{C_L}{C_D}\right)}(\cos \gamma_* - \cos \gamma_e) = 0 \qquad\qquad \textbf{7.61}$$

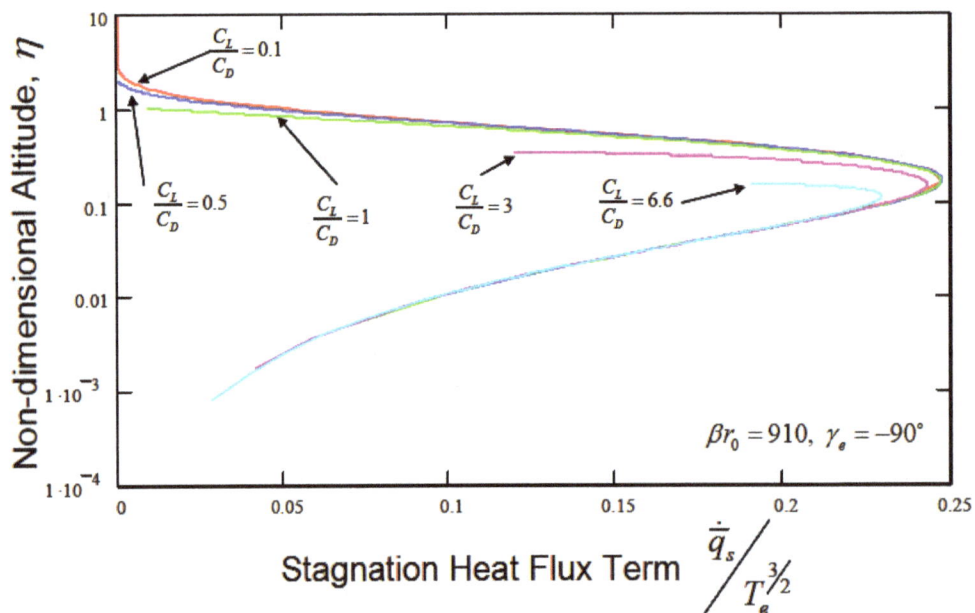

Figure 7-8: *Stagnation Heat Flux for Steep Gliding Entry at Near Circular Speeds (High-Altitude Entry)*

Or, similar to what we found earlier, we can solve for γ_*:

$$\gamma_* = 2\tan^{-1}\left[\frac{\left(\dfrac{C_L}{C_D}\right) - \sqrt{\left(\dfrac{C_L}{C_D}\right)^2 + 36\sin^2\gamma_e}}{6\left(1+\cos\gamma_e\right)}\right] \qquad \textbf{7.62}$$

(Again, this shows us the proper value of the flight-path angle is negative.) The maximum occurs at an altitude of:

$$\eta_* = -\frac{\sin\gamma_*}{6} \qquad \textbf{7.63}$$

The corresponding expressions for the kinetic energy and velocity at this point,

$$T_* = T_e\exp\left[-\frac{2\left(\gamma_* - \gamma_e\right)}{\left(\dfrac{C_L}{C_D}\right)}\right] \qquad \textbf{7.64}$$

$$\frac{{}^R V_*}{{}^R V_e} = \exp\left[-\frac{\left(\gamma_* - \gamma_e\right)}{\left(\dfrac{C_L}{C_D}\right)}\right] \qquad \textbf{7.65}$$

appear to be the same as Eqs. (7.55) and (7.56). However, the *value* for γ_* in these equations will be different when identifying the point of $\dot{\bar{q}}_{w_{max}}$ (Eqs. (7.54) - (7.57)) and $\dot{\bar{q}}_{s_{max}}$ (Eqs. (7.62) - (7.65)).

Using these expressions in Eq. (7.59), we can express the maximum stagnation heat flux as

$$\dot{q}_{s_{max}} = \dot{q}_{s_*} = T_e^{3/2} \left(-\frac{\sin \gamma_*}{6} \right)^{1/2} \exp\left[-\frac{3(\gamma_* - \gamma_e)}{\left(\dfrac{C_L}{C_D} \right)} \right] \qquad \textbf{7.66}$$

once we've found γ_*.

Comparing Eqs. (7.57) and (7.63) does not readily tell us which heat flux occurs first because the value of γ_* is computed differently in each equation. The relative magnitude of the maximum fluxes isn't evident from Eqs. (7.59) and (7.66) either. These comparisons are left as an exercise.

The maximum heating fluxes and where they occur during medium and steep gliding entry are summarized in Table 7-2.

7.6.3   Skip Entry

We examined skip entry in Section 5.5. In that section, we found the kinetic energy in terms of the flight-path angle and the entry conditions:

$$T = T_e \exp\left[\frac{-2(\gamma - \gamma_e)}{\left(\dfrac{C_L}{C_D} \right)} \right] \qquad \textbf{7.67}$$

Similarly, we found the relationship between flight-path angle and altitude:

$$\cos \gamma - \cos \gamma_e = \frac{C_L}{C_D}(\eta - \eta_e) \qquad \textbf{7.68}$$

Table 7-2: Points of Maximum Heat Flux for Medium and Steep Gliding Entry

	$\dot{\bar{q}}_{w_*}$	$\dot{\bar{q}}_{s_*}$
Maximum Value	$-\dfrac{T_e^{3/2}\sin\gamma_*}{3}\exp\left[-\dfrac{3(\gamma_*-\gamma_e)}{\left(\dfrac{C_L}{C_D}\right)}\right]$	$T_e^{3/2}\left(-\dfrac{\sin\gamma_*}{6}\right)^{1/2}\exp\left[-\dfrac{3(\gamma_*-\gamma_e)}{\left(\dfrac{C_L}{C_D}\right)}\right]$
T_*	$T_e\exp\left[-\dfrac{2(\gamma_*-\gamma_e)}{\left(\dfrac{C_L}{C_D}\right)}\right]$	$T_e\exp\left[-\dfrac{2(\gamma_*-\gamma_e)}{\left(\dfrac{C_L}{C_D}\right)}\right]$
η_*	$-\dfrac{\sin\gamma_*}{3}$	$-\dfrac{\sin\gamma_*}{6}$
$\dfrac{^RV_*}{^RV_e}$	$\exp\left[-\dfrac{(\gamma_*-\gamma_e)}{\left(\dfrac{C_L}{C_D}\right)}\right]$	$\exp\left[-\dfrac{(\gamma_*-\gamma_e)}{\left(\dfrac{C_L}{C_D}\right)}\right]$
γ_* **given by**	$\sin\gamma_*+\dfrac{3}{\left(\dfrac{C_L}{C_D}\right)}(\cos\gamma_*-\cos\gamma_e)=0$	$\sin\gamma_*+\dfrac{6}{\left(\dfrac{C_L}{C_D}\right)}(\cos\gamma_*-\cos\gamma_e)=0$

At the end of a skip, we found $\gamma_f=-\gamma_e$, so we can write the final kinetic energy (after one skip) as:

$$T_f = T_e\exp\left[\dfrac{4\gamma_e}{\left(\dfrac{C_L}{C_D}\right)}\right]$$

7.69

Thus, using Eq. (7.24), the total heat transfer during each skip is:

$$\bar{Q}_f = -\left(\frac{c_f A}{C_D S}\right) T_e \left\{ \exp\left[\frac{4\gamma_e}{\left(\frac{C_L}{C_D}\right)}\right] - 1 \right\}$$

7.70

To study the heat transfer during multiple skips, let T_{e_i} and T_{f_i} be the initial (entry) and final (exit) kinetic energies, respectively, during the ith skip. Then, the total heat transfer during n passes into the atmosphere is the sum of that generated in each skip:

$$\bar{Q}_{f_{total}} = -\left(\frac{c_f A}{C_D S}\right) \sum_{i=1}^{n} \left(T_{f_i} - T_{e_i}\right)$$

7.71

Using the equations in Chapter 2 for the orbital portion between atmospheric encounters, we can find the relationship between $T_{e_{i+1}}$ and T_{f_i}. Or, we can accept from the symmetry of the orbit that

$$T_{e_{i+1}} = T_{f_i}$$

7.72

and:

$$\begin{aligned}
\bar{Q}_{f_{total}} &= -\left(\frac{c_f A}{C_D S}\right) \sum_{i=1}^{n} \left(T_{e_{i+1}} - T_{e_i}\right) \\
&= -\left(\frac{c_f A}{C_D S}\right) \left(T_{e_{n+1}} - T_{e_1}\right) \\
&= -\left(\frac{c_f A}{C_D S}\right) \left(T_{f_n} - T_{e_1}\right)
\end{aligned}$$

7.73

Thus, the overall heat transfer during multiple skips looks identical to what we found for shallow gliding entry (Eq. (7.27)) -- simply proportional to the difference in initial and final kinetic energy. If the skip entry is followed by a shallow glide, we can "steal" the result for the limiting (maximum) total heat transfer from Section 7.6.1:

$$\bar{Q}_{f_{\text{limit}}} = \left(\frac{c_f A}{C_D S} \right) T_{e_1} \qquad \textbf{7.74}$$

Or, when the initial entry speed is near circular:

$$\bar{Q}_{f_{\text{limit}}} = \frac{1}{2} \left(\frac{c_f A}{C_D S} \right) \qquad \textbf{7.75}$$

Looking at this equation, we can see a blunt vehicle will minimize the total heat absorbed.

The heat flux equations are easily found at this point. First, solve Eq. (7.68) for η:

$$\eta = \frac{\cos \gamma - \cos \gamma_e}{\left(\dfrac{C_L}{C_D} \right)} + \eta_e \qquad \textbf{7.76}$$

Then, compare our solution skip entry (Eqs. (7.67) and (7.76)) with what we found for steep gliding entry (Eqs. (7.45) and (7.46)). The equations are the same (albeit, for dramatically different reasons); therefore, we'll have the same results for the flux rates, maximum flux rates, and critical altitudes! There is no need to repeat the derivations.

Plots of the heat fluxes are similar to those in the previous section, except that the flight-path angle can be assumed to run the range $\gamma_e \rightarrow -\gamma_e$ during the skip. Also, since there is no requirement for the entry to be steep, we need to look at the heating at various entry angles in addition to with various C_L/C_D values. Figure 7-9 shows the heat fluxes for an extremely steep ($\gamma_e = -90°$) skip entry with three lift-to-drag ratios. Note that for $C_L/C_D = 0.1$, there isn't enough lift to "pull out" of the dive. Figure 7-10 shows the same information for a shallow

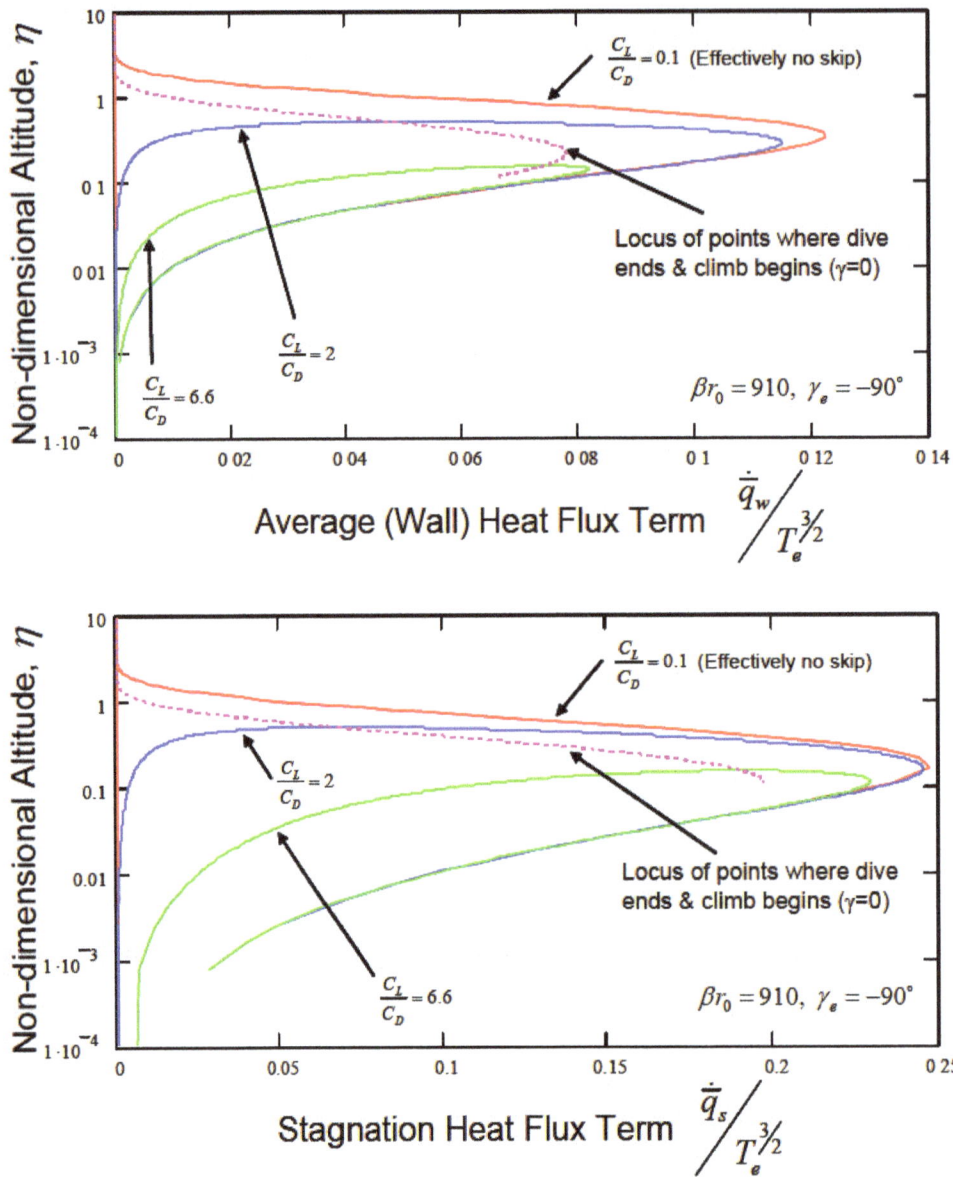

Figure 7-9: Heat Flux for a Steep Skip Entry (High-Altitude Entry)

($\gamma_e = -5°$) skip entry. In it, all of the calculated trajectories are able to complete the skip. By comparing the two plots, we can observe that more lift and less steep entries result in lower heat fluxes. (What isn't shown, however, is the total heat

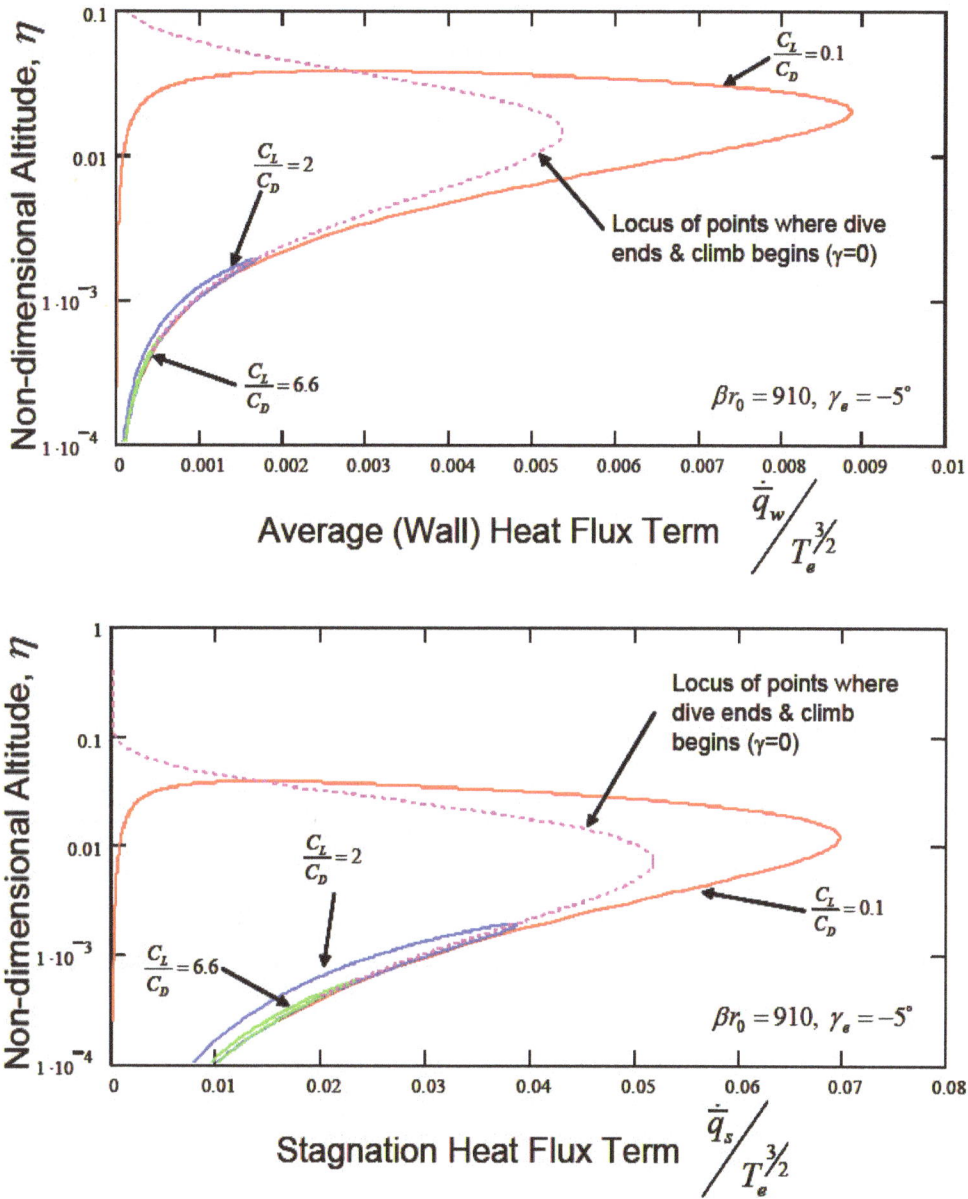

Figure 7-10: Heat Flux for a Shallow Skip Entry (High-Altitude Entry)

transfer to the vehicle and the final kinetic energy remaining to be dissipated! Both may be important in the trajectory selection.) Table 7-3 summarizes the critical values for heat fluxes.

Table 7-3: Points of Maximum Heat Flux for Skip Entry

	$\dot{q}_{w_*}$	$\dot{q}_{s_*}$
Maximum Value	$-\dfrac{T_e^{3/2}\sin\gamma_*}{3}\exp\left[-\dfrac{3(\gamma_*-\gamma_e)}{\left(\dfrac{C_L}{C_D}\right)}\right]$	$T_e^{3/2}\left(-\dfrac{\sin\gamma_*}{6}\right)^{1/2}\exp\left[-\dfrac{3(\gamma_*-\gamma_e)}{\left(\dfrac{C_L}{C_D}\right)}\right]$
T_*	$T_e\exp\left[-\dfrac{2(\gamma_*-\gamma_e)}{\left(\dfrac{C_L}{C_D}\right)}\right]$	$T_e\exp\left[-\dfrac{2(\gamma_*-\gamma_e)}{\left(\dfrac{C_L}{C_D}\right)}\right]$
η_*	$-\dfrac{\sin\gamma_*}{3}$	$-\dfrac{\sin\gamma_*}{6}$
$\dfrac{^R V_*}{^R V_e}$	$\exp\left[-\dfrac{(\gamma_*-\gamma_e)}{\left(\dfrac{C_L}{C_D}\right)}\right]$	$\exp\left[-\dfrac{(\gamma_*-\gamma_e)}{\left(\dfrac{C_L}{C_D}\right)}\right]$
γ_* **given by**	$\sin\gamma_*+\dfrac{3}{\left(\dfrac{C_L}{C_D}\right)}(\cos\gamma_*-\cos\gamma_e)=0$	$\sin\gamma_*+\dfrac{6}{\left(\dfrac{C_L}{C_D}\right)}(\cos\gamma_*-\cos\gamma_e)=0$

7.6.4 Steep Ballistic Entry

The next trajectory type of interest is that of steep ballistic entry. We first examined this type of entry back in Section 5.6. Before looking at the more precise solution we found, it is helpful to look at a somewhat simplified analysis.

7.6.4.1 Steep Ballistic Entry, Ignoring Gravity

We've already shown (in Section 5.6.2) that we can find a somewhat "simplified" approximation for steep ballistic entry by ignoring gravity. As a first analysis, we will use that solution to solve for the heat transfer during these types of entries. The solution was

$$\cos \gamma = \cos \gamma_e \qquad \textbf{7.77}$$

(more precisely, $\gamma = \gamma_e$) and:

$$T = T_e \exp\left[\frac{2(\eta - \eta_e)}{\sin \gamma_e}\right] \qquad \textbf{7.78}$$

In this case, we'll write the total heat transfer slightly differently by using Eq. (7.25)

$$\overline{Q}_f = -\left(\frac{c_f A}{C_D S}\right)\left(\frac{g_0 r_0}{{}^R V_0^2}\right)\left\{T_e \exp\left[\frac{2(\eta_f - \eta_e)}{\sin \gamma_e}\right] - T_e\right\} \qquad \textbf{7.79}$$

which, with some simplification, this becomes:

$$\overline{Q}_f = -\frac{1}{2}\left(\frac{c_f A}{C_D S}\right)\left(\frac{{}^R V_e}{{}^R V_0}\right)^2\left\{\exp\left[\frac{2(\eta_f - \eta_e)}{\sin \gamma_e}\right] - 1\right\} \qquad \textbf{7.80}$$

When entry is from a high altitude, $\overline{Q}_f$ can be reduced further to:

$$\overline{Q}_f = -\frac{1}{2}\left(\frac{c_f A}{C_D S}\right)\left(\frac{{}^R V_e}{{}^R V_0}\right)^2\left[\exp\left(\frac{2\eta_f}{\sin \gamma_e}\right) - 1\right] \qquad \textbf{7.81}$$

Two different and distinct "classes" of steep ballistic entry are represented by Eq. (7.81).

When the vehicle has a "relatively" small mass,

$$\eta_f = \frac{\rho_f SC_D}{2m\beta} \gg 1 \qquad\qquad \textbf{7.82}$$

after it has descended into the appreciable atmosphere. Thus, for "light" vehicles,

$$\left| \frac{2\eta_f}{\sin \gamma_e} \right| \gg 1 \qquad\qquad \textbf{7.83}$$

and the exponential term in Eq. (7.81) approaches zero and the total heat transfer is simply:

$$\bar{Q}_f \approx \frac{1}{2} \left(\frac{c_f A}{C_D S} \right) \left(\frac{{}^R V_e}{{}^R V_0} \right)^2 \qquad\qquad \textbf{7.84}$$

Note that this is mathematically equivalent to assuming $T_f \ll T_e$ or $T_f \approx 0$ in Eq. (5.26). Both of these conditions are representative of what actually happens for light entry vehicles. Equation (7.84) indicates the total heat transfer for these vehicles is minimized with blunt geometries.

For a "dense" (and likely more massive) vehicle,

$$\eta_f = \frac{\rho_f SC_D}{2m\beta} \ll 1 \qquad\qquad \textbf{7.85}$$

and:

$$\left| \frac{2\eta_f}{\sin \gamma_e} \right| \ll 1 \qquad\qquad \textbf{7.86}$$

In this situation, the exponential term in Eq. (7.81) can be approximated with the first few terms of a Taylor series expansion:

$$\exp\left(\frac{2\eta_f}{\sin\gamma_e}\right) \approx 1 + \frac{2\eta_f}{\sin\gamma_e} \qquad \textbf{7.87}$$

In turn, the expression for total heat transfer during the entry becomes

$$\bar{Q}_f \approx -\frac{1}{2}\left(\frac{c_f A}{C_D S}\right)\left(\frac{{}^R V_e}{{}^R V_0}\right)^2 \frac{2\eta_f}{\sin\gamma_e}$$

$$\approx -\frac{\rho_f c_f A}{2m\beta\sin\gamma_e}\left(\frac{{}^R V_e}{{}^R V_0}\right)^2 \qquad \textbf{7.88}$$

for dense vehicles. Equation (7.88) indicates the total heat transfer for these vehicles is minimized when the surface friction c_f and "wetted" area A are small and the mass is large. ("Small" surface area and "large" mass are seemingly logical characteristics of "dense" vehicles.)

Substituting Eq. (7.78) into the definition of the average wall heat flux, we can get an expression for the flux at any point in the trajectory:

$$\dot{\bar{q}}_w = \eta T^{3/2} = T_e^{3/2}\eta\exp\left[\frac{3(\eta-\eta_e)}{\sin\gamma_e}\right] \qquad \textbf{7.89}$$

Or, when entry is from a high altitude, this becomes:

$$\dot{\bar{q}}_w = T_e^{3/2}\eta\exp\left(\frac{3\eta}{\sin\gamma_e}\right) \qquad \textbf{7.90}$$

Figure 7-11 plots this relationship.

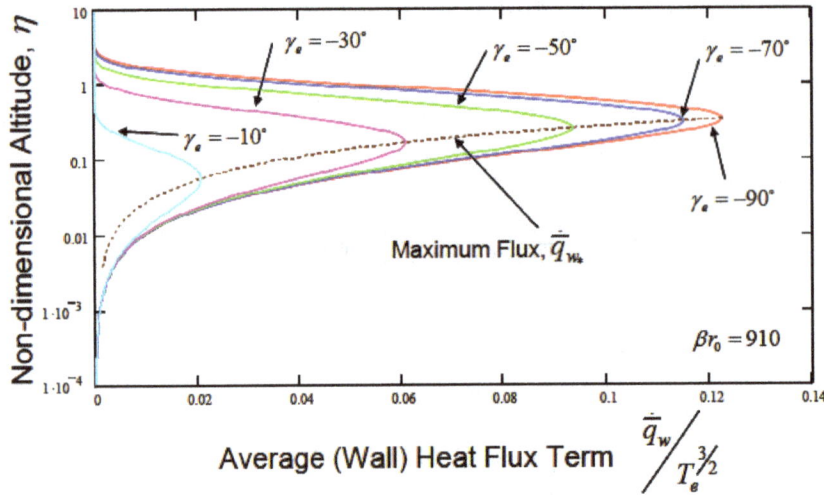

Figure 7-11: Average (Wall) Heat Flux for Steep Ballistic Entry Ignoring Gravity (High-Altitude Entry)

When this expression is maximized with respect to (non-dimensional) altitude,

$$\frac{d\overline{\dot{q}}_w}{d\eta} = T_e^{3/2}\left[\exp\left(\frac{3\eta_*}{\sin\gamma_e}\right) + \frac{3\eta_*}{\sin\gamma_e}\exp\left(\frac{3\eta_*}{\sin\gamma_e}\right)\right] = 0 \qquad \textbf{7.91}$$

and solved, we find the altitude for maximum average wall heat flux to be:

$$\eta_* = -\frac{\sin\gamma_e}{3} \qquad \textbf{7.92}$$

Of course, this is achieved only if $\eta_* < \eta_s$; i.e., the vehicle does not impact the planet first! The corresponding kinetic energy at the critical altitude given in Eq. (7.92) is found by evaluating Eq. (7.78):

$$T_* = \frac{T_e}{e^{2/3}} \qquad \textbf{7.93}$$

Notice that this kinetic energy (and, hence, the velocity) is completely independent of vehicle characteristics! Finally, the maximum flux rate can be evaluated using the values in Eqs. (7.92) and (7.93):

$$\dot{\bar{q}}_{w_*} = \frac{-T_e^{3/2} \sin \gamma_e}{3e} \qquad \textbf{7.94}$$

This maximum flux (and its corresponding altitude) was shown in Figure 7-11.

Turning to the stagnation heat flux, we can get an expression for the flux at any point in the trajectory by using Eq. (7.78):

$$\dot{\bar{q}}_s = \eta^{1/2} T^{3/2} = T_e^{3/2} \eta^{1/2} \exp\left[\frac{3(\eta - \eta_e)}{\sin \gamma_e}\right] \qquad \textbf{7.95}$$

Or, when entry is from a high altitude, this becomes

$$\dot{\bar{q}}_s = T_e^{3/2} \eta^{1/2} \exp\left(\frac{3\eta}{\sin \gamma_e}\right) \qquad \textbf{7.96}$$

and can be seen in Figure 7-12. When this expression is maximized with respect to (non-dimensional) altitude,

$$\frac{d\dot{\bar{q}}_s}{d\eta} = T_e^{3/2} \left[\frac{1}{2\eta_*^{1/2}} \exp\left(\frac{3\eta_*}{\sin \gamma_e}\right) + \frac{3\eta_*^{1/2}}{\sin \gamma_e} \exp\left(\frac{3\eta_*}{\sin \gamma_e}\right)\right] = 0 \qquad \textbf{7.97}$$

and solved, we find the altitude for maximum stagnation heat flux to be:

$$\eta_* = -\frac{\sin \gamma_e}{6} \qquad \textbf{7.98}$$

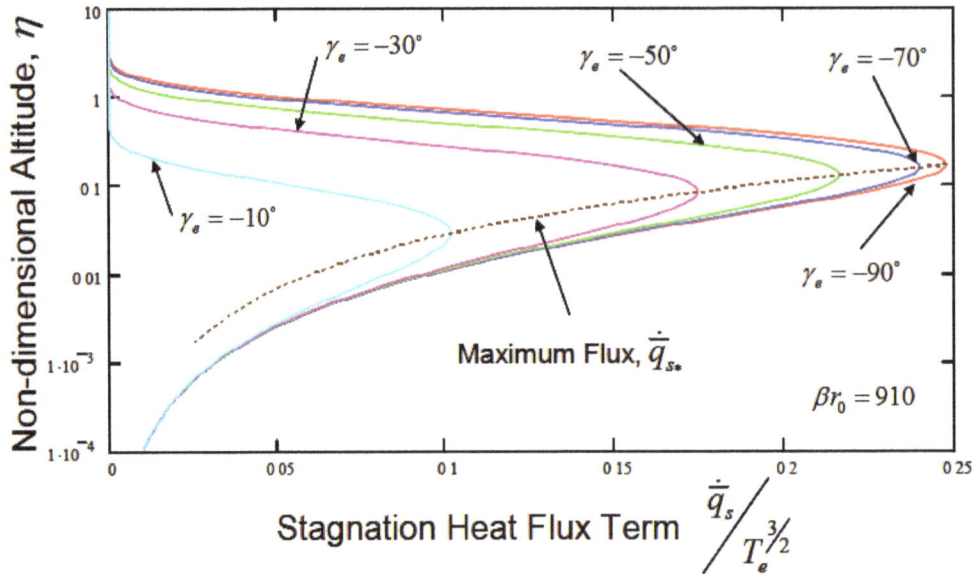

Figure 7-12: Stagnation Heat Flux for Steep Ballistic Entry Ignoring Gravity (High-Altitude Entry)

(Again, this is possible only if $\eta_* < \eta_s$.) The corresponding kinetic energy at the critical altitude given in Eq. (7.98) is found by evaluating Eq. (7.78):

$$T_* = \frac{T_e}{e^{1/3}} \qquad\qquad \textbf{7.99}$$

As with the critical kinetic energy for the average wall flux, this critical kinetic energy is completely independent of vehicle characteristics! The maximum stagnation flux rate can be evaluated using the values in Eqs. (7.98) and (7.99):

$$\dot{\bar{q}}_{s_*} = T_e^{3/2} \sqrt{\frac{-\sin\gamma_e}{6e}} \qquad\qquad \textbf{7.100}$$

The maximum stagnation heat flux (and its corresponding altitude) was shown in Figure 7-12. The critical heat flux values for this type of entry are summarized in Table 7-4.

Table 7-4: Points of Maximum Heat Flux for Steep Ballistic Entry Ignoring Gravity

	$\dot{\bar{q}}_{w_\star}$	$\dot{\bar{q}}_{s_\star}$
Maximum Value	$\dfrac{-T_e^{3/2}\sin\gamma_e}{3e}$	$T_e^{3/2}\sqrt{\dfrac{-\sin\gamma_e}{6e}}$
$T_\star$	$\dfrac{T_e}{e^{2/3}}$	$\dfrac{T_e}{e^{1/3}}$
$\eta_\star$	$-\dfrac{\sin\gamma_e}{3}$	$-\dfrac{\sin\gamma_e}{6}$
$\dfrac{^R V_\star}{\sqrt{g_0 r_0}}$	$\dfrac{\sqrt{2T_e}}{e^{1/3}}$	$\dfrac{\sqrt{2T_e}}{e^{1/6}}$

7.6.4.2 Steep Ballistic Entry, Including Gravity

The previous section looked at steep ballistic entry while ignoring gravity. In Section 5.6.1, however, we found a "better" solution for steep ballistic entry which included gravity. (It was better in the sense it used one less simplifying assumption.)

It was:

$$T(\alpha) = \frac{e^{-\alpha}}{\beta r_0}\left[T_e \beta r_0 e^{\alpha_e} - Ei(\alpha_e) + Ei(\alpha)\right] \qquad \textbf{7.101}$$

where:

$$\alpha = \frac{-2\eta}{\sin\gamma_e} \qquad \textbf{7.102}$$

We can use this solution to examine this type of entry again, but with gravity included this time. As before, we'll write the total heat transfer by using Eq. (7.25):

$$\bar{Q}_f = -\left(\frac{c_f A}{C_D S}\right)\left(\frac{g_0 r_0}{{}^R V_0^2}\right)\left\{\frac{e^{-\alpha_f}}{\beta r_0}\left[T_e \beta r_0 e^{\alpha_e} - Ei(\alpha_e) + Ei(\alpha_f)\right] - T_e\right\} \qquad \textbf{7.103}$$

Unfortunately, we can't simply assume entry is from a high altitude and blindly substitute $\eta_e \approx 0$ this time because $\eta_e \approx 0$ means $\alpha_e \approx 0$ and $Ei(0) = -\infty$! To proceed, we need to take a look at the exponential integral more closely for high altitudes.

In the range of (about) $0 < \alpha_e < 0.37$, $Ei(\alpha_e)$ rapidly becomes finite and small; indeed, $|Ei(0.1)| \leq 1.65$. ($Ei(\alpha_e)$ becomes positive for $\alpha_e \geq 0.373$.) At the other end of the trajectory, $Ei(\alpha_f)$ will become large compared to $Ei(\alpha_e)$. By the time $\alpha_f \geq 3.8$, $Ei(\alpha_f) > 17$. In light of these observations, it is reasonable assume for entry from high altitude we can make the approximation:

$$-Ei(\alpha_e) + Ei(\alpha_f) \approx Ei(\alpha_f) \qquad \textbf{7.104}$$

Implicit in this approximation is the requirement that the initial entry isn't from a "hard vacuum" or else $\alpha_e = 0$ and $Ei(\alpha_e) = -\infty$. This is not overly restrictive since very little (relatively speaking) aerodynamic heating occurs in the thinnest parts of the atmosphere for most planets. Equation (7.103) is then reduced to

$$\bar{Q}_f = -\left(\frac{c_f A}{C_D S}\right)\left(\frac{g_0 r_0}{{}^R V_0^2}\right)\left\{T_e\left[e^{-(\alpha_f - \alpha_e)} - 1\right] + \frac{e^{-\alpha_f}}{\beta r_0}Ei(\alpha_f)\right\} \qquad \textbf{7.105}$$

after some rearranging. This can be simplified even more.

The last term in the bracket is very small compared to T_e. The portion due

to altitude $\dfrac{Ei(\alpha_f)}{e^{\alpha_f}}$ is small as shown in Figure 7-13. Further, $\dfrac{1}{\beta r_0} \ll 1$ for most

atmospheres of concern, so the entire term can be safely ignored for our purposes.
Replacing T_e with its definition, we can simplify our expression for the total heat
transfer during entry:

$$\bar{Q}_f = \left(\frac{1}{2}\right)\left(\frac{c_f A}{C_D S}\right)\left(\frac{{}^R V_e}{{}^R V_0}\right)^2 \left[1 - e^{-(\alpha_f - \alpha_e)}\right] \qquad \textbf{7.106}$$

If α is replaced by its definition in Eq. (7.106) and a negative sign factored out,
we have:

$$\bar{Q}_f = -\left(\frac{1}{2}\right)\left(\frac{c_f A}{C_D S}\right)\left(\frac{{}^R V_e}{{}^R V_0}\right)^2 \left\{\exp\left[\frac{2\left(\eta_f - \eta_e\right)}{\sin \gamma_e}\right] - 1\right\} \qquad \textbf{7.107}$$

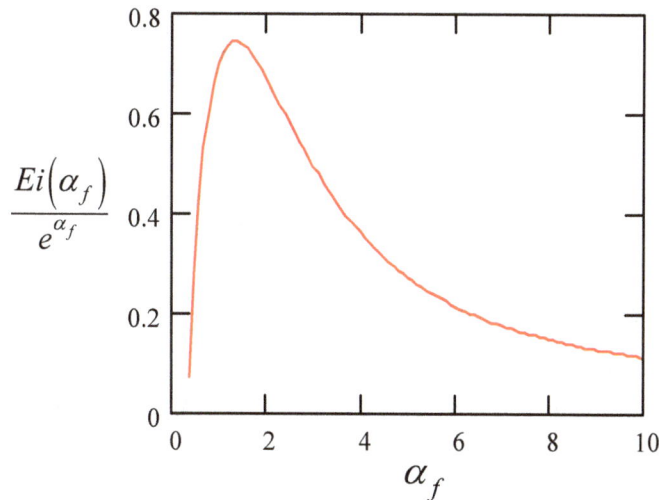

Figure 7-13: Relative Magnitude of Exponential Terms

This is identical to what we found for the total heating when we ignored gravity (Eq. (7.81)), so we can immediately conclude

$$\bar{Q}_f \approx \frac{1}{2}\left(\frac{c_f A}{C_D S}\right)\left(\frac{^R V_e}{^R V_0}\right)^2 \qquad \textbf{7.108}$$

for "light" vehicles and

$$\bar{Q}_f \approx -\frac{\rho_f c_f A}{2m\beta \sin\gamma_e}\left(\frac{^R V_e}{^R V_0}\right)^2 \qquad \textbf{7.109}$$

for dense vehicles.

The average wall heat flux for this type of entry is simplest when written in terms of α since we have both $\eta(\alpha)$ and $T(\alpha)$:

$$\dot{\bar{q}}_w = \eta T^{3\!/\!2} = \left(\frac{-\alpha \sin\gamma_e}{2}\right)\left\{\frac{e^{-\alpha}}{\beta r_0}\left[T_e \beta r_0 e^{\alpha_e} - Ei(\alpha_e) + Ei(\alpha)\right]\right\}^{3\!/\!2} \qquad \textbf{7.110}$$

But, for those that like consistency, $\dot{\bar{q}}_w$ can be written in terms of η also:

$$\dot{\bar{q}}_w = \eta\left\{\frac{\exp\left(\dfrac{2\eta}{\sin\gamma_e}\right)}{\beta r_0}\left[T_e \beta r_0 \exp\left(\frac{-2\eta_e}{\sin\gamma_e}\right) - Ei\left(\frac{-2\eta_e}{\sin\gamma_e}\right) + Ei\left(\frac{-2\eta}{\sin\gamma_e}\right)\right]\right\}^{3\!/\!2} \qquad \textbf{7.111}$$

Unlike for the other entries studied in this chapter, we cannot easily remove the requirement to pick an initial kinetic energy by plotting $\dfrac{\dot{\bar{q}}_w}{T_e^{3\!/\!2}}$ versus η. (Notice how it doesn't factor out of the brackets in Eq. (7.111).) It can be shown, however, that plots of $\dfrac{\dot{\bar{q}}_w}{T_e^{3\!/\!2}}$ versus η are relatively insensitive to the specific value of T_e used for any given (steep) entry angle γ_e. Similarly, we must select a nonzero value for η_e to avoid the singularity in Eq. (7.111). Figure 7-14 shows the typical trends of the wall heat flux for steep ballistic entry with gravity included.

Maximizing $\dot{\bar{q}}_w$ with respect to α (using Eq. (7.110)) leads to solving

$$\left[T_e \beta r_0 e^{\alpha_e} - Ei\left(\alpha_e\right) + Ei\left(\alpha_*\right) \right]\left(1 - \frac{3\alpha_*}{2} \right) + \frac{3}{2}e^{\alpha_*} = 0 \qquad \textbf{7.112}$$

to find α_*. In turn, $\eta_* = \dfrac{-\alpha_* \sin\gamma_e}{2}$ gives the corresponding altitude. (Note, this altitude could have been found directly by maximizing Eq. (7.111) with respect to

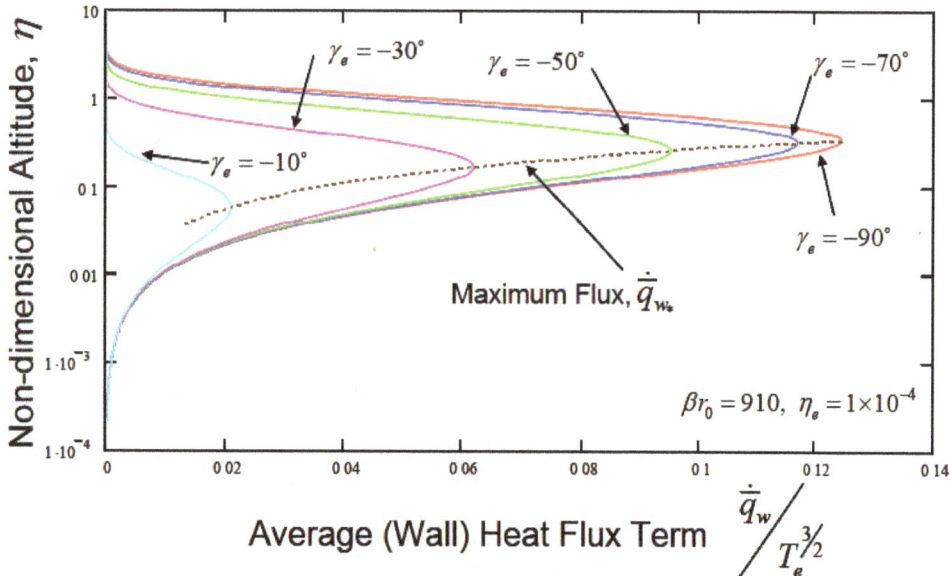

Figure 7-14: Average (Wall) Heat Flux for Steep Ballistic Entry Including Gravity (High-Altitude Entry)

η, but that is left as an exercise.) Evaluating $T(\alpha_*)$ and simplifying with the aid of Eq. (7.112) gives us the kinetic energy

$$T_* = \frac{1}{\beta r_0 \left(\alpha_* - \frac{2}{3} \right)} \qquad \textbf{7.113}$$

and velocity:

$$\frac{^R V_*}{^R V_0} = \left[\frac{2}{\beta r_0 \left(\alpha_* - \frac{2}{3} \right)} \right]^{\frac{1}{2}} \qquad \textbf{7.114}$$

In writing Eqs. (7.113) and (7.114), we have assumed $\alpha_* > \frac{2}{3}$. Interestingly, for a large range of initial conditions, α_* is very close to violating this mathematical constraint! Based on these two equations, that would imply this heating maximum is very near a velocity maximum. Finally, the maximum flux $\dot{\bar{q}}_{w_*}$ can be assembled

$$\dot{\bar{q}}_{w_*} = \eta_* T_*^{\frac{3}{2}} = \left(\frac{-\alpha_* \sin \gamma_e}{2} \right) \left[\frac{1}{\beta r_0 \left(\alpha_* - \frac{2}{3} \right)} \right]^{\frac{3}{2}} \qquad \textbf{7.115}$$

and was shown in Figure 7-14.

Finding and maximizing the stagnation heat flux is only slightly more difficult. Beginning with

$$\dot{\bar{q}}_s = \eta^{1/2} T^{3/2} = \left(\frac{-\alpha \sin \gamma_e}{2} \right)^{1/2} \left\{ \frac{e^{-\alpha}}{\beta r_0} \left[T_e \beta r_0 e^{\alpha_e} - Ei(\alpha_e) + Ei(\alpha) \right] \right\}^{3/2} \qquad \textbf{7.116}$$

and maximizing with respect to α leads to solving

$$\left[T_e\beta r_0 e^{\alpha_e} - Ei(\alpha_e) + Ei(\alpha_*)\right](1-3\alpha_*)+3e^{\alpha_*}=0 \qquad \textbf{7.117}$$

to find $\alpha_* = \dfrac{-2\eta_*}{\sin\gamma_e}$ where $\dot{\bar{q}}_s$ reaches its peak value. Representative plots for the stagnation heat flux are shown in Figure 7-15.

Equations (7.117) and (7.101) combine to give the corresponding kinetic energy at the point of maximum heat flux:

$$T_* = \frac{1}{\beta r_0\left(\alpha_* - \dfrac{1}{3}\right)} \qquad \textbf{7.118}$$

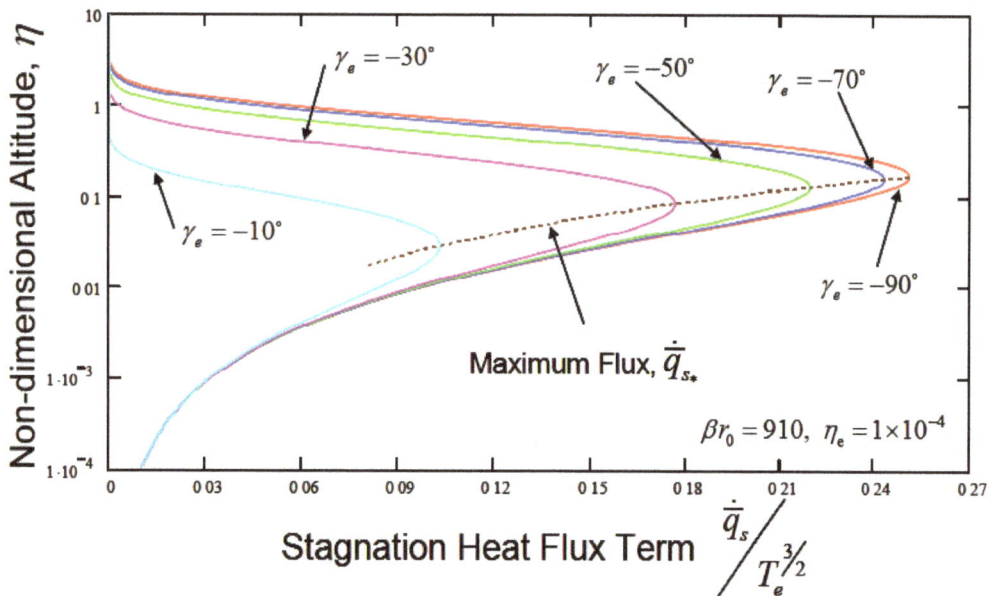

Figure 7-15: Stagnation Heat Flux for Steep Ballistic Entry Including Gravity (High-Altitude Entry)

Similarly, the velocity is:

$$\frac{^RV_*}{^RV_0} = \left[\frac{2}{\beta r_0 \left(\alpha_* - \dfrac{1}{3} \right)} \right]^{1/2}$$

7.119

Note, Eqs. (7.118) and (7.119) require $\alpha_* > \frac{1}{3}$ in order to be valid. The maximum value itself is:

$$\dot{\bar{q}}_{s*} = \eta_*^{1/2} T_*^{3/2} = \left(\frac{-\alpha_* \sin \gamma_e}{2} \right)^{1/2} \left[\frac{1}{\beta r_0 \left(\alpha_* - \dfrac{1}{3} \right)} \right]^{3/2}$$

7.120

Table 7-5 summarizes the critical values for wall and stagnation heat fluxes for the steep ballistic entry when gravity is included in the analysis.

Table 7-5: Points of Maximum Heat Flux for Steep Ballistic Entry Including Gravity

	$\dot{\overline{q}}_{w_*}$	$\dot{\overline{q}}_{s_*}$
Maximum Value	$\left(\dfrac{-\alpha_* \sin\gamma_e}{2}\right)\left[\dfrac{1}{\beta r_0\left(\alpha_* - \dfrac{2}{3}\right)}\right]^{3/2}$	$\left(\dfrac{-\alpha_* \sin\gamma_e}{2}\right)^{1/2}\left[\dfrac{1}{\beta r_0\left(\alpha_* - \dfrac{1}{3}\right)}\right]^{3/2}$
T_*	$\dfrac{1}{\beta r_0\left(\alpha_* - \dfrac{2}{3}\right)}$	$\dfrac{1}{\beta r_0\left(\alpha_* - \dfrac{1}{3}\right)}$
η_*	$\dfrac{-\alpha_* \sin\gamma_e}{2}$	$\dfrac{-\alpha_* \sin\gamma_e}{2}$
$\dfrac{^RV_*}{^RV_e}$	$\left[\dfrac{2}{\beta r_0\left(\alpha_* - \dfrac{2}{3}\right)}\right]^{1/2}$	$\left[\dfrac{2}{\beta r_0\left(\alpha_* - \dfrac{1}{3}\right)}\right]^{1/2}$
α_* **given by**	$\left[T_e\beta r_0 e^{\alpha_e} - Ei(\alpha_e) + Ei(\alpha_*)\right]$ $\cdot\left(1 - \dfrac{3\alpha_*}{2}\right) + \dfrac{3}{2}e^{\alpha_*} = 0$	$\left[T_e\beta r_0 e^{\alpha_e} - Ei(\alpha_e) + Ei(\alpha_*)\right]$ $\cdot(1 - 3\alpha_*) + 3e^{\alpha_*} = 0$

7.7 Problems

Material Understanding:

1. Starting with

$$\frac{d}{dt}\left(^R V\right) = -\frac{\rho C_D S\, ^R V^2}{2m}$$

   show that:

$$\frac{dt}{dT} = -\frac{1}{2\beta\sqrt{2g_0 r_0}}\frac{1}{\eta T^{3/2}}$$

2. Equation (7.48) gives the total heat absorbed by the vehicle during gliding entry at medium and steep entry. If you are free to tailor the vehicle, what can be done to minimize $\bar{Q}_f$? Identify at least two characteristics of the vehicle and/or the entry conditions that *appear* to result in less overall heat transfer. Why do you suspect they reduce $\bar{Q}_f$?

3. Starting with Eq. (7.52), show that the solution for the flight-path angle at which the maximum wall heat flux $\dot{\bar{q}}_{w_*}$ for medium/steep gliding entry is given by Eq. (7.54).

4. Is it possible for $\dot{\bar{q}}_{w_*}$ to be larger than $\dot{\bar{q}}_{s_*}$ for any given steep ballistic entry? Use the simplified solution ignoring gravity (Section 7.6.4.1) to justify your answer.

5. In Section 7.6.4.2, we found the altitude η_* corresponding to the maximum average heating flux $\dot{\bar{q}}_w$ for a steep ballistic entry (with gravity included). We did this by maximizing Eq. (7.110) with respect to α and then finding α_* to get eventually get η_*. Show that you can find this "critical" altitude directly by maximizing

$$\dot{\bar{q}}_w = \eta \left\{ \frac{\exp\left(\dfrac{2\eta}{\sin\gamma_e}\right)}{\beta r_0} \left[T_e \beta r_0 \exp\left(\frac{-2\eta_e}{\sin\gamma_e}\right) - Ei\left(\frac{-2\eta_e}{\sin\gamma_e}\right) + Ei\left(\frac{-2\eta}{\sin\gamma_e}\right) \right] \right\}^{3/2}$$

with respect to η and solving for η_* directly.

6. Beginning with

$$\dot{\bar{q}}_s = \eta^{1/2} T^{3/2} = \left(\frac{-\alpha \sin\gamma_e}{2} \right)^{1/2} \left\{ \frac{e^{-\alpha}}{\beta r_0} \left[T_e \beta r_0 e^{\alpha_e} - Ei(\alpha_e) + Ei(\alpha) \right] \right\}^{3/2}$$

show that maximizing $\dot{\bar{q}}_s$ with respect to α leads to solving

$$\left[T_e \beta r_0 e^{\alpha_e} - Ei(\alpha_e) + Ei(\alpha_*) \right](1 - 3\alpha_*) + 3e^{\alpha_*} = 0$$

for α_*.

Computational Insights:

7. In Problem 1 of Chapter 6, a gliding entry was solved using first-order, second-order, and numerical integration methods. Those solutions can be used to plot the heat flux rates for the trajectories. Plot altitude η versus $\dot{\bar{q}}_w \big/ T_e^{3/2}$ and $\dot{\bar{q}}_s \big/ T_e^{3/2}$ for each of the three solutions.

8. Over a range of steep entry angles, compare the timing and magnitude of the maximum heat fluxes experienced during gliding entries. Use $C_L/C_D = 1$ for simplicity.

9. Sections 7.6.4.1 and 7.6.4.2 derived solutions for heat flux rates for steep ballistic entry without and with gravity (respectively) in the analyses. Using at least four different sets of initial conditions (T_e, γ_e), make plots of $\dot{\bar{q}}_s$ versus η. (Assume an Earth-like atmosphere with $\beta r_0 \approx 910$.) Show the solution curves for both analyses on each plot. How similar are they in peak values? How similar are the estimates of the altitude of that peak?

10. Demonstrate the assertion on page 210 that plots of $\dfrac{\dot{\bar{q}}_w}{T_e^{3/2}}$ versus η are relatively insensitive to the specific value of T_e used for a given entry angle in steep, ballistic entry (including gravity). Pick at least four values of T_e and plot $\dfrac{\dot{\bar{q}}_w}{T_e^{3/2}}$ versus η for results. (Assume an Earth-like atmosphere with $\beta r_0 \approx 910$.)

11. Is the assertion in Problem 10 valid for Mars ($\beta r_0 \approx 350$) and Jupiter ($\beta r_0 \approx 3000$)?

This page intentionally blank.

Chapter 8

Entry Corridor

8.1 Introduction

For a given vehicle (or object), there is a fairly restrictive set of conditions which must be met to ensure it survives atmospheric entry. The two most obvious restrictions (deceleration and heating limits) were discussed in Chapter 5 and Chapter 7. The entry conditions and, in most situations, the vehicle design will dictate the maximum deceleration and heating that will be encountered. If the entry angle is too steep at its entry speed, the vehicle will be destroyed by dynamic and/or thermal loads. If it is too shallow, the vehicle may simply pass through the upper atmosphere and continue on past the planet forever. Between these two extremes is an *entry corridor* where the vehicle can be guaranteed to not only to be "captured" by the atmosphere but also to stay within its design limits.

Before a vehicle begins its atmospheric entry, we can assume it is on an orbital conic (trajectory) with respect to the planet. In Figure 8-1, a point on the orbit (outside the atmosphere) is labeled with a "1." Within the plane of the motion and using $\vec{r}_1$ as a reference direction, the *inertial* velocity V_1, flight-path

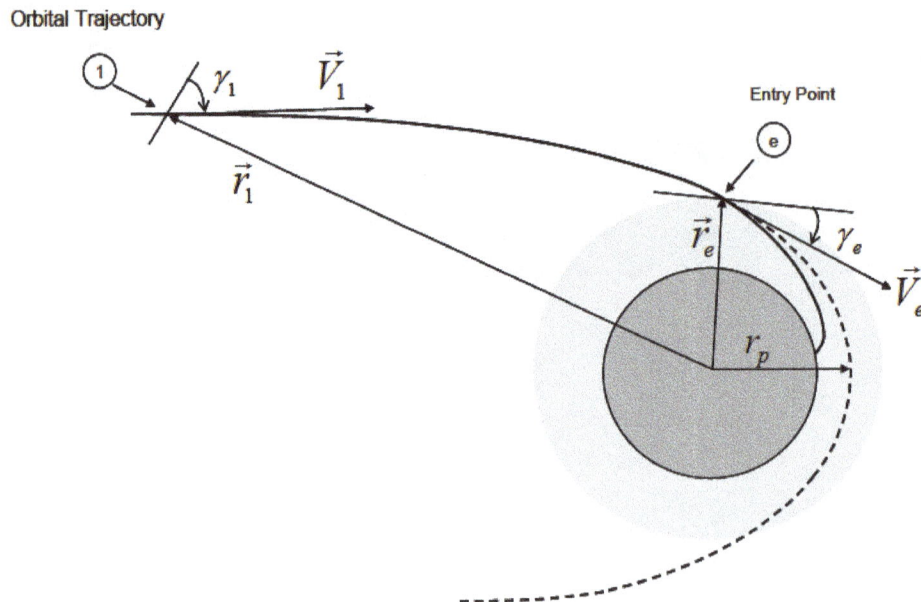

Figure 8-1: Orbital Parameters and Associated Entry Conditions

angle γ_1, and radius r_1 completely describe the orbit. (These do not, however, specify the *orientation* of the orbital plane in space.) That orbit is fixed and the conditions later at entry (V_e, γ_e, and r_e) are already ordained. However, if the vehicle has the ability to maneuver, it can change the orbit and "target" an entry corridor. This chapter is concerned with defining and finding the entry corridor.

8.2 Finding Entry Conditions from Approach Conic

If V_1, γ_1, and r_1 are known values (measured with respect to an inertial point) well outside the atmosphere on an orbital trajectory (a "conic"), then we can find the entry conditions when the atmosphere is entered at r_e. In theory, there is some arbitrariness in setting the value of r_e because the radius to the "top of the atmosphere" is poorly defined. In practice, though, the value can be selected based on some criterion such as when the atmospheric drag force reaches a certain fraction of the gravity force. For our purposes, we'll assume an acceptable value of r_e has been dictated.

The equations found in Chapter 2 can be used to relate the given orbital parameters (at "1") to those at atmospheric entry (at "e"). For any orbit, the vis-viva equation

$$V^2 = \mu \left(\frac{2}{r} - \frac{1}{a} \right)$$

8.1

can be solved for the semimajor axis from the position and velocity at any point on the orbit. Thus, given the conditions at "1," we have

$$a = \frac{r_1}{2 - \left(\dfrac{r_1 V_1^2}{\mu} \right)}$$

8.2

for this orbit. Then, at entry (where r_e is known), the velocity can be calculated:

$$V_e = \sqrt{\mu \left(\frac{2}{r_e} - \frac{1}{a} \right)}$$

8.3

The flight-path angle is found by noting specific momentum is conserved along the orbit

$$h = r_1 V_1 \cos \gamma_1 = r_e V_e \cos \gamma_e$$

8.4

and solving:

$$\gamma_e = \cos^{-1} \left[\frac{r_1 V_1 \cos \gamma_1}{r_e V_e} \right]$$

8.5

In this equation, the solution with $\gamma_e < 0$ should be selected since the radius is decreasing as point "e" is approached.

As you might imagine, this problem can be turned around. Given an entry radius, velocity, and flight-path angle, the approach conic can be found. It is this "backwards" solution that tends to be used to find the entry corridor and the associated entry conics. The next sections illustrate the process.

Before continuing, however, a word of caution is required. The velocities on the conic are all *inertial* and care must be taken not to confuse their values with those *relative* to a rotating planet; i.e., $V_e \neq {}^R V_e$ in general.

8.3 Finding the Approach Conic from Entry Conditions

When the entry radius r_e, velocity V_e (or kinetic energy T_e), and flight-path angle γ_e are known, it is possible to determine the approach conic (e.g., the orbit) the vehicle followed to arrive at the atmosphere. (More correctly, it is possible to determine the orbit *shape* within the orbital plane – three scalar values are insufficient to define the orientation of the orbital plane in space.)

Given the entry velocity and radius, the vis-viva equation can be solved for the semimajor axis of the orbit:

$$ a = \frac{r_e}{2 - \left(\dfrac{r_e V_e^2}{\mu} \right)} \qquad\qquad \textbf{8.6} $$

The eccentricity and specific angular momentum are related by

$$ \frac{h^2}{\mu} = a\left(1 - e^2\right) \qquad\qquad \textbf{8.7} $$

for all conics. (In the case of parabolic orbits, $a = \infty$ and $e = 1$, so some care may needed in certain applications.) The constant value of h is known from:

$$h = r_e V_e \cos \gamma_e \qquad \qquad \textbf{8.8}$$

Combining Eqs. (8.6) - (8.8) and solving for eccentricity, we find

$$e = \sqrt{1 - \frac{r_e V_e^2}{\mu} \left[2 - \left(\frac{r_e V_e^2}{\mu} \right) \right] \cos^2 \gamma_e} \qquad \qquad \textbf{8.9}$$

Or, equivalently:

$$e = \sqrt{\sin^2 \gamma_e + \left[1 - \left(\frac{r_e V_e^2}{\mu} \right) \right]^2 \cos^2 \gamma_e} \qquad \qquad \textbf{8.10}$$

The semimajor axis and eccentricity found with Eqs. (8.6) and (8.9) or (8.10), respectively, define the approach orbit. If more detail is required, the conic equation can be solved

$$r_e = \frac{\left(\dfrac{h^2}{\mu} \right)}{1 + e \cos \nu_e} \qquad \qquad \textbf{8.11}$$

for the true anomaly at entry ν_e.

8.4 Approach Periapsis

We will find it helpful to describe these approach conics in terms of the periapsis radius *that would exist* if the atmosphere (or planet surface) did not change the two-body motion. Figure 8-1 shows periapsis radius r_p relative to the "initial" point and entry point. To find a general expression for r_p, we will need to find the orbital eccentricity e first.

For two-body orbital motion, eccentricity and specific angular momentum are related by

$$\frac{h^2}{\mu} = a\left(1 - e^2\right)$$

8.12

for all conics. Using Eqs. (8.2) and (8.4), this can be solved for eccentricity in terms of given orbital parameters:

$$e = \sqrt{1 - \frac{r_1 V_1^2}{\mu}\left[2 - \left(\frac{r_1 V_1^2}{\mu}\right)\right]\cos^2 \gamma_1}$$

8.13

(This equation correctly gives $e = 1$ for parabolic orbits.) For non-parabolic orbits, the periapsis can be written entirely in terms of the given conditions:

$$r_p = a(1 - e) = \frac{r_1}{2 - \left(\frac{r_1 V_1^2}{\mu}\right)}\left[1 - \sqrt{1 - \frac{r_1 V_1^2}{\mu}\left[2 - \left(\frac{r_1 V_1^2}{\mu}\right)\right]\cos^2 \gamma_1}\right]$$

8.14

For parabolic orbits, the expression is somewhat simpler:

$$r_p = \frac{h^2}{2\mu} = \frac{r_1^2 V_1^2 \cos^2 \gamma_1}{2\mu}$$

8.15

As shown in Figure 8-2, there is a maximum periapsis radius which will still allow the vehicle to be "captured" in one pass past the planet. Similarly, there is a minimum radius below which the vehicle encounters too much drag. These radii define the "overshoot" and "undershoot" boundaries, respectively. The difference between the radii is Δr_p and all of the conics (orbits) with periapsis values between these boundaries lie in the "entry corridor."

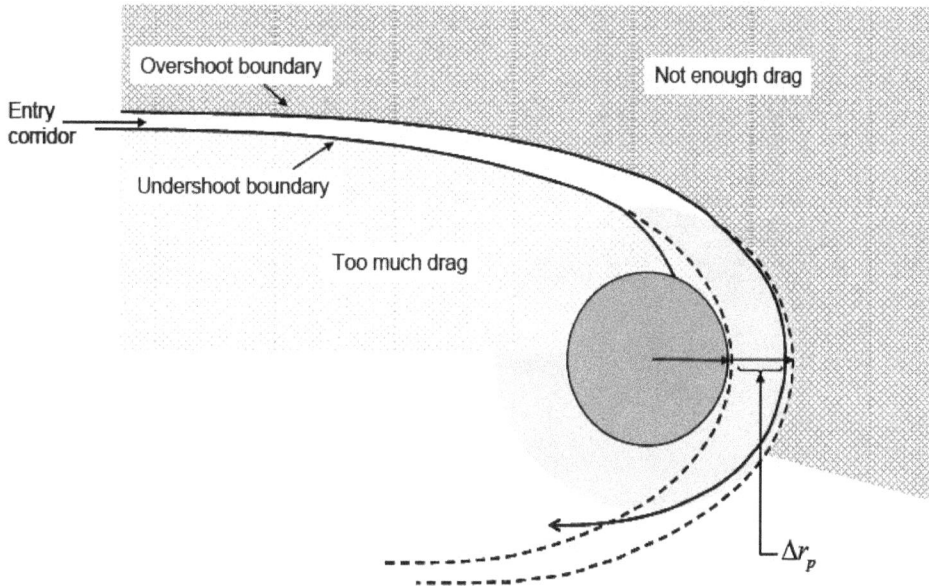

Figure 8-2: Entry Corridor

8.4.1 Undershoot Boundary

The undershoot boundary is the limiting entry path which results in a "safe" atmospheric entry. For illustration, we will assume the limiting factor is a maximum allowed deceleration, say $\dfrac{a_{decel}}{g_0} \leq \left(\dfrac{a_{decel}}{g_0} \right)_{failure}$. (The limit could just as easily be defined by a maximum heating rate.) If we restrict ourselves to a *planar entry*, we can use the second-order solution from Chapter 6 to find the entry trajectory which "just hits" the maximum deceleration. (Using a second-order solution frees us from the necessity of deciding which *specific* first-order solution is appropriate.) To simplify the analysis (without affecting the qualitative results), we can also assume the planet's *atmosphere does not rotate*; i.e.,

$V_e = {}^R V_e$. Setting the maximum deceleration on an arbitrary entry trajectory to our maximum allowed value, we have

$$\left(\frac{a_{decel}}{g_0} \right)_{failure} = -\beta r_0 T_* \sin \gamma_* \sqrt{1 + \left(\frac{C_L}{C_D} \right)^2} \qquad 8.16$$

$$\gamma_* = \gamma_e - \frac{1}{2} \ln \left(\frac{T_*}{T_e} \right) \left[\frac{C_L}{C_D} + \frac{2}{\beta r_0 \sin \gamma_*} \left(\frac{1}{2T_*} - 1 \right) \cos \gamma_* \right] \qquad 8.17$$

$$\cos \gamma_* = \frac{\cos \gamma_e - \left(\dfrac{\sin \gamma_*}{2} \right) \left(\dfrac{C_L}{C_D} \right)}{1 + \dfrac{1}{\beta r_0} \left(\dfrac{1}{2T_*} - 1 \right)} \qquad 8.18$$

where, as usual, r_0 is a reference radius (which may or may not be the same as r_e). In these three equations, there are four unknowns: γ_e, T_e, γ_*, and T_*. Two (γ_* and T_*) are "fixed" by dynamics and cannot be "selected" and two (γ_e and T_e) are "free" variables. Only one of the two free variables can be arbitrarily selected since there are just three equations in the four unknowns. Typically, it is harder to change the magnitude of the orbital velocity (i.e., T_e) than it is to change its orientation slightly (i.e., γ_e), so it is normally assumed to be known. Given T_e, Eqs. (8.16) - (8.18) can be solved for γ_e, γ_*, and T_*. (Note: there is nothing to preclude selecting a value for γ_e and solving for T_e, γ_*, and T_* instead.)

Armed with T_e and γ_e, we can solve for the approach conic defining the undershoot boundary. Rewriting Eq. (8.6) and Eq. (8.9) in terms of T_e in place of

V_e we can define the approach conic for the undershoot boundary:

$$a = \frac{r_e}{2\left[1 - \left(\dfrac{r_e T_e g_0 r_0}{\mu}\right)\right]}$$ 8.19

$$e = \sqrt{1 - \frac{4 r_e T_e g_0 r_0}{\mu}\left[1 - \left(\frac{r_e T_e g_0 r_0}{\mu}\right)\right]\cos^2 \gamma_e}$$ 8.20

In a homework problem in Chapter 4, it was shown for the current assumptions

$$\mu = g_0 r_0^2$$ 8.21

so Eqs. (8.19) and (8.20) can be simplified:

$$a = \frac{r_e}{2\left[1 - T_e\left(\dfrac{r_e}{r_0}\right)\right]}$$ 8.22

$$e = \sqrt{1 - 4 T_e\left(\frac{r_e}{r_0}\right)\left[1 - T_e\left(\frac{r_e}{r_0}\right)\right]\cos^2 \gamma_e}$$ 8.23

Notice these equations simplify even more when the reference radius r_0 is selected to be the entry radius r_e:

$$a = \frac{r_e}{2(1 - T_e)}$$ 8.24

$$e = \sqrt{1 - 4 T_e (1 - T_e)\cos^2 \gamma_e}$$ 8.25

Finally, Eq. (8.14) gives the corresponding periapsis radius for an undershoot boundary (for $r_0 = r_e$):

$$r_{P_{under}} = a(1-e) = \frac{r_e}{2(1-T_e)}\left[1 - \sqrt{1 - 4T_e(1-T_e)\cos^2\gamma_e}\right]$$

8.26

Varying the assumed value of T_e (or, alternatively, the presumed value of γ_e), Eqs. (8.16) - (8.18) and (8.24) - (8.26) can be repeatedly solved to find a "family" of conics which all meet the criterion:

$$\left(\frac{a_{decel}}{g_0}\right)_{max} = \left(\frac{a_{decel}}{g_0}\right)_{failure}$$

8.27

8.4.2 Overshoot Boundary

At the other end of the problem, there is a boundary above which the vehicle will not be captured. On the overshoot boundary, the vehicle encounters just enough atmospheric drag to avoid exiting the atmosphere. Traditionally, assuming the velocity on the overshoot boundary is reduced to circular orbital velocity (at a radius of r_e) is considered adequate for defining the boundary. (Note that this assumes there is enough energy and/or lift for the vehicle to "skip" back to the edge of the atmosphere. When skip does not occur, we will be forced to define the overshoot boundary differently.)

It is tempting to use the second-order equations from Chapter 6 like we did for the undershoot boundary. However, Loh's equations (and the first-order skip solution from Chapter 5) do not give realistic results for an important entry scenario. Specifically, negative lift can be used to "hold" the vehicle in the atmosphere while it decelerates. As a simple validation of this assertion, consider an entry at T_e and γ_e with $r_e = r_0$. Then, both Loh's solution and the first-order solution become

$$T_e = \frac{1}{2}\exp\left(\frac{-4\gamma_e}{\dfrac{C_L}{C_D}}\right) \qquad\qquad \textbf{8.28}$$

for the overshoot boundary. For entry, $\gamma_e < 0$ so with negative lift $\left(\dfrac{-4\gamma_e}{\dfrac{C_L}{C_D}}\right) < 0$.

It is apparent Eq. (8.28) *cannot be satisfied* for negative lift when $T_e > \dfrac{1}{2}$ (which is the only case of interest when looking for an overshoot boundary). Thus, for negative lift, we are forced to abandon the solutions we have already derived if we wish to define the overshoot boundary. Another alternative would be to directly solve the differential equations for planar entry (Eqs. (5.1) and (5.2)) subject to the known boundary conditions.

Rewriting these with kinetic energy as the independent variable gives:

$$\frac{d\eta}{dT} = \frac{\beta r_0 \eta \sin\gamma}{2T\beta r_0 \eta + \sin\gamma} \qquad\qquad \textbf{8.29}$$

$$\frac{d\gamma}{dT} = -\left[\frac{C_L}{C_D} + \frac{1}{\beta r_0 \eta}\left(1 - \frac{1}{2T}\right)\cos\gamma\right]\left(\frac{\beta r_0 \eta}{2T\beta r_0 \eta + \sin\gamma}\right) \qquad\qquad \textbf{8.30}$$

Equations (8.29) and (8.30) can be integrated as a two-point boundary value problem (TPBVP) with the boundary conditions

$$\eta = \eta_e \quad \text{at} \quad T = T_e \tag{8.31}$$

$$\eta = \eta_e \quad \text{at} \quad T_f = \frac{{}^R V_f^2}{2 g_o r_o} = \frac{r_0}{2 r_e} \tag{8.32}$$

where it has been assumed T_e is "fixed" and the flight-path angle throughout is unknown (but will be found when the TPBVP is solved).

This TPBVP is, mathematically at least, well-posed and straight-forward to set up for solution. Once solved,

$$r_{P_{over}} = a(1-e) = \frac{r_e}{2(1-T_e)} \left[1 - \sqrt{1 - 4 T_e (1 - T_e) \cos^2 \gamma_e} \right] \tag{8.33}$$

gives the periapsis radius for the overshoot boundary (for $r_0 = r_e$). However, numerically solving the problem can be exceedingly difficult. Both γ and η are very small at entry and exit. Further complicating the numerics, γ probably remains small but changes sign while η may vary over several orders of magnitude!

Because of these difficulties (and other less obvious reasons), traditional texts often introduce another non-dimensional form of the equations better suited for studying the entry corridor, specifically Chapman's theory (14-17; 39:169-177; 58:178-225). Rather than introduce an entirely new *approximation* theory (complete with new variables and a requirement to numerically integrate), we will simply accept and present Chapman's results for the overshoot boundary. (In Section 9.4 we will revisit the overshoot boundary with a new theory related to Chapman's.)

Chapman, in his classic analyses, chose to characterize the entry conic with a parameter more convenient to his particular equations than our periapsis radius r_p (Refs. 14-17). This "periapsis parameter" F_p is defined by

$$F_p = \frac{\rho_p S C_D}{2m} \sqrt{\frac{r_p}{\beta}} = \eta_p \sqrt{\beta r_p} \qquad \textbf{8.34}$$

where the "p" subscript again denotes the conditions at the periapsis of the hypothetical two-body conic. Note, since η_p is a function of r_p, there is a one-to-one correspondence between Chapman's F_p parameter and the periapsis radius r_p we introduced earlier. His results for the overshoot boundary are presented in Figure 8-3.

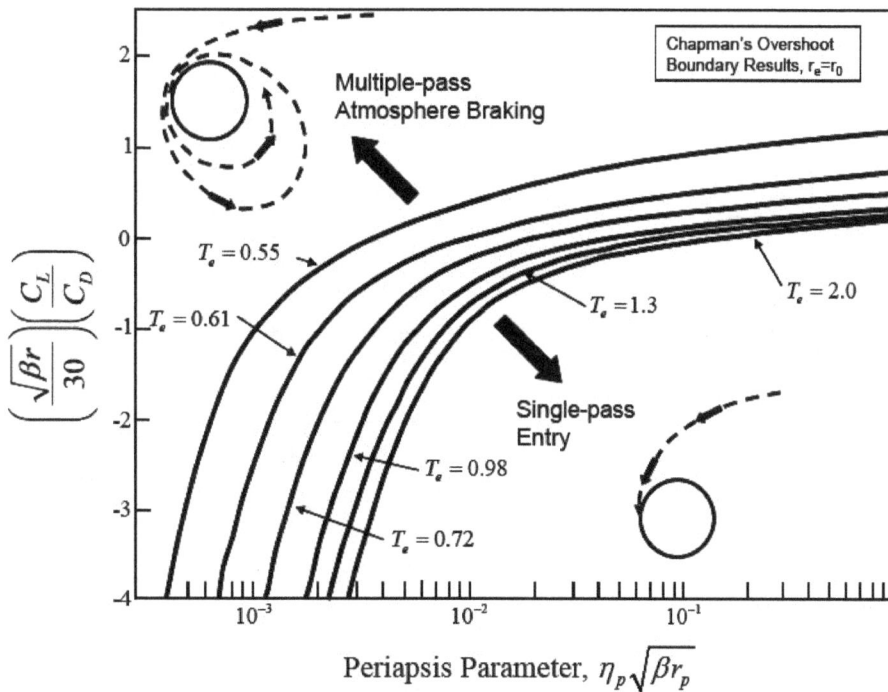

Figure 8-3: Chapman's Overshoot Boundaries for Single Pass Entries (15:18; 39:170; 58:220)

233

In this figure, each curve represents an overshoot boundary for specific entry kinetic energy over a range of entry flight-path angles (as reflected in the range of periapsis parameters). Above the curve, the vehicle fails to remain in the atmosphere or, at best, requires multiple passes to complete the entry. Below the curve, entry occurs on the first pass. The term $\sqrt{\beta r}$ on the y-axis (and Eq. (8.34)) should be replaced with the appropriate *average* value for whatever atmosphere is of interest to be consistent with Chapman's formulation. (For Earth, Chapman used the value $\sqrt{\beta r} = 30$.)

8.4.2.1 Special Case of Positive Lift

When there is positive lift, either Loh's solution or the first-order skip solution from Section 5.5 can be used to directly solve for the overshoot boundary. Both approaches relate the entry kinetic energy to the entry flight-path angle on the overshoot boundary by

$$T_e = \frac{1}{2}\exp\left(\frac{-4\gamma_e}{\dfrac{C_L}{C_D}}\right)$$

8.35

when $r_e = r_0$. With this same condition, the corresponding approach radius is:

$$r_{P_{over}} = \frac{r_e}{2\left(1-T_e\right)}\left\{1 - \sqrt{1 - 4T_e\left(1-T_e\right)\cos^2\left[\frac{1}{4}\left(\frac{C_L}{C_D}\right)\ln\left(2T_e\right)\right]}\right\}$$

8.36

The derivation of Eq. (8.36) is left as an exercise.

8.4.2.2 Special Case for Unobtainable Overshoot Boundaries

Certain entries may exclude the possibility for a skip-type trajectory based on physically realizable entry conditions. Consider three examples:

- *Steep ballistic entry:* The vehicle *may* never reapproach the upper edge of the atmosphere before impacting the planet, so the overshoot boundary (as defined earlier) doesn't exist.

- *Entry with small positive lift or large negative lift:* Mathematically, it *might* be possible to find a combination of T_e and γ_e to define an overshoot boundary. However, other considerations (such as maneuvering fuel) might prevent the vehicle from actually realizing such an orbital approach. For all intents and purposes, the overshoot boundary in this case might as well be undefined.

- *Entry at less than circular speed:* If the vehicle's entry speed is already less than what is required to maintain a circular orbit, no overshoot boundary needs to be defined. (Of course, if T_e is free to change, an overshoot boundary *can be* found.)

In situations such as these, the convention is to use the radius of the planet $R_\oplus$ in lieu of an overshoot radius $r_{p_{over}}$.

8.5 Entry Corridor Width

Earlier, $\Delta r_p = r_{p_{over}} - r_{p_{under}}$ was introduced as a "measure" of the entry corridor. The undershoot radius $r_{p_{under}}$ is easily obtained directly from Eq. (8.26).

When defined, the overshoot radius $r_{p_{over}}$ can be found by solving the TPBVP defined by Eqs. (8.29) - (8.32) or by using Chapman's solution and solving for the radius from the periapsis parameter. (When lift is positive, Eq. (8.36) also provides the value.) When undefined, the overshoot radius is replaced by $R_\oplus$ (39:173). Thus, the entry corridor can be defined by

Entry Corridor $= \Delta r_p = r_{p_{over}} - r_{p_{under}}$

Entry Corridor $= R_\oplus - r_{p_{under}}$ (when the overshoot boundary cannot be realized)

Just as easily, the entry corridor could be defined in terms of Chapman's periapsis parameter

$$\Delta F_p = F_{p_{under}} - F_{p_{over}} \qquad \text{8.37}$$

(with the obvious changes when the overshoot boundary cannot be realized).

Chapman's original works contain a wide range of calculated overshoot and undershoot boundaries and those should be referenced for more detailed discussions (Refs. 14-17). Loh and Vinh provide excellent summaries of his work (39:170-177; 58:178-225).

8.6 Problems

Material Understanding:

1. Prove that Eq. (8.13)

$$e = \sqrt{1 - \frac{r_1 V_1^2 \cos^2 \gamma_1}{\mu^2}\left(2\mu - r_1 V_1^2\right)}$$

reduces to $e = 1$ for parabolic orbits.

2. For an overshoot boundary, show that Loh's equations and first order theory both reduce to Eq. (8.28)

$$T_e = \frac{1}{2}\exp\left(\frac{-4\gamma_e}{\dfrac{C_L}{C_D}}\right)$$

when $r_e = r_0$.

3. Show that the equations of motion for planar entry from Chapter 5 (Eqs. (5.1) and (5.2)) can be written in terms of kinetic energy as the independent variable:

$$\frac{d\eta}{dT} = \frac{\beta r_0 \eta \sin \gamma}{2T\beta r_0 \eta + \sin \gamma}$$

$$\frac{d\gamma}{dT} = -\left[\frac{C_L}{C_D} + \frac{1}{\beta r_0 \eta}\left(1 - \frac{1}{2T}\right)\cos \gamma\right]\left(\frac{\beta r_0 \eta}{2T\beta r_0 \eta + \sin \gamma}\right)$$

4. Show that for the case where $\dfrac{C_L}{C_D} > 0$ and $r_e = r_0$, the approach radius for the overshoot boundary can be written as:

$$r_{P_{over}} = \frac{r_e}{2(1-T_e)}\left\{1-\sqrt{1-4T_e(1-T_e)\cos^2\left[\frac{1}{4}\left(\frac{C_L}{C_D}\right)\ln(2T_e)\right]}\right\}$$

5. For the situation where $\dfrac{C_L}{C_D} = 0.4$ and $\gamma_e = -0.2$ radians, the *overshoot boundary* is well-defined using our first-order skip equations (from Section 5.5). Assume the reference radius is equal to the entry radius (i.e., $r_0 = r_e$) and answer the following:

   a. Find the required entry kinetic energy T_e which places this vehicle exactly on the overshoot boundary.

   b. Find the overshoot radius $r_{P_{over}}$.

Chapter 9

Unified Theory

9.1 Introduction

In earlier chapters, classical theories for planar entry were presented. Chapter 5 derived several closed-form solutions that were each limited to one specific type of entry trajectory. In its range of validity, each solution is accurate and helpful for identifying trends (such as the benefits of adding lift to reduce deceleration or heating rates). Loh's theory in Chapter 6, while empirical, alleviated most of the restrictions and is applicable to a broad range of entries. It is, in fact, actually quite accurate also (58:226).

The next level of complexity is to remove the restriction of planar entry. In doing so, it is also desirable to retain the "universality" of the equations. That is to say, whenever possible, explicit knowledge of the vehicle mass, shape, size, etc. should not be required; "universal" variables similar to η and T should be used if possible. Vinh and Brace derived a set of "universal equations" of this form valid for three-dimensional planetary entry as well as orbital motion (40:9-11; 57:295-299; 58:227-232; 59). This chapter will derive and analyze their universal equations.

9.2 Universal Equations for Three-Dimensional Entries

For a non-thrusting entry near a spherical, non-rotating planet, the kinematic equations (Eqs. (3.35) - (3.37)) and force equations (Eqs. (3.65) - (3.67)) become:

$$\dot{r} = {}^R V \sin \gamma \qquad \text{9.1}$$

$$\dot{\theta} = \frac{{}^R V \cos \gamma \cos \psi}{r \cos \phi} \qquad \text{9.2}$$

$$\dot{\phi} = \frac{{}^R V \cos \gamma \sin \psi}{r} \qquad \text{9.3}$$

$$ {}^R \dot{V} = -\frac{D}{m} - g \sin \gamma \qquad \text{9.4}$$

$$ {}^R V \dot{\gamma} = \frac{L}{m} \cos \sigma - g \cos \gamma + \frac{{}^R V^2}{r} \cos \gamma \qquad \text{9.5}$$

$$ {}^R V \dot{\psi} = \frac{L \sin \sigma}{m \cos \gamma} - \frac{{}^R V^2}{r} \cos \gamma \cos \psi \tan \phi \qquad \text{9.6}$$

Introducing our usual assumptions for the forms of lift and drag,

$$L = \frac{\rho C_L S}{2} {}^R V^2 \qquad \text{9.7}$$

$$D = \frac{\rho C_D S}{2} {}^R V^2 \qquad \text{9.8}$$

the last three equations of motion can be rewritten as:

$$^{R}\dot{V} = -\frac{\rho C_{D} S\,^{R}V^{2}}{2m} - g\sin\gamma \qquad \textbf{9.9}$$

$$^{R}V\dot{\gamma} = \left(\frac{\rho C_{L} S\,^{R}V^{2}}{2m}\right)\cos\sigma - g\cos\gamma + \frac{^{R}V^{2}}{r}\cos\gamma \qquad \textbf{9.10}$$

$$^{R}V\dot{\psi} = \left(\frac{\rho C_{L} S\,^{R}V^{2}}{2m}\right)\left(\frac{\sin\sigma}{\cos\gamma}\right) - \frac{^{R}V^{2}}{r}\cos\gamma\cos\psi\tan\phi \qquad \textbf{9.11}$$

Vinh found that changing the independent variable from time to

$$s = \int_{0}^{t} \frac{^{R}V}{r}\cos\gamma\,dt \qquad \textbf{9.12}$$

facilitated the problem formulation in universal variables. This variable is monotonically increasing when $\cos\gamma > 0$, so it is a reasonable replacement for time in this application (58:229). Note, however, this substitution will make it impossible to examine perfectly vertical trajectories $(\gamma = -\pi/2)$. Dividing Eqs. (9.1) - (9.3) and (9.9) - (9.11) by $\dfrac{ds}{dt} = \dfrac{^{R}V}{r}\cos\gamma$, the equations of motion become:

$$\frac{dr}{ds} = r\tan\gamma \qquad \textbf{9.13}$$

$$\frac{d\theta}{ds} = \frac{\cos\psi}{\cos\phi} \qquad \textbf{9.14}$$

$$\frac{d\phi}{ds} = \sin\psi \qquad\qquad \textbf{9.15}$$

$$\frac{d\,^{R}V^{2}}{ds} = -\frac{r\rho C_{D}S\,^{R}V^{2}}{m\cos\gamma} - 2gr\tan\gamma \qquad\qquad \textbf{9.16}$$

$$\frac{d\gamma}{ds} = \left(\frac{r\rho C_{L}S}{2m\cos\gamma}\right)\cos\sigma + \left(1 - \frac{gr}{^{R}V^{2}}\right) \qquad\qquad \textbf{9.17}$$

$$\frac{d\psi}{ds} = \left(\frac{r\rho C_{L}S}{2m\cos^{2}\gamma}\right)\sin\sigma - \cos\psi\tan\phi \qquad\qquad \textbf{9.18}$$

The gravity term in these (universal equations) should be replaced by an inverse square, central gravity term

$$g = g(r) = \frac{\mu}{r^{2}} \qquad\qquad \textbf{9.19}$$

where μ is the gravitational constant for the planet. To be consistent with our previous derivations, the atmospheric density will be considered to vary exponentially (at least locally) according to:

$$\frac{d\rho}{\rho} = \frac{d\rho(r)}{\rho(r)} = -\beta dr \qquad\qquad \textbf{9.20}$$

(Vinh expands the equations to include more general isothermal atmospheres as well and his works should be reviewed for details if required (58:228-230; 59).)

Vinh also introduced a pair of dependent variable changes

$$u = \frac{^{R}V^{2}\cos^{2}\gamma}{gr} \qquad\qquad \textbf{9.21}$$

$$Z = \frac{\rho C_{D}S}{2m}\sqrt{\frac{r}{\beta}} \qquad\qquad \textbf{9.22}$$

to replace velocity and radius, respectively (16; 58:229). (u is a measure of kinetic energy along perpendicular to the radius and Z is a measure of altitude.) Vinh refers to these as "modified" Chapman variables because of their similarity to those used by Chapman in his classic papers (Refs. 14-17). The derivatives of these new variables (for a strictly exponential atmosphere where β is a constant) are straight-forward to calculate. They are

$$\frac{dZ}{ds} = \beta r \left(\frac{1}{2\beta r} - 1 \right) Z \tan \gamma \qquad \textbf{9.23}$$

$$\frac{du}{ds} = -\frac{2Zu\sqrt{\beta r}}{\cos \gamma} \left(1 + \frac{C_L}{C_D} \tan \gamma \cos \sigma + \frac{\sin \gamma}{2Z\sqrt{\beta r}} \right) \qquad \textbf{9.24}$$

after simplifying. Strictly speaking, r in Eqs. (9.23) and (9.24) should be replaced in favor of Z; however, Vinh (and Chapman before him) shows replacing βr with an *average* value $\overline{\beta r}$ greatly simplifies the mathematics and does not significantly change the results when studying trajectories into most atmospheres of interest (15:6-7; 58:230). Further, since $\overline{\beta r}$ is typically large, Eqs. (9.23) and (9.24) can be simplified to

$$\frac{dZ}{ds} = -\overline{\beta r} Z \tan \gamma \qquad \textbf{9.25}$$

$$\textbf{9.26}$$
$$\frac{du}{ds} = -\frac{2Zu\sqrt{\overline{\beta r}}}{\cos \gamma} \left(1 + \frac{C_L}{C_D} \tan \gamma \cos \sigma + \frac{\sin \gamma}{2Z\sqrt{\overline{\beta r}}} \right)$$

with little loss in accuracy.

The differential equations for γ and ψ still contain r and ^RV, so they need to be rewritten with modified Chapman variables:

$$\frac{d\gamma}{ds} = \frac{Z\sqrt{\overline{\beta}r}}{\cos\gamma}\left[\frac{C_L}{C_D}\cos\sigma + \frac{\cos\gamma}{Z\sqrt{\overline{\beta}r}}\left(1 - \frac{\cos^2\gamma}{u}\right)\right] \qquad \textbf{9.27}$$

$$\frac{d\psi}{ds} = \frac{Z\sqrt{\overline{\beta}r}}{\cos^2\gamma}\left(\frac{C_L}{C_D}\sin\sigma - \frac{\cos^2\gamma}{Z\sqrt{\overline{\beta}r}}\cos\psi\tan\phi\right) \qquad \textbf{9.28}$$

The differential equations for θ and ϕ do not contain r or ^RV so they do not need to be changed and we can now assemble a complete set of the six three-dimensional equations of motion:

$$\frac{dZ}{ds} = -\overline{\beta}rZ\tan\gamma \qquad \textbf{9.29}$$

$$\frac{du}{ds} = -\frac{2Zu\sqrt{\overline{\beta}r}}{\cos\gamma}\left(1 + \frac{C_L}{C_D}\cos\sigma\tan\gamma + \frac{\sin\gamma}{2Z\sqrt{\overline{\beta}r}}\right) \qquad \textbf{9.30}$$

$$\frac{d\theta}{ds} = \frac{\cos\psi}{\cos\phi} \qquad \textbf{9.31}$$

$$\frac{d\phi}{ds} = \sin\psi \qquad \textbf{9.32}$$

$$\frac{d\gamma}{ds} = \frac{Z\sqrt{\overline{\beta}r}}{\cos\gamma}\left[\frac{C_L}{C_D}\cos\sigma + \frac{\cos\gamma}{Z\sqrt{\overline{\beta}r}}\left(1 - \frac{\cos^2\gamma}{u}\right)\right] \qquad \textbf{9.33}$$

$$\frac{d\psi}{ds} = \frac{Z\sqrt{\overline{\beta}r}}{\cos^2\gamma}\left(\frac{C_L}{C_D}\sin\sigma - \frac{\cos^2\gamma}{Z\sqrt{\overline{\beta}r}}\cos\psi\tan\phi\right) \qquad \textbf{9.34}$$

Equations (9.29) - (9.34) are forms of those first derived by Vinh and Brace (59). They are valid for a constant drag coefficient. (The definition of Z in Eq. (9.22) introduces this requirement.) Like the equations derived in Section 4.4, they are free from the characteristics of mass, size, and shape of the vehicle. Once the atmosphere has been specified with a characteristic average $\overline{\beta r}$, these equations can be integrated just *one time* for each set of specified entry conditions, constant lift-to-drag ratio C_L / C_D, and constant bank angle σ. Solutions, given in "Z-Tables" and graphs have been calculated and published for Z (7; 18:106-277; 58:241-246). With modern desktop computers, we can easily solve these equations in a few seconds, so we will not rely on tables for the solutions to the differential equations. (In addition, finding our own numerical solutions frees us to vary the bank angle rather than hold it constant.)

9.3 Reduction to Other Solutions

In Chapter 6, we developed Loh's Second Order-Solution and demonstrated it could be reduced to classic first-order solutions when the same simplifying assumptions were introduced. In much the same way, we can show Eqs. (9.29) - (9.34) are consistent with all known first- and second-order solutions (58:227). The text by Vinh, et al. details four of these reductions (58:232-241). Here, we will only cover two: the Keplerian solution and Loh's Second-Order solution.

9.3.1 Keplerian Solution

If Eqs. (9.29) - (9.34) represent a unified solution for all atmospheric entries, then they should model two-body orbital motion outside of the atmosphere. From the definition of Z, it is clear $Z = 0$ in orbital problems. Taking the limit as $Z \rightarrow 0$, the differential equations become:

$$\frac{dZ}{ds} = 0 \qquad \text{9.35}$$

$$\frac{du}{ds} = -u \tan \gamma \qquad \text{9.36}$$

$$\frac{d\theta}{ds} = \frac{\cos \psi}{\cos \phi} \qquad \text{9.37}$$

$$\frac{d\phi}{ds} = \sin \psi \qquad \text{9.38}$$

$$\frac{d\gamma}{ds} = \left(1 - \frac{\cos^2 \gamma}{u} \right) \qquad \text{9.39}$$

$$\frac{d\psi}{ds} = -\cos \psi \tan \phi \qquad \text{9.40}$$

Solving these begins with dividing Eq. (9.36) by Eq. (9.13) to get:

$$\frac{du}{dr} = -\frac{u}{r} \qquad \text{9.41}$$

This is readily integrated as

$$\ln (ur) = k_1 \qquad \text{9.42}$$

where k_1 is a constant of integration. Putting this in a slightly different form and renaming the constant to p will aid us later:

$$u = \frac{p}{r} \qquad \text{9.43}$$

Notice this relates velocity (through u) to the radius, so we can expect to use it later in finding the vis-viva and the classic conics equations first introduced in Chapter 2.

Dividing Eq. (9.38) by Eq. (9.40) yields

$$\frac{d\phi}{d\psi} = -\frac{\tan\psi}{\tan\phi} \qquad \textbf{9.44}$$

which separates and integrates to give:

$$\ln(\cos\phi\cos\psi) = k_2 \qquad \textbf{9.45}$$

k_2 is a constant of integration. Rewriting this slightly in terms of a new integration constant i:

$$\cos\phi\cos\psi = \cos i \qquad \textbf{9.46}$$

Similarly, dividing Eq. (9.37) by Eq. (9.40) gives:

$$\frac{d\theta}{d\psi} = -\frac{1}{\sin\phi} \qquad \textbf{9.47}$$

The $\sin\phi$ term can be eliminated in favor of an expression containing only constants and ψ. From Eq. (9.46):

$$\sin\phi = \begin{cases} +\sqrt{1-\left(\dfrac{\cos i}{\cos\psi}\right)^2} & \text{for } 0\le\phi\le\dfrac{\pi}{2} \\[3ex] -\sqrt{1-\left(\dfrac{\cos i}{\cos\psi}\right)^2} & \text{for } \dfrac{-\pi}{2}\le\phi\le 0 \end{cases} \qquad \textbf{9.48}$$

For compactness, this will be written as

$$\sin\phi = \pm\sqrt{1 - \left(\frac{\cos i}{\cos\psi}\right)^2}$$

9.49

where the "upper" sign is for $0 \le \phi \le \dfrac{\pi}{2}$ and the "lower" sign is for $\dfrac{-\pi}{2} \le \phi \le 0$.

Noting $-\dfrac{\pi}{2} \le \psi \le \dfrac{\pi}{2}$, this can be written:

$$\sin\phi = \pm\frac{\sqrt{\cos^2\psi - \cos^2 i}}{\cos\psi}$$

9.50

Therefore,

$$\frac{d\theta}{d\psi} = -\frac{1}{\sin\phi} = \mp\frac{\cos\psi}{\sqrt{\cos^2\psi - \cos^2 i}}$$

9.51

With a simple variable change,

$$x = \sin\psi$$

9.52

this can be put in a form that is easily integrated:

$$\int d\theta = \mp\int \frac{dx}{\sqrt{\sin^2 i - x^2}}$$

9.53

Doing so yields

$$\theta + k_3 = \mp\sin^{-1}\left(\frac{x}{|\sin i|}\right)$$

9.54

where k_3 is a constant of integration. Changing back to the original variables and simplifying, this can be written

$$|\sin i|\sin(\theta + k_3) = \mp\sin\psi$$

9.55

A new constant of integration, $k_4 = k_3 - \dfrac{\pi}{2}$, can be introduced to change the sine to a cosine:

$$\left|\sin i\right|\cos\left(\theta + k_4\right) = \mp\sin\psi \qquad \text{9.56}$$

With two more simplifications, Eq. (9.56) can be put into the form found by Vinh. First, assume $0 \le i \le \pi$ so that $\sin i \ge 0$. (When the physical meaning of i is established, this assumption will be clearly justified.) Second, assume the constant k_4 is replaced with a new constant so that the positive sign on the right-hand side of Eq. (9.56) is valid:

$$\sin i\cos\left(\theta - \Omega\right) = \sin\psi \qquad \text{9.57}$$

Dividing Eq. (9.39) by Eq. (9.36) relates the flight-path angle γ and kinetic energy variable u:

$$\frac{d\gamma}{du} = -\frac{1}{u\tan\gamma}\left(1 - \frac{\cos^2\gamma}{u}\right) \qquad \text{9.58}$$

A change of variables with $y = \dfrac{1}{\cos^2\gamma}$, allows this to be written as a linear first-order differential equation:

$$\frac{dy}{du} + \frac{2}{u}y = \frac{2}{u^2} \qquad \text{9.59}$$

The solution this form of differential equations is

$$y = \frac{2u + k_5}{u^2} \qquad \text{9.60}$$

where k_5 is a constant of integration (19:43-44). Substituting for y, the solution relating flight-path angle and the kinetic energy parameter becomes:

$$\cos^2 \gamma = \frac{u^2}{2u - (1 - e^2)}$$

9.61

The constant k_5 has been replaced with $k_5 = -(1 - e^2)$ in Eq. (9.61) to simplify later discussions.

At this point, it is helpful to introduce some spherical trigonometry to aid in identifying the physical significance of some of the integration constants. Consider the right spherical triangle shown in Figure 9-1. Using the notation shown, we can write down several identities to help us in the derivations:

$$\sin a = \sin c \sin \alpha$$

9.62

$$\cos c = \cos a \cos b$$

9.63

$$\sin b = \tan a \cot \alpha$$

9.64

$$\cos \beta = \tan a \cot c$$

9.65

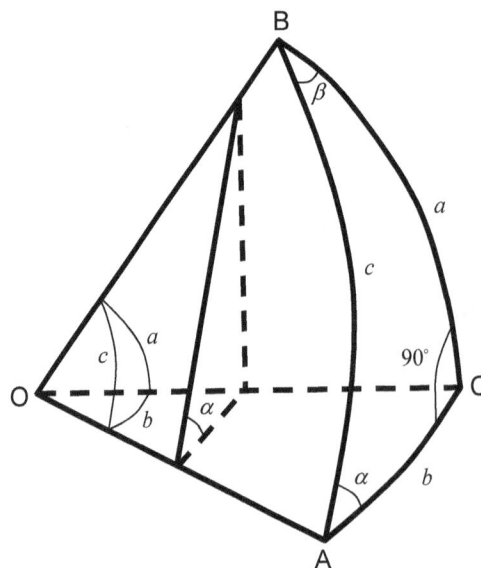

Figure 9-1: General Right Spherical Triangle

These are well-known and will not be proven here (6:145-148; 8:8-9). Figure 9-2 shows this geometry superimposed on our entry trajectory. (This figure illustrates the case for $\theta > \Omega$. The situation for $\Omega > \theta$ is similar and will not be shown here.) From the figure, it's obvious that, if we prove this motion is orbital motion, the constant Ω will be the right ascension of the ascending node. Similarly, the angle labeled α is the orbital inclination if we are modeling orbital motion. The angle c is an angular measure "travel" about a normal to the plane of motion.

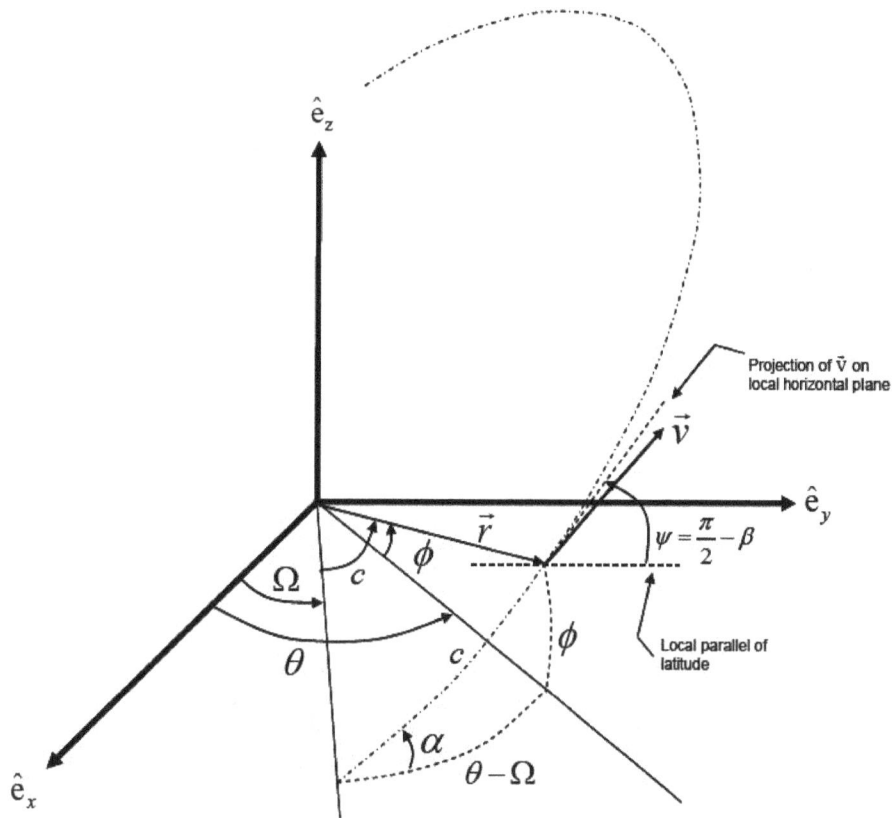

Figure 9-2: Spherical Geometry of Vehicle Trajectory

Continuing on with proving our equations model Keplerian motion, we rewrite Eqs. (9.62) - (9.65) to reflect the variables we are using (and those shown in Figure 9-2):

$$\sin\phi = \sin c \sin\alpha \qquad\qquad \textbf{9.66}$$

$$\cos c = \cos\phi\cos(\theta - \Omega) \qquad\qquad \textbf{9.67}$$

$$\sin(\theta - \Omega) = \tan\phi\cot\alpha \qquad\qquad \textbf{9.68}$$

$$\sin\psi = \tan\phi\cot c \qquad\qquad \textbf{9.69}$$

We can show the angle α in Figure 9-2 and the constant of integration i in Eq. (9.57) are one in the same. Begin by squaring and combining Eqs. (9.46) and (9.57):

$$\left(\frac{\cos i}{\cos\phi}\right)^2 + \left[\sin i\cos(\theta - \Omega)\right]^2 = \cos^2\psi + \sin^2\psi \qquad\qquad \textbf{9.70}$$

After recognizing the right-hand-side is equal to one and simplifying, this becomes:

$$\sin^2(\theta - \Omega) = \tan^2\phi\cot^2 i \qquad\qquad \textbf{9.71}$$

Taking the square root gives

$$\sin(\theta - \Omega) = \tan\phi\cot i \qquad\qquad \textbf{9.72}$$

where the positive root has been taken. (The negative root is a valid solution and would imply different values for the constants Ω and i or a situation where $\Omega > \theta$. While valid, we do not need to discuss such situations since the geometry is similar and the final results are the same.) Comparing Eqs. (9.68) and (9.72) it is evident $\alpha = i$, although we have yet to show this is orbital motion and i is, therefore, the inclination.

Combining Eqs. (9.57) and (9.67) and differentiating with respect to s gives:

$$\left(\frac{dc}{ds}\right) = \left(\frac{\sin\phi}{\sin c}\right)\left(\frac{\sin\psi}{\sin i}\right)\left(\frac{d\phi}{ds}\right) - \left(\frac{\cos\phi}{\sin c}\right)\left(\frac{\cos\psi}{\sin i}\right)\left(\frac{d\psi}{ds}\right)$$

9.73

Equations (9.38) and (9.40) eliminate $\left(\dfrac{d\phi}{ds}\right)$ and $\left(\dfrac{d\psi}{ds}\right)$, respectively:

$$\left(\frac{dc}{ds}\right) = \left(\frac{1}{\sin c \sin i}\right)\left[\sin\phi\sin\psi\left(\sin\psi\right) - \cos\phi\cos\psi\left(-\cos\psi\tan\phi\right)\right]$$

$$= \left(\frac{1}{\sin c \sin i}\right)\left(\sin\phi\sin^2\psi + \sin\phi\cos^2\psi\right)$$

$$= \frac{\sin\phi}{\sin c \sin i}$$

9.74

If Eq. (9.66) is substituted into this (and $\alpha = i$ is recognized), the equation becomes

$$\left(\frac{dc}{ds}\right) = 1$$

9.75

or, rewritten with the aid of $\dfrac{du}{ds} = -u\tan\gamma$:

$$\left(\frac{dc}{du}\right) = \frac{-1}{u\tan\gamma}$$

9.76

Equation (9.61) can be solved for $\cos\gamma$ and $\sin\gamma$:

$$\cos\gamma = \frac{u}{\sqrt{2u-(1-e^2)}} \qquad \textbf{9.77}$$

$$\sin\gamma = \pm\sqrt{\frac{2u-(1-e^2)-u^2}{2u-(1-e^2)}} \qquad \textbf{9.78}$$

Since $\dfrac{-\pi}{2} \leq \gamma \leq \dfrac{\pi}{2}$, there is no sign ambiguity in the cosine, but the sine term may be either positive or negative. These two equations can be used to eliminate $\tan\gamma$ in Eq. (9.76):

$$\left(\frac{dc}{du}\right) = \mp\frac{1}{\sqrt{2u-(1-e^2)-u^2}} \qquad \textbf{9.79}$$

Changing the independent variable to $x = 1-u$ puts this into a form easily found in integral tables:

$$\left(\frac{dc}{dx}\right) = \pm\frac{1}{\sqrt{e^2-x^2}} \qquad \textbf{9.80}$$

Integrating and simplifying, this becomes

$$u = 1 \pm e\cos(c-k_6) \qquad \textbf{9.81}$$

where k_6 is a constant of integration. As we've done before, we can select the constant (and call it ω) such that the positive sign in Eq. (9.81) is valid:

$$u = 1 + e\cos(c-\omega) \qquad \textbf{9.82}$$

Eliminating u with the solution we found in Eq. (9.43), we (finally) get an expression relating the radius to the polar angle $c - \omega$:

$$r = \frac{p}{1 + e\cos(c - \omega)}$$ 9.83

This is the well-known equation for a conic. For two-body orbital motion, it is normally written as

$$r = \frac{p}{1 + e\cos v}$$ 9.84

when true anomaly $v = c - \omega$ is used to describe the polar angle. Similarly, Eq. (9.82) can be rewritten as:

$$u = 1 + e\cos v$$ 9.85

This form will be useful in a later section.

Equation (9.84) shows that our universal equations (Eqs. (9.29) - (9.34)) give a solution for Keplerian motion when $Z \to 0$. Having established this as orbital motion, the constants of integration we found along the way are easily identified with their orbital counterparts: p is the semilatus rectum; i is the inclination; Ω is the right ascension of the ascending node; e is the eccentricity; and ω is the argument of periapsis.

9.3.2 Loh's Solution

Equations (9.29) - (9.34) can be shown to be consistent with Loh's solutions we found in Chapter 6. Those solutions were for planar motion, so we set $\sigma = 0°$. Equations (9.29), (9.30), and (9.33) decouple from the rest to give equations of planar motion:

$$\frac{dZ}{ds} = -\overline{\beta r} Z \tan \gamma \qquad \text{9.86}$$

$$\frac{du}{ds} = -\frac{2Zu\sqrt{\overline{\beta r}}}{\cos \gamma}\left(1 + \frac{C_L}{C_D}\tan \gamma + \frac{\sin \gamma}{2Z\sqrt{\overline{\beta r}}}\right) \qquad \text{9.87}$$

$$\frac{d\gamma}{ds} = \frac{Z\sqrt{\overline{\beta r}}}{\cos \gamma}\left[\frac{C_L}{C_D} + \frac{\cos \gamma}{Z\sqrt{\overline{\beta r}}}\left(1 - \frac{\cos^2 \gamma}{u}\right)\right] \qquad \text{9.88}$$

Recall from Chapter 6, Loh made the empirical observation that the group of terms

$$G = \frac{-1}{\beta r_0 \eta}\left(1 - \frac{1}{2T}\right)\cos \gamma \qquad \text{9.89}$$

was constant for the purposes of integration. In terms of this chapter's universal variables

$$\eta = \frac{Z}{\sqrt{\overline{\beta r}}} \qquad \text{9.90}$$

and:

$$T = \frac{u}{2\cos^2 \gamma} \qquad \text{9.91}$$

(In getting these two relationships, it has been assumed $\beta r_0 = \overline{\beta r}$ and $g_0 r_0 = gr$ during the entry.) Thus, G can be rewritten as:

$$G = \frac{-1}{\sqrt{\overline{\beta r}}Z}\left(1 - \frac{\cos^2 \gamma}{u}\right)\cos \gamma \qquad \text{9.92}$$

Substituting for Loh's constant G and changing the independent variable to Z in Eq. (9.88) by dividing by (9.86), we get an easily integrated equation:

$$\frac{d\gamma}{dZ} = \frac{-1}{\sqrt{\beta r} \sin \gamma} \left(\frac{C_L}{C_D} - G \right) \qquad \textbf{9.93}$$

Or, after integrating,

$$\cos \gamma - \cos \gamma_e = \frac{\left(\dfrac{C_L}{C_D} - G \right)}{\sqrt{\beta r}} (Z - Z_e) \qquad \textbf{9.94}$$

where γ_e and Z_e are the entry conditions. Replacing Z with η this exactly matches Eq. (6.11):

$$\cos \gamma - \cos \gamma_e = \left(\frac{C_L}{C_D} - G \right)(\eta - \eta_e) \qquad \textbf{9.95}$$

This is one-half of Loh's Second-Order Solution (and is one-half of Loh's Unified Solution as well).

To find the other half of Loh's solution, we begin by dividing Eq. (9.87) by Eq. (9.88):

$$\frac{du}{d\gamma} = \frac{-2u}{\left(\dfrac{C_L}{C_D} - G \right)} \left(1 + \frac{C_L}{C_D} \tan \gamma + \frac{\sin \gamma}{2Z\sqrt{\beta r}} \right) \qquad \textbf{9.96}$$

Rewriting this slightly, it becomes:

$$\frac{du}{d\gamma} = \frac{-2u}{\left(\dfrac{C_L}{C_D} - G\right)} \left(1 + \frac{C_L}{C_D}\tan\gamma + \frac{\sin\gamma}{Z\sqrt{\beta r}} - \frac{\sin\gamma\cos^2\gamma}{Zu\sqrt{\beta r}} + \frac{\sin\gamma\cos^2\gamma}{Zu\sqrt{\beta r}} - \frac{\sin\gamma}{2Z\sqrt{\beta r}}\right)$$

$$= \frac{-2u}{\left(\dfrac{C_L}{C_D} - G\right)} \left[1 + \tan\gamma\left(\frac{C_L}{C_D} - G\right) - \frac{\sin\gamma}{2Z\sqrt{\beta r}}\left(1 - \frac{2\cos^2\gamma}{u}\right)\right] \qquad \textbf{9.97}$$

Rearranging and grouping terms, this becomes:

$$\left(\frac{du}{d\gamma} + 2u\tan\gamma\right) + \frac{2u}{\left(\dfrac{C_L}{C_D} - G\right)} = -\frac{\sin\gamma\left(1 - \dfrac{2\cos^2\gamma}{u}\right)}{2Z\sqrt{\beta r}\left(\dfrac{C_L}{C_D} - G\right)} \qquad \textbf{9.98}$$

Eliminating Z with Eq. (9.94) and assuming a high-altitude entry ($Z_e = 0$):

$$\left(\frac{du}{d\gamma} + 2u\tan\gamma\right) + \frac{2u}{\left(\dfrac{C_L}{C_D} - G\right)} = -\frac{\sin\gamma\left(1 - \dfrac{2\cos^2\gamma}{u}\right)}{2\overline{\beta r}\left(\cos\gamma - \cos\gamma_e\right)} \qquad \textbf{9.99}$$

Or, with a little more creative manipulation, we "simplify" it as

$$\frac{d}{d\gamma}\left(\frac{u}{2\cos^2\gamma}\right) + \frac{2\left(\dfrac{u}{2\cos^2\gamma}\right)}{\left(\dfrac{C_L}{C_D} - G\right)} = -\frac{\sin\gamma\left(1 - \dfrac{2\cos^2\gamma}{u}\right)}{4\cos^2\gamma\,\overline{\beta r}\left(\cos\gamma - \cos\gamma_e\right)} \qquad \textbf{9.100}$$

which makes the substitution of T straightforward:

$$\frac{dT}{d\gamma} + \frac{2T}{\left(\dfrac{C_L}{C_D} - G\right)} = -\left[\frac{\sin\gamma}{\overline{\beta r}\left(\cos\gamma - \cos\gamma_e\right)}\right]\left[\frac{\left(1 - \dfrac{1}{T}\right)}{4\cos^2\gamma}\right]$$

9.101

Equation (9.101) is very similar to, but not exactly, the differential equation we solved in Chapter 6 for Loh's Universal Theory (Eq. (6.19)). However, if we make the same assumption as we did for his second-order solution, $\overline{\beta r} \gg 1$, then we can exactly match Eq. (6.31)

$$\frac{dT}{d\gamma} + \frac{2T}{\left(\dfrac{C_L}{C_D} - G\right)} = 0$$

9.102

and develop the same solution as in Eq. (6.35):

$$\gamma = \gamma_e - \frac{1}{2}\ln\left(\frac{T}{T_e}\right)\left[\frac{C_L}{C_D} + \frac{1}{\beta\, r_0\eta}\left(1 - \frac{1}{2T}\right)\cos\gamma\right]$$

9.103

Finding Eqs. (9.95) and (9.103) from Vinh and Brace's universal equations (Eqs. (9.29) - (9.34)) shows they are consistent with Loh's Second-Order Solution. Since we have already shown examples (in Chapter 6) where Loh's Second-Order Solution reduces to first-order solutions, we can state with confidence Vinh and Brace's universal equations also reduce to known first-order solutions.

Before leaving this section, we can revisit finding the second of Loh's equations without relying on the assumption $\overline{\beta r} \gg 1$. Vinh, et al. began with an expression for $\dfrac{du}{dZ}$ which we can get by dividing Eq. (9.87) by Eq. (9.86)

$$\frac{du}{dZ} = \frac{2u}{\sqrt{\overline{\beta r}} \sin \gamma} \left(1 + \frac{C_L}{C_D} \tan \gamma + \frac{\sin \gamma}{2Z \sqrt{\overline{\beta r}}} \right) \qquad \textbf{9.104}$$

(58:240). Rearranging it slightly, we will begin with:

$$\frac{1}{u}\frac{du}{dZ} - \frac{1}{Z\overline{\beta r}} = \frac{2}{\sqrt{\overline{\beta r}} \sin \gamma} \left(1 + \frac{C_L}{C_D} \tan \gamma \right) \qquad \textbf{9.105}$$

The left-hand-side is a perfect derivative, so it simplifies to:

$$\frac{d}{dZ}\left(\ln u - \frac{1}{\overline{\beta r}} \ln Z \right) = \frac{2}{\sqrt{\overline{\beta r}} \sin \gamma} \left(1 + \frac{C_L}{C_D} \tan \gamma \right) \qquad \textbf{9.106}$$

If this is divided by Eq. (9.93), the independent variable is changed to γ

$$\frac{d}{d\gamma}\left(\ln u - \frac{1}{\overline{\beta r}} \ln Z \right) = \frac{-2}{\left(\dfrac{C_L}{C_D} - G \right)} \left(1 + \frac{C_L}{C_D} \tan \gamma \right) \qquad \textbf{9.107}$$

and is easily integrated (with the aid of a basic integral table):

$$\ln\left(\frac{u}{u_e} \right) - \frac{1}{\overline{\beta r}} \ln\left(\frac{Z}{Z_e} \right) = \frac{-2}{\left(\dfrac{C_L}{C_D} - G \right)} \left[(\gamma - \gamma_e) - \frac{C_L}{C_D} \ln\left(\frac{\cos \gamma}{\cos \gamma_e} \right) \right] \qquad \textbf{9.108}$$

The entry conditions have been used in place of a constant of integration. When rearranged, it takes on a more familiar looking form:

$$\gamma - \gamma_e = \frac{C_L}{C_D} \ln\left(\frac{\cos\gamma}{\cos\gamma_e}\right) - \frac{\left(\frac{C_L}{C_D} - G\right)}{2} \ln\left(\frac{u}{u_e}\right) + \frac{\left(\frac{C_L}{C_D} - G\right)}{2} \frac{1}{\beta r} \ln\left(\frac{Z}{Z_e}\right) \qquad \textbf{9.109}$$

Equations (9.94) and (9.109) constitute Loh's Unified Solution in the current variables (58:240). As was the case in Chapter 6, we have two equations in three variables and by choosing one, we can solve for the other two at any point along the trajectory.

9.4 Entry Corridor Revisited

In Chapter 8 we introduced the concept of an entry corridor. Calculating the undershoot boundary was fairly simple; however, the overshoot boundary proved to be a numerically difficult task. To that end, we simply introduced Chapman's results without proof and moved on. With the current unified theory, we can now revisit the problem to obtain a robust solution technique. In this formulation, we will follow the method described by Vinh, <u>et al.</u> and approach the problem from a consideration of decelerations experienced by the vehicle (58:244:252).

If we, once again, assume the drag force is the dominant term in the tangential direction and lift is dominant in the normal direction, we can write the total non-dimensional deceleration as

$$\frac{a_{decel}}{g_0} = \frac{Zu\sqrt{\beta r}}{\cos^2\gamma} \sqrt{1 + \left(\frac{C_L}{C_D}\right)^2} \qquad \textbf{9.110}$$

Since these are all proportional, the maxima (or minima) occurs at the same point in the trajectory. The point of the extrema ("critical") deceleration is given by

$$u \cos \gamma \left(\frac{dZ}{ds} \right) + Z \cos \gamma \left(\frac{du}{ds} \right) + 2Zu \sin \gamma \left(\frac{d\gamma}{ds} \right) = 0 \qquad \textbf{9.111}$$

Using Eqs. (9.86) - (9.88) to simplify, this "optimality condition" becomes

$$2\sqrt{\overline{\beta r}} Z_* + \left(\overline{\beta r} - 1 \right) \sin \gamma_* + \frac{2 \sin \gamma_* \cos^2 \gamma_*}{u_*} = 0 \qquad \textbf{9.112}$$

where the "*" subscript denotes conditions at the critical point. From this equation, it is obvious the maximum (or minimum) can only be satisfied when $\gamma_* < 0$; in-other-words, during an inbound portion of a trajectory. Of course, at this point, the total deceleration is

$$\left(\frac{a_{decel}}{g_0} \right)_* = \frac{Z_* u_* \sqrt{\overline{\beta r}}}{\cos^2 \gamma_*} \sqrt{1 + \left(\frac{C_L}{C_D} \right)^2} \qquad \textbf{9.113}$$

and it remains to be determined if this is a minimum or maximum value along the trajectory.

9.4.1 Overshoot Boundary

If we assume atmospheric entry occurs when the deceleration reaches some small fraction f of the gravity acceleration, then along the overshoot boundary $\frac{a_{decel}}{g_0}$ should never drop below f. (This definition is different from the one used in Section 8.4.2, but produces similar results.) Put another way, the *minimum* value of deceleration along the overshoot boundary is f and occurs at the point satisfying Eq. (9.112). The value of f is somewhat arbitrary; we'll follow Vinh's lead and use $f = 0.05$ (58:251). If we choose a Z_* and solve

$$2\sqrt{\overline{\beta r}}\, Z_* + \left(\overline{\beta r} - 1\right)\sin\gamma_* + \frac{2\sin\gamma_* \cos^2\gamma_*}{u_*} = 0 \qquad \textbf{9.114}$$

and

$$\left(\frac{a_{decel}}{g_0}\right)_* = \frac{Z_* u_* \sqrt{\overline{\beta r}}}{\cos^2\gamma_*}\sqrt{1 + \left(\frac{C_L}{C_D}\right)^2} = f \qquad \textbf{9.115}$$

for u_* and γ_*, we can find a point along an overshoot boundary. Since the entry corridor is considered a "planar problem," integrating Eqs. (9.29), (9.30), and (9.33) "backwards" to the point where $\left(\dfrac{a_{decel}}{g_0}\right) = f$ again is all that is required to get the entry conditions Z_e, u_e, and γ_e.

As in Section 8.4, we can find the periapsis of non-parabolic approach orbits with the following equation:

$$r_{P_{over}} = \frac{r_e}{2 - \left(\dfrac{u_e}{\cos^2\gamma_e}\right)}\left\{1 - \sqrt{1 - u_e\left[2 - \left(\dfrac{u_e}{\cos^2\gamma_e}\right)\right]}\right\} \qquad \textbf{9.116}$$

Equivalently (and perhaps more conveniently in this situation), we can form the Chapman periapsis parameter

$$F_p = \frac{\rho_p S C_D}{2m}\sqrt{\frac{r_p}{\beta}} \qquad \textbf{9.117}$$

where, by observation, we note $F_p = Z_p$. Z_p is easily found by solving

$$\frac{u_e^2}{\cos^2 \gamma_e} - 2u_e = u_p^2 - 2u_p \qquad\qquad \textbf{9.118}$$

and

$$Z_p = Z_e \sqrt{\frac{u_e}{u_p}} \exp\left[\overline{\beta r}\left(1 - \frac{u_e}{u_p} \right) \right] \qquad\qquad \textbf{9.119}$$

simultaneously. Equations (9.118) and (9.119) were found from the definition of Z and by evaluating the orbital constants p and e at both the entry radius and hypothetical periapsis. The derivation/proof is left as an exercise. Once found, this Z_p is the Chapman periapsis parameter for the overshoot boundary; i.e., $F_{P_{over}} = Z_{P_{over}}$.

By scanning through values of Z_*, a range of overshoot boundaries can be found, each with different entry conditions. Figure 9-3 shows the solutions for ballistic entries into Earth's atmosphere for entry speeds between circular and 2.2 times circular. This was generated by scanning $1.68 \times 10^{-3} \le Z_* \le 1.91 \times 10^{-3}$ and repeatedly solving the problem described above.

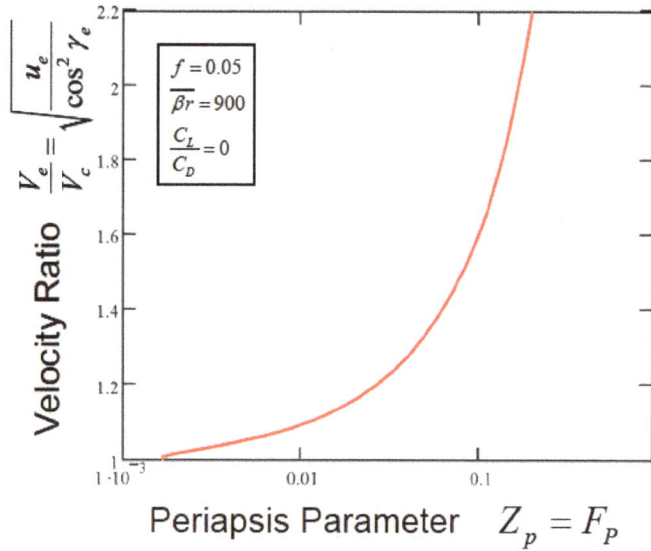

Figure 9-3: Overshoot Boundary for Ballistic Entry into Earth's Atmosphere

To summarize this section, a "cookbook" approach to finding overshoot boundaries is given below.

Step 1: Pick a value for Z_*.

Step 2: Solve Eqs. (9.114) and (9.115) for u_* and γ_*. This is a point of minimum deceleration on the trajectory.

Step 3: Integrate Eqs. (9.29), (9.30), and (9.33) "backwards" until $\left(\dfrac{a_{decel}}{g_0} \right) = f$ again to find the entry conditions $Z = Z_e$, $u = u_e$, and

$\gamma = \gamma_e$. This is a two-point boundary value problem (TPBVP), defined by solving the three differential equations subject to the boundary conditions

$$Z = Z_*, \ u = u_*, \ \gamma = \gamma_* \ \text{at} \ s = s_*$$

$$\frac{Z_e u_e \sqrt{\bar{\beta} r}}{\cos^2 \gamma_e} \sqrt{1 + \left(\frac{C_L}{C_D}\right)^2} = f \ \text{at} \ s = 0$$

where s_* is an unknown value corresponding to the point of minimum deceleration.

Step 4: Solve Eqs. (9.118) and (9.119) simultaneously to find $F_{P_{over}} = Z_{P_{over}}$. (Equivalently, the initial conditions from *Step 3* could be used to calculate a dimensional radius $r_{P_{over}}$.)

Step 5: Select another value for Z_* and repeat.

9.4.2 Undershoot Boundary

On the undershoot boundary, the vehicle experiences a maximum deceleration equal to some value dictated by structural, thermal, or some other limitation. In this case, we are looking for a deceleration *maximum* found by satisfying Eq. (9.112). Similar to the overshoot boundary, we solve for the "critical" point by simultaneously solving Eqs. (9.112) and (9.113), this time setting the maximum deceleration to the maximum allowed:

$$\left(\frac{a_{decel}}{g_0}\right)_* = \frac{Z_* u_* \sqrt{\bar{\beta} r}}{\cos^2 \gamma_*} \sqrt{1 + \left(\frac{C_L}{C_D}\right)^2} = \left(\frac{a_{decel}}{g_0}\right)_{failure} \qquad 9.122$$

This solution for Z_*, u_* and γ_* represents a point along an undershoot boundary. This point can be used as the starting point for a *backwards* integration to the point given by $\left(\dfrac{a_{decel}}{g_0}\right) = f$ to get the entry conditions Z_e, u_e, and γ_e. Then, just as in finding the overshoot boundary, the undershoot periapsis radius $r_{p_{under}}$ or periapsis parameter $F_{p_{under}} = Z_{p_{under}}$ can be calculated. Figure 9-4 shows the undershoot boundaries for ballistic entry with six different deceleration limits. Using these equations, it can be shown the minimum peak deceleration is in the range $5.5 < \left(\dfrac{a_{decel}}{g_0}\right)_* < 6.55$ and occurs at slightly hyperbolic entry speeds.

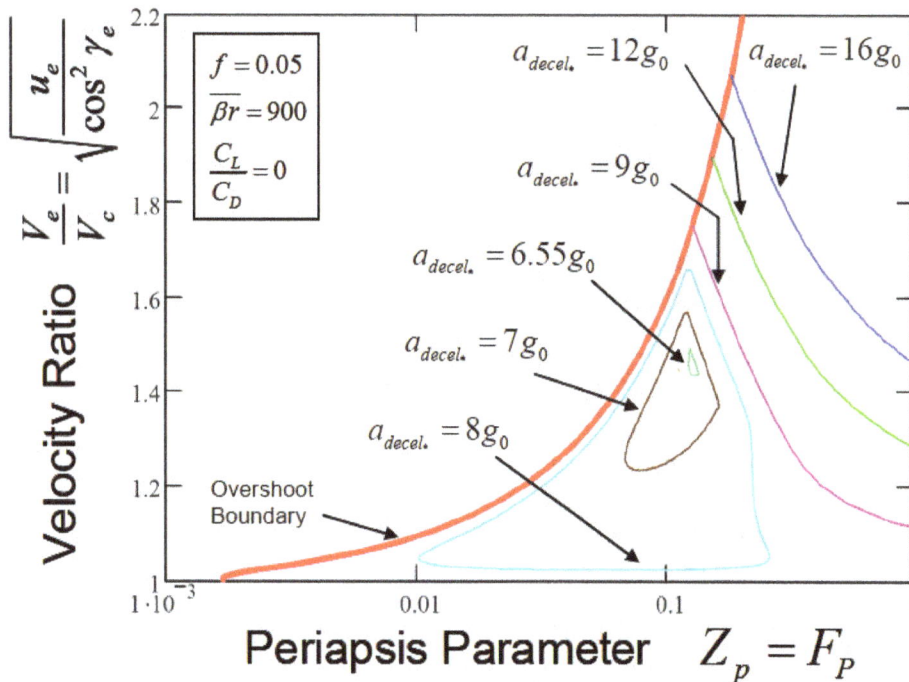

Figure 9-4: Overshoot and Undershoot Boundaries for Ballistic Entry into Earth's Atmosphere

To summarize this section, a "cookbook" approach to finding undershoot boundaries is given below.

Step 1: Define a limiting deceleration $\left(\dfrac{a_{decel}}{g_0}\right)_{failure}$.

Step 2: Pick a value for Z_*.

Step 3: Solve Eqs. (9.114) and (9.122) for u_* and γ_*. This is a point of maximum deceleration on the trajectory.

Step 4: Integrate Eqs. (9.29), (9.30), and (9.33) "backwards" until $\left(\dfrac{a_{decel}}{g_0}\right) = f$ to find the entry conditions $Z = Z_e$, $u = u_e$, and $\gamma = \gamma_e$. This is a TPBVP, defined by solving the three differential equations subject to the boundary conditions

$$Z = Z_*, \ u = u_*, \ \gamma = \gamma_* \ \text{at} \ s = s_*$$

$$\frac{Z_e u_e \sqrt{\beta r}}{\cos^2 \gamma_e} \sqrt{1 + \left(\frac{C_L}{C_D}\right)^2} = f \ \text{at} \ s = 0$$

where s_* is an unknown value corresponding to the point of maximum deceleration.

Step 5: Solve Eqs. (9.118) and (9.119) simultaneously to find $F_{P_{under}} = Z_{P_{under}}$. (Equivalently, the initial conditions from *Step 4* could be used to calculate a dimensional radius $r_{P_{under}}$.)

Step 6: Select another value for Z_* and repeat Steps 2-5 until the entire undershoot boundary is defined.

Step 7: Define a new $\left(\dfrac{a_{decel}}{g_0}\right)_{failure}$ (if desired) and repeat Steps 2-6.

9.4.3 Entry Corridor Width

In Section 8.5, one definition of the entry corridor width was given as:

$$\Delta F_p = F_{P_{under}} - F_{P_{over}} \qquad\qquad \textbf{9.125}$$

Or, in the notation of this chapter:

$$\Delta Z_p = Z_{P_{under}} - Z_{P_{over}} \qquad\qquad \textbf{9.126}$$

In the case of ballistic entry, these differences can be read directly from Figure 9-4 once an entry speed is selected. Similar plots can be made for lifting entry.

As an example, consider a ballistic parabolic entry. Figure 9-5 shows the line of constant entry velocity along which the entry corridor width can be measured. Figure 9-6 replots the information along that line to show the maximum deceleration experienced as a function of Z_p. The figure also shows the entry corridor for a maximum deceleration of $10g_0$ ($a_{decel_*} = 10g_0$) is given by $0.065 \le Z_p \le 0.295$ or $\Delta Z_p = 0.23$. For Earth, this represents a width of about 10.8 km.

It is possible to increase the corridor width by using lift. *Negative lift raises the overshoot boundary radius* (i.e., $Z_{P_{over}}$ is reduced) by enabling the vehicle to approach at a more shallow angle (or faster speed) and still be captured. This, along with the opposite effect with positive lift, is shown in Figure 9-7.

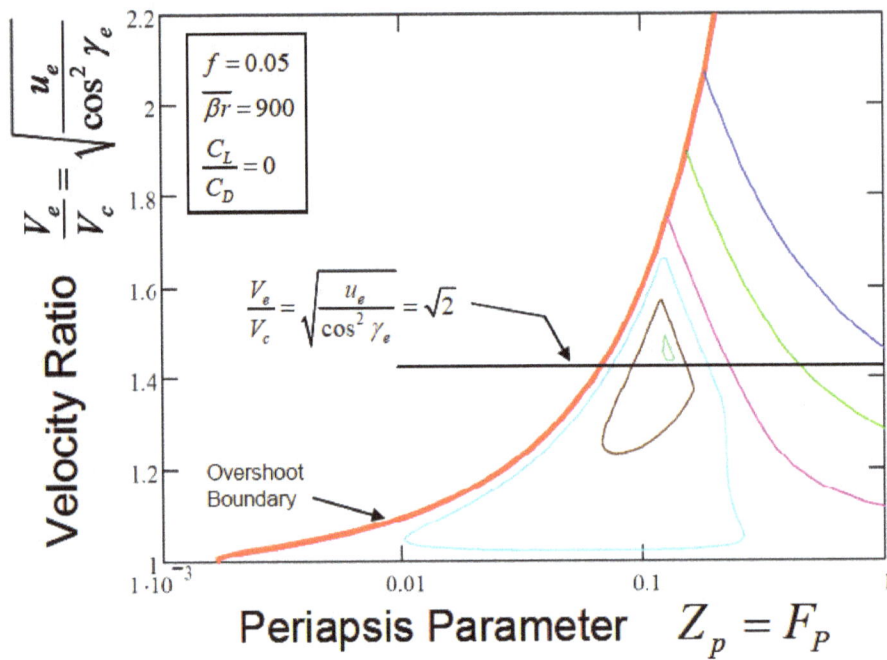

Figure 9-5: Measuring the Entry Corridor Width for Parabolic Ballistic Entry

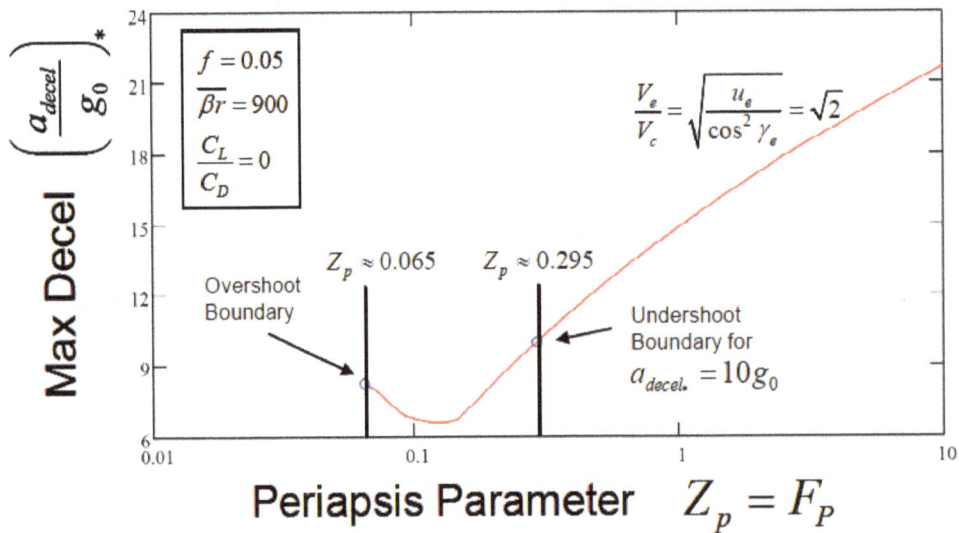

Figure 9-6: Measuring the Entry Corridor Width for $a_{decel.} = 10g_0$

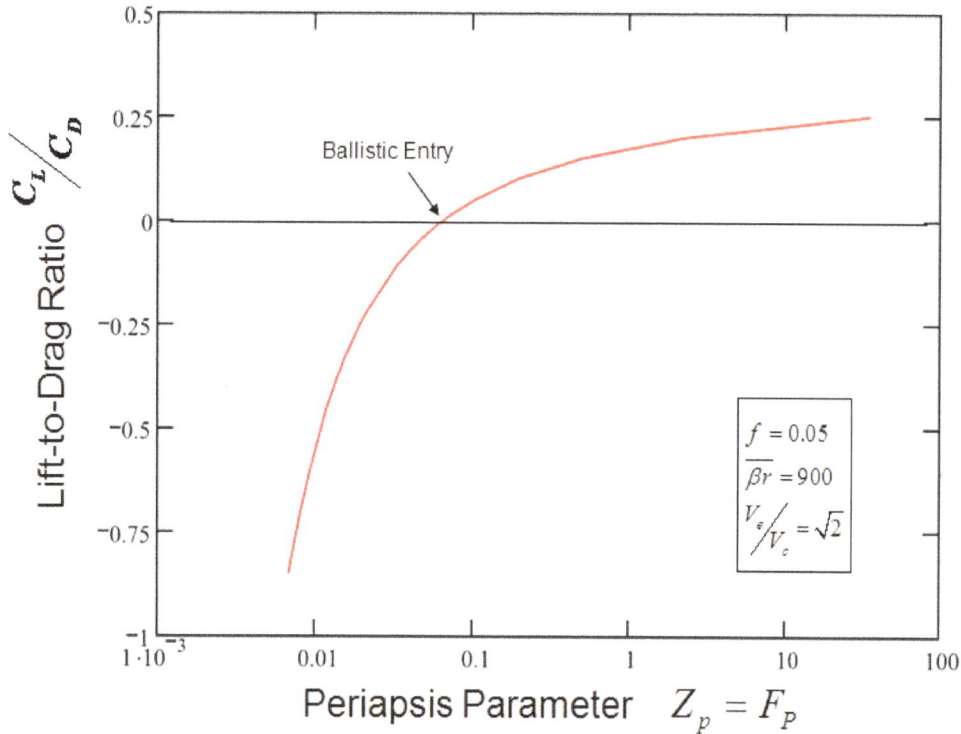

Figure 9-7: Overshoot Boundary for Parabolic Entry with Lift

Likewise, *positive lift* can allow steeper (or faster) entries for a given maximum deceleration. This *lowers the undershoot boundary radius* (i.e., $Z_{p_{under}}$ is increased). Figure 9-8 shows this for a parabolic entry with $a_{decel_*} = 10g_0$ and a range of lift capability. For example, a vehicle with $-0.34 \leq \dfrac{C_L}{C_D} \leq 0.34$ can expand its entry corridor to $0.015 \leq Z_p \leq 15.020$, or $\Delta Z_p \approx 15$. For Earth, this represents about 49.4 km – almost 4.5 times as large as ballistic entry alone!

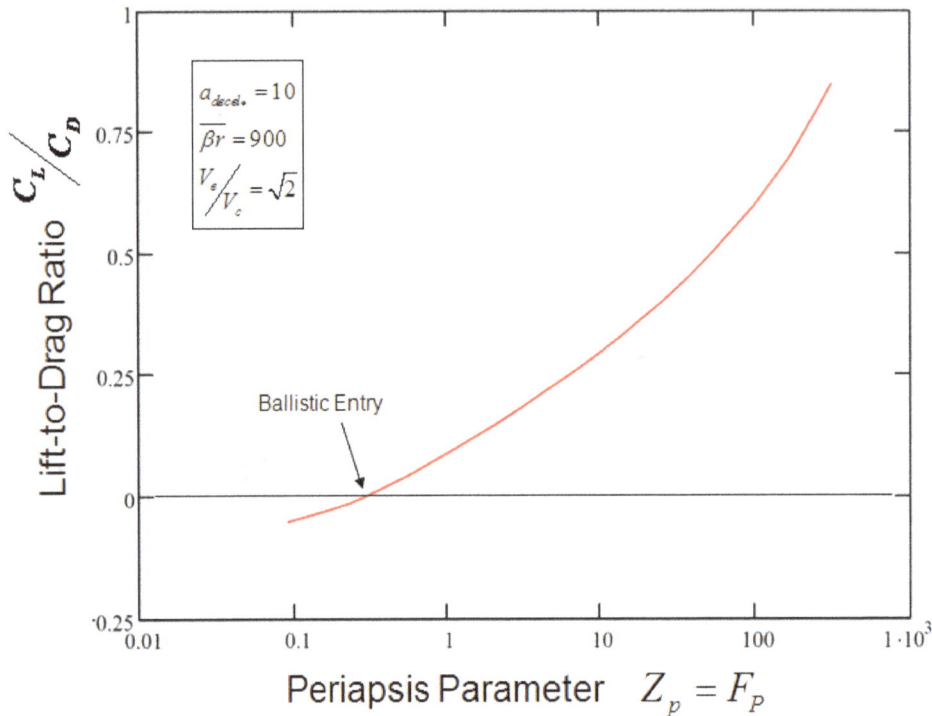

Figure 9-8: *Undershoot Boundary for* $a_{decel_*} = 10 g_0$ *Parabolic Entry with Lift*

9.5 Multipass Entry

As we've already shown, Eqs. (9.29) - (9.34) are valid for both entry and orbital trajectories. With these, we can examine multipass entries like the one in Figure 9-9 by integrating a single set of equations forward from a set of initial conditions until impact with the planet. There are, however, some practical aspects which need to be addressed before running to the computer. Specifically, to handle the portions of the trajectory where $Z \to 0$, a few changes and warnings will make the solution much easier to numerically find.

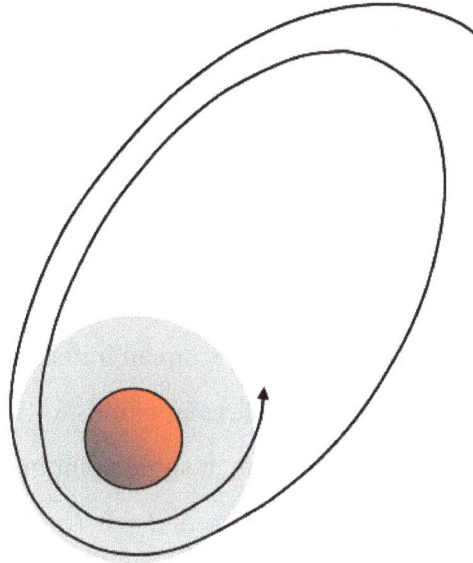

Figure 9-9: Multipass Entry

As written, Eqs. (9.30), (9.33), and (9.34) have an apparent singularity at $Z = 0$. A slight rearranging of the equations will eliminate it. These equations (and the other three for completeness) become:

$$\frac{dZ}{ds} = -\overline{\beta r} Z \tan \gamma \qquad\qquad \textbf{9.127}$$

$$\frac{du}{ds} = -\frac{2Zu\sqrt{\overline{\beta r}}}{\cos \gamma}\left(1 + \frac{C_L}{C_D}\cos \sigma \tan \gamma\right) - u \tan \gamma \qquad\qquad \textbf{9.128}$$

$$\frac{d\theta}{ds} = \frac{\cos \psi}{\cos \phi} \qquad\qquad \textbf{9.129}$$

$$\frac{d\phi}{ds} = \sin \psi \qquad\qquad \textbf{9.130}$$

273

$$\frac{d\gamma}{ds} = \frac{Z\sqrt{\beta r}}{\cos\gamma}\frac{C_L}{C_D}\cos\sigma + \left(1 - \frac{\cos^2\gamma}{u}\right)$$

9.131

$$\frac{d\psi}{ds} = \frac{Z\sqrt{\beta r}}{\cos^2\gamma}\frac{C_L}{C_D}\sin\sigma - \cos\psi\tan\phi$$

9.132

Assuming we begin our integration from an initial point where $Z \neq 0$, a perfect computer could easily integrate these equations to find trajectories. Unfortunately, no computer is perfect and, at some point in the integration, Z might take on the value of *exactly* zero. When $Z = 0$ (or the computer *thinks* $Z = 0$), Eq. (9.127) yields $dZ/ds = 0$ and the value of Z remains fixed at $Z = 0$ forever more. (This statement assumes we are integrating "forward" in s and find our value of Z becoming "numerically" zero.) Fortunately, there is a method to handle this situation.

As the trajectory leaves the atmosphere and Z becomes sufficiently small, two-body orbital motion can be assumed. (We can assume Z is very small without assuming it to be exactly zero.) If the vehicle returns to the atmosphere, the orbit is elliptical. Figure 9-10 illustrates the situation. Symmetry lets us immediately relate three of the exit conditions to three of the conditions at the next entry:

$$Z_2 = Z_1$$

9.133

$$u_2 = u_1$$

9.134

$$\gamma_2 = -\gamma_1$$

9.135

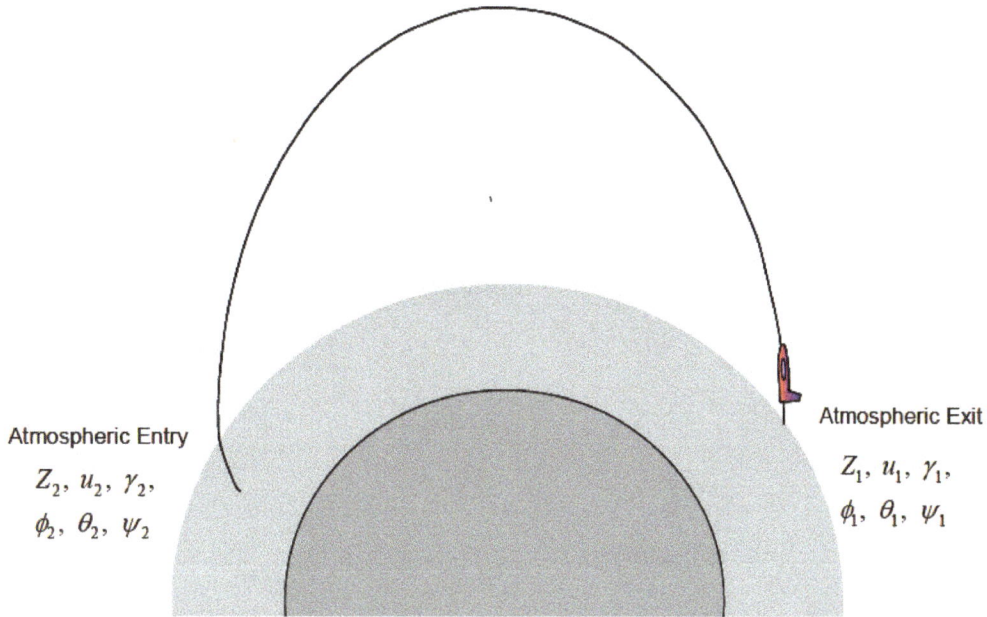

Atmospheric Entry

Z_2, u_2, γ_2,

ϕ_2, θ_2, ψ_2

Atmospheric Exit

Z_1, u_1, γ_1,

ϕ_1, θ_1, ψ_1

Figure 9-10: Keplerian Portion of the Trajectory

During the orbit (just after moving beyond the "exit point"), these parameters are easily related by

$$Z = 0 \qquad\qquad \textbf{9.136}$$

and

$$\cos^2 \gamma = \frac{u^2}{2u - (1 - e^2)} \qquad\qquad \textbf{9.137}$$

where the (constant) eccentricity is found by evaluating Eq. (9.137) at u_1 and γ_1.

The other three trajectory parameters can be found by integrating

$$\frac{d\theta}{ds} = \frac{\cos \psi}{\cos \phi} \qquad\qquad \textbf{9.138}$$

$$\frac{d\phi}{ds} = \sin \psi \qquad\qquad \textbf{9.139}$$

$$\frac{d\psi}{ds} = -\cos\psi \tan\phi \qquad \qquad \textbf{9.140}$$

forward from ϕ_1, θ_1, and ψ_1 once the value of s_2 is known. (These are just Eqs. (9.129), (9.130), and (9.132) rewritten with $Z = 0$.)

To calculate s_2, consider the definition of s given in Eq. (9.12):

$$s = \int_0^t \frac{{}^R V}{r} \cos\gamma\, dt \qquad \qquad \textbf{9.141}$$

Between s_1 and s_2, the trajectory is simple two-body motion and, since the planet isn't rotating, ${}^R V = {}^I V$. Thus, we can simplify this expression by introducing the (constant) angular momentum to eliminate the flight-path angle during our time of interest:

$$s_2 - s_1 = h \int_{t_1}^{t_2} \frac{1}{r^2}\, dt \qquad \qquad \textbf{9.142}$$

The variable of integration can be changed by using the two-body relationship

$$\frac{dv}{dt} = \frac{na^2 \sqrt{1-e^2}}{r^2} = \frac{\sqrt{\mu a}\sqrt{1-e^2}}{r^2} \qquad \qquad \textbf{9.143}$$

to replace time in favor of true anomaly:

$$s_2 - s_1 = \frac{h}{\sqrt{\mu a}\sqrt{1-e^2}} \int_{v_1}^{v_2} dv \qquad \qquad \textbf{9.144}$$

Since the motion must be elliptical during this time, $p = h^2/\mu = a(1 - e^2)$ and Eq. (9.144) simplifies to:

$$s_2 - s_1 = \int_{\nu_1}^{\nu_2} d\nu = \nu_2 - \nu_1 \qquad \text{9.145}$$

Figure 9-11 illustrates the symmetry in the orbital portion of the problem. With the aid of the graphic, it is obvious $\nu_2 = 2\pi - \nu_1$; therefore, we have:

$$s_2 - s_1 = 2(\pi - \nu_1) \qquad \text{9.146}$$

The exit true anomaly is found by evaluating Eq. (9.85) at the exit point

$$\nu_1 = \cos^{-1}\left(\frac{u_1 - 1}{e}\right) \qquad \text{9.147}$$

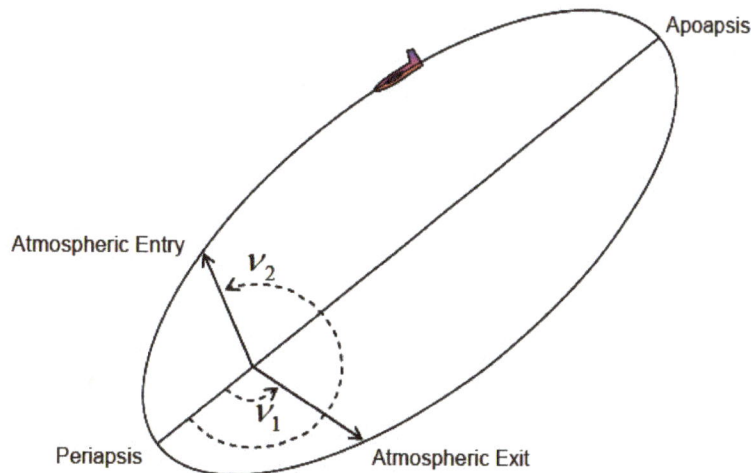

Figure 9-11: True Anomaly for Exit and Reentry Points

where the eccentricity e is similarly calculated from the conditions at the exit point by solving Eq. (9.137):

$$e = \sqrt{\frac{u_1^2}{\cos^2 \gamma_1} - 2u_1 + 1}$$

9.148

When the vehicle reenters the atmosphere, $Z_2 = Z_1 \neq 0$ and the numerical integration of Eqs. (9.127) - (9.132) can continue from s_2 to the next atmospheric exit (if one exists).

Figure 9-12 shows a typical multipass entry as viewed in the $u - \gamma$ plane. The approximate points of entry into and out of the atmosphere are marked and can be distinguished by the abrupt changes in the ellipse-like shape of the curve. In this case, the vehicle clearly enters and exits the atmosphere twice. On the third entry, it is captured.

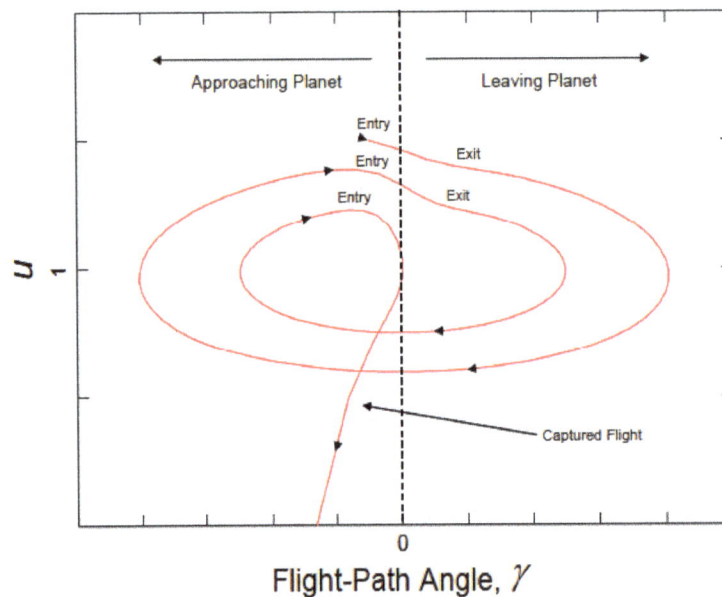

Figure 9-12: Ballistic Multipass Entry on the $u - \gamma$ Plane

When the vehicle is outside the atmosphere, its trajectory is Keplerian and its motion in the $u - \gamma$ plane is given Eq. (9.137). As the vehicle passes through the atmosphere, the instantaneous value of the eccentricity rapidly changes and the path in the $u - \gamma$ plane changes. When it exits the atmosphere, the eccentricity becomes constant again and the vehicle follows a new curve given by Eq. (9.137) for the new value of eccentricity. To illustrate, Figure 9-13 plots the eccentricity corresponding to the same multipass entry. Notice how the eccentricity changes once the vehicle encounters the atmosphere; its motion is no longer Keplerian and eccentricity is not even approximately constant.

Before leaving this section, it's a simple matter to look at the effect of lift on multipass entries. As we would expect, positive lift (i.e., lift pulling the vehicle "up" away from the planet) increases the number of passes required to be "captured." Figure 9-14 compares the typical impact on eccentricity for a purely

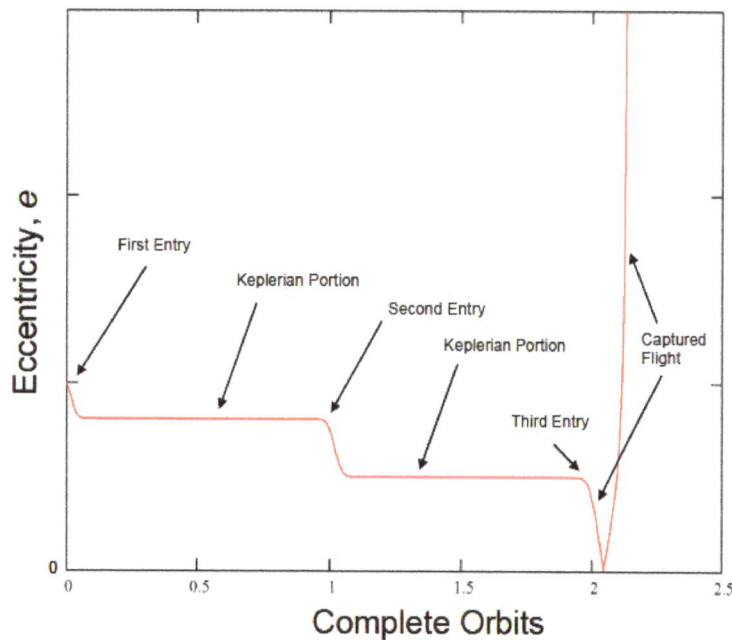

Figure 9-13: Eccentricity Change in Ballistic Multipass Entry

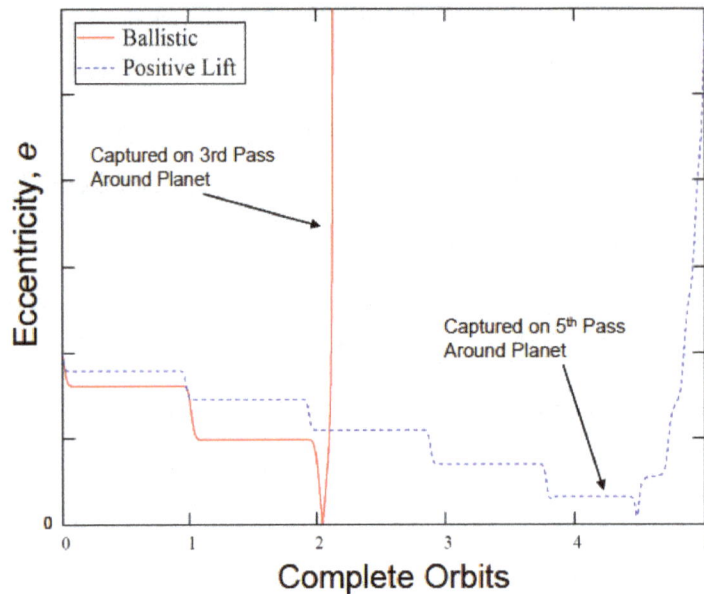

Figure 9-14: Effect of Lift on Multipass Entries

ballistic entry and one with positive lift. For completeness, the $u - \gamma$ plane view of entry with lift is shown in Figure 9-15. The climb and dive "porpoising" motion caused by the lift is evident in the final portion of the entry. As the lift tries to pull the vehicle up, γ becomes positive (or less negative) and the vehicle continues to lose speed (u decreases). Then, as it loses speed, it loses lift and begins to fall (or fall faster). Falling, it builds up more lift, only to repeat the cycle. Finally, gravity wins and the vehicle never generates enough lift to significantly alter the flight-path angle again.

9.6 Applied Skip Entry: NASA's Orion Crew Exploration Vehicle

NASA's Crew Exploration Vehicle (CEV) will use a skip entry to increase its flexibility. When returning from low-Earth orbit (LEO), the CEV will use a

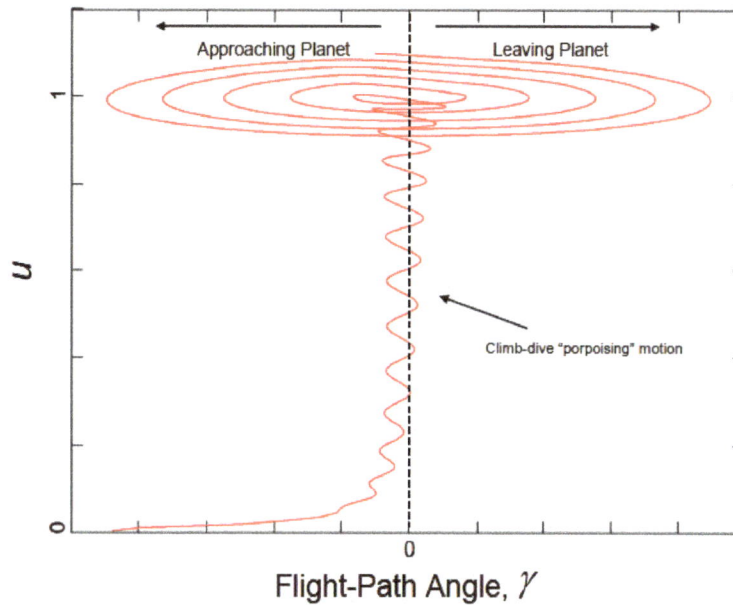

Figure 9-15: Multipass Entry with Lift on the $u - \gamma$ Plane

skip to extend the in-track range (if required). When returning from the moon, a skip followed by a partial Keplerian orbit gives the vehicle the freedom to leave the moon at any time and still land at any point on the Earth. Multiple authors have examined various vehicle configurations and the corresponding skip performance (2; 25; 34; 44; 45; 67). Of particular interest to us, Kaya used a form of Vinh and Brace's universal equations (as given in Eqs. (9.127) - (9.132)) to study the "current" (as of late 2007) configuration of the Orion CEV for lunar return trajectories (34:29, 38-42).

Kaya studied planar trajectories with a fixed lift-to-drag ratio. He developed software to choose the entry flight-path-angle and latitude required to land at any selected point on the Earth without exceeding a predetermined deceleration limit. A typical trajectory is shown in Figure 9-16 with the scale of the atmosphere and skip greatly exaggerated for clarity. As the figure shows, the

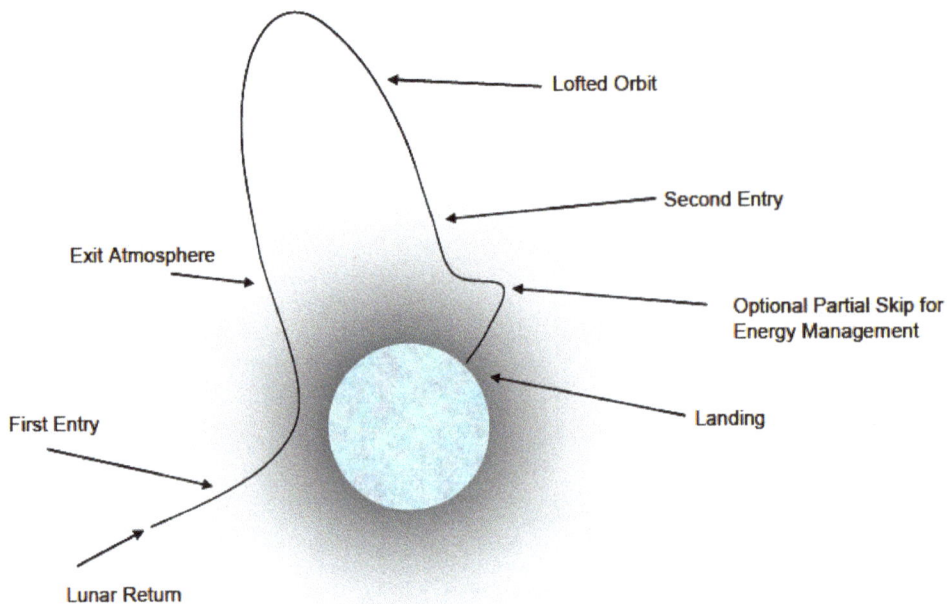

Figure 9-16: Notional Orion CEV Skip Entry

CEV performs an initial skip, completely exiting the atmosphere. At the apogee of the ensuing orbital portion, the vehicle *could* perform a thrusting maneuver and circularize its orbit. Assuming it does not, the CEV returns to the atmosphere again. On this second entry, the lift vector *could* be adjusted to correct position errors and better "target" the landing point.

After solving the problem in non-dimensional universal variables, Kaya converted the results to physical parameters for analysis. Typical results in physical parameters are shown in Figure 9-17. Kaya found the baseline Orion configuration (a capsule reminiscent of Apollo) could leave the moon at any time and land at any point on the Earth if it employed a skip trajectory of this type (34:64). Adding to the flexibility, a reasonable thrusting maneuver at the apogee

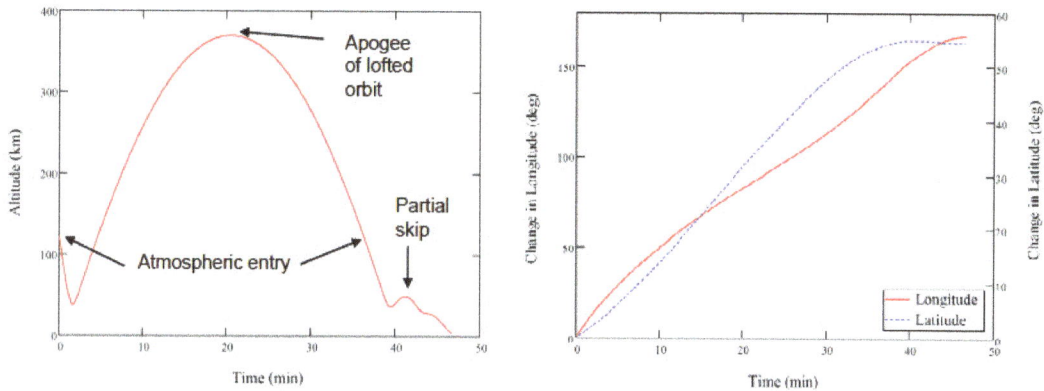

Figure 9-17: Typical Orion CEV Skip Entry

of the skip could enable the capsule to enter an Earth orbit for an extended number of orbits.

9.7 A (Very) Quick Look at Non-planar Entries

By its definition, the analysis of the entry corridor Section 9.4 involved looking at planar entries. Strictly for convenience, the multipass examples in Section 9.5 were computed as planar problems (even though nothing *required* them to be planar). Kaya, to limit the scope of his study, also limited his work to planar entries. The universal equations allow us to look at using lift to modify the impact (or landing) point of a reentry vehicle to some point not in the original entry plane.

In Figure 9-18, three trajectories are shown. Each enters the atmosphere with exactly the same flight-path angle and velocity in a plane aligned with the equator. The ballistic entry continues with planar motion until impact. The two with lift and bank move out of the entry plane and impact the planet with cross-track and in-track range differences. Figure 9-19 shows these changes in ground tracks (and impact points) as well as flight times.

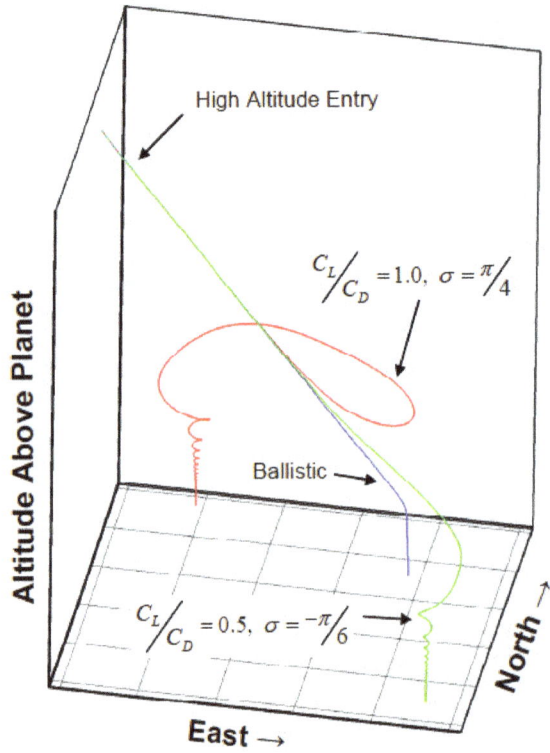

Figure 9-18: Effect of Lift Vector on Three-Dimensional Trajectory

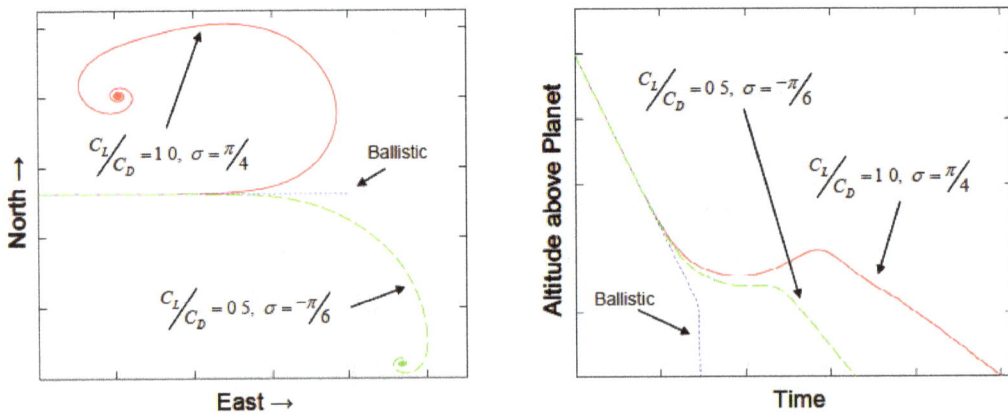

Figure 9-19: Corresponding Ground Tracks and Altitude Profiles

The two figures demonstrate the value of using a lift vector to modify the entry path. While these examples used constant lift vectors for the entire entry, it's easy to see that a varying lift vector could be used to truly customize the trajectory. The topic of "maneuvering entry vehicles" is worthy of at least a chapter by itself and will not be addressed here.

9.8 Summary

In this chapter, we removed the restriction of planar entry in our analyses. To do this, we derived Vinh and Brace's set of motion equations and verified they are valid during all non-thrusting phases of entry. We maintained the "universality" of the equations so we do not need detailed knowledge of the vehicle mass, shape, size, etc. (We did, however, change our non-dimensional variables somewhat.) These universal equations can be used for more detailed analyses (such as extending Kaya's work to include out-of-plane lift). Chapter 10 will further generalize these equations to include variable lift and bank.

9.9 Problems

Material Understanding:

1. The expression for $\dfrac{dZ}{ds}$ in Eq. (9.23) is for a strictly exponential atmosphere.

    Other models of planetary atmospheres still have the exponential characteristic

    $$\frac{d\rho}{\rho} = -\beta dr$$

    but β may not be a constant. (E.g., for an *isothermal* model, β is proportional to the gravity of the planet; i.e., $\beta = kg$.) Show that a more general expression for $\dfrac{dZ}{ds}$ is given by

    $$\frac{dZ}{ds} = \beta r \left(\frac{1}{2\beta^2} \frac{d\beta}{dr} + \frac{1}{2\beta r} - 1 \right) Z \tan \gamma$$

2. Show how to find Eq. (9.110):

    $$\frac{a_{decel}}{g_0} = \frac{Zu\sqrt{\beta r}}{\cos^2 \gamma} \sqrt{1 + \left(\frac{C_L}{C_D} \right)^2}$$

    Make use of the *appropriate* equations already derived in other chapters.

3. Prove that satisfying Eq. (9.112)

$$2\sqrt{\overline{\beta r}}Z_* + \left(\overline{\beta r} - 1\right)\sin\gamma_* + \frac{2\sin\gamma_*\cos^2\gamma_*}{u_*} = 0$$

is *necessary* (but not *sufficient*) to identify the point of maximum deceleration.

4. Show that the entry conditions are related to the hypothetical periapsis conditions by Eqs. (9.118) and (9.119):

$$\frac{u_e^2}{\cos^2\gamma_e} - 2u_e = u_p^2 - 2u_p$$

$$Z_p = Z_e\sqrt{\frac{u_e}{u_p}}\exp\left[\overline{\beta r}\left(1 - \frac{u_e}{u_p}\right)\right]$$

5. Using a value of $\beta = 0.14\,\mathrm{km^{-1}}$, show $0.065 \leq Z_p \leq 0.295$ represents a periapsis altitude range of about 10.8 km.

6. The discussion on multipass entries in Section 9.5 addressed how to numerically integrate the universal equations across portions of the trajectory that pass out of the atmosphere and then return. It did not, however, address how to begin the problem when the initial conditions are outside the atmosphere. Discuss how you would solve for the trajectory from an initial point (at $s_i = 0$) given by $Z_i = 0$, u_i, γ_i, ϕ_i, θ_i, and ψ_i to the atmospheric entry point (at s_e) given by $Z_e \neq 0$, u_e, γ_e, ϕ_e, θ_e, and ψ_e. You can assume you have one additional piece of information, such as the initial radius r_i or eccentricity e_i (if required).

Computational Insights:

7. Negative lift (i.e., $\sigma = 180°$ to give lift that pulls the vehicle down towards the planet) can be used to "hold" a vehicle in the atmosphere at speeds greater than the ballistic overshoot speed. This expands the width of the entry corridor. NASA's Crew Exploration Vehicle (CEV) can generate a lift-to-drag ratio of $\dfrac{C_L}{C_D} = 0.34$. Calculate and plot the overshoot boundary for a CEV with negative lift for the range $1 < \sqrt{\dfrac{u_e}{\cos^2 \gamma_e}} \leq 2.2$. Show that the overshoot boundary for parabolic entry is shifted to the left to $Z_{P_{over}} \approx 0.015$ when CEV uses negative lift. When "paired" with a ballistic undershoot boundary, show that the corresponding corridor width for a maximum deceleration of $10 g_0$ expands to about 21.3 km. (Caution: depending on the accuracy of your TPBVP solver, this answer may vary somewhat; i.e., $Z_{P_{over}} \approx 0.013 - 0.016$.)

Chapter 10

Entry with Variable Lift and Bank

10.1 Introduction

We've been examining situations where the lift-to-drag ratio and bank angle have been held constant. Indeed, the universal equations in Chapter 9 were derived in a form that could be integrated just once for a given atmosphere, set of entry conditions, constant lift-to-drag ratio, and constant bank angle σ. Such solutions have been calculated and published in "Z-Tables" and graphs (7; 18:106-277; 58:241-246). With modern desktop computers, it's usually quicker to integrate the differential equations than it is to refer to the Z-Tables, but the solutions are still in our handy non-dimensional variables, freeing us from the need to integrate for every possible vehicle configuration.

There are several interesting classes of entry where variable lift and bank can be studied without knowing too much about the design of the entry vehicle. (In-other-words, we can keep the results as general as possible and not require any detailed knowledge of the vehicle mass, surface area, etc.) This chapter will introduce the equations necessary for study and look at the control profiles a vehicle might use to follow several classes of entry. It is intended only as an introduction to the maneuvering entry vehicle problem.

10.2 Equations of Motion for Variable Lift and Bank

The non-dimensional universal equations derived in Chapter 9 allow us to examine situations where the lift varies as long as the drag coefficient C_D is constant. For reference, these equations were

$$\frac{dZ}{ds} = -\overline{\beta r} Z \tan \gamma \qquad\qquad \textbf{10.1}$$

$$\frac{du}{ds} = -\frac{2Zu\sqrt{\beta r}}{\cos \gamma} \left(1 + \frac{C_L}{C_D} \cos \sigma \tan \gamma + \frac{\sin \gamma}{2Z\sqrt{\beta r}} \right) \qquad\qquad \textbf{10.2}$$

$$\frac{d\theta}{ds} = \frac{\cos \psi}{\cos \phi} \qquad\qquad \textbf{10.3}$$

$$\frac{d\phi}{ds} = \sin \psi \qquad\qquad \textbf{10.4}$$

$$\frac{d\gamma}{ds} = \frac{Z\sqrt{\beta r}}{\cos \gamma} \left[\frac{C_L}{C_D} \cos \sigma + \frac{\cos \gamma}{Z\sqrt{\beta r}} \left(1 - \frac{\cos^2 \gamma}{u} \right) \right] \qquad\qquad \textbf{10.5}$$

$$\frac{d\psi}{ds} = \frac{Z\sqrt{\beta r}}{\cos^2 \gamma} \left(\frac{C_L}{C_D} \sin \sigma - \frac{\cos^2 \gamma}{Z\sqrt{\beta r}} \cos \psi \tan \phi \right) \qquad\qquad \textbf{10.6}$$

where $s = \int\limits_{0}^{t} \dfrac{{}^{R}V}{r}\cos\gamma\, dt$ was used as the independent variable instead of time

(with flight-path angle constrained such that $\cos\gamma > 0$). The modified Chapman variables

$$u = \frac{{}^{R}V^{2}\cos^{2}\gamma}{gr} \qquad\qquad \textbf{10.7}$$

$$Z = \frac{\rho C_{D}S}{2m}\sqrt{\frac{r}{\beta}} \qquad\qquad \textbf{10.8}$$

related the non-dimensional kinetic energy u and altitude Z to their physical counterparts.

The assumption of a constant drag coefficient may be a good approximation in some cases; however, it's generally more accurate to model the increase in drag that accompanies a reentry vehicle's increase in lift (47:240-246; 53:236-237; 58:308-311). The dimensional equations we derived in Chapter 3 can certainly be used, but we will stick with non-dimensional equations and "universal variables" similar to u and altitude Z (or T and η before them). Vinh, et al. derived several variations to Eqs. (10.1) - (10.6) to allow C_{D} (as well as C_{L}) to vary during entry (62:1617-1618; 58:308-313). We will derive one set of their equations for use here.

When a entry vehicle develops lift, there is a corresponding increase in the drag. A simple, yet fairly general, relationship between the drag C_{D} and lift C_{L} is known as the *drag polar* and can be written as

$$C_{D} = C_{D_{0}} + KC_{L}{}^{n} \qquad\qquad \textbf{10.9}$$

where C_{D_0} is the zero-lift drag coefficient; K is the induced drag parameter; and n is the drag polar parameter. In our application, C_{D_0}, K, and n are constants and we've assumed the minimum drag occurs at the zero lift condition. (A similar expression can be given when the minimum drag is at nonzero lift (68:269). The *induced drag* caused by lift is represented by the term $KC_L^{\ n}$.

With this relationship, the lift-to-drag ratio can be written:

$$\frac{C_L}{C_D} = \frac{C_L}{C_{D_0} + KC_L^{\ n}}$$
10.10

To find the maximum ratio, C_L can be treated as the independent variable and

$$\frac{d}{dC_L}\left(\frac{C_L}{C_D}\right) = \frac{C_{D_0} - KC_L^{\ n}(n-1)}{\left(C_{D_0} + KC_L^{\ n}\right)^2} = 0$$
10.11

solved to find the constant K:

$$K = \frac{C_{D_0}}{(n-1)C_{L*}^{\ n}}$$
10.12

C_{L*} is the lift coefficient corresponding to the maximum lift-to-drag ratio and *not* necessarily the maximum value of C_L. (Similarly, the drag coefficient corresponding to the maximum lift-to-drag ratio is C_{D*}.) Substituting into Eq. (10.10) gives the corresponding maximum

$$\left(\frac{C_L}{C_D}\right)_* = \frac{C_{L*}}{C_{D*}} = \frac{(n-1)C_{L*}}{nC_{D_0}}$$
10.13

which readily identifies

$$C_{D*} = \left(\frac{n}{n-1}\right) C_{D_0} \qquad \text{10.14}$$

as the drag coefficient corresponding to the maximum lift-to-drag ratio.

Rescaled lift and drag coefficients can be defined as:

$$C_L = C_{L*}\lambda \qquad \text{10.15}$$

$$C_D = C_{D*}f(\lambda) \qquad \text{10.16}$$

Notice they take on their "critical" values (C_{L*} and C_{D*}) when $\lambda = \lambda_* = 1$.

Because of this, these two definitions are introduced as "normalized" lift and drag coefficients in some formulations:

$$\lambda = \bar{C}_L = \frac{C_L}{C_{L*}} \qquad \text{10.17}$$

$$f(\lambda) = \bar{C}_D = \frac{C_D}{C_{D*}} \qquad \text{10.18}$$

Note that there will be a maximum value λ_{max} based on the maximum lift a vehicle can generate. At this point, we do not need to worry about that value but we should know it exists. (A similar limit exists for maximum *negative* lift.)

Regardless of how λ and $f(\lambda)$ are introduced, comparing Eqs. (10.9), (10.12), (10.14), (10.15) gives $f(\lambda)$:

$$f(\lambda) = \frac{(n-1) + \lambda^n}{n} \qquad \text{10.19}$$

Using $L = \dfrac{C_L}{C_D}$ as shorthand, a corresponding "normalized" lift-to-drag ratio can also be introduced:

$$\overline{L} = \frac{\left(\dfrac{C_L}{C_D}\right)}{\left(\dfrac{C_L}{C_D}\right)_*} = \frac{L}{L_*} \qquad\qquad \textbf{10.20}$$

With these definitions, we will now turn to rewriting Eqs. (10.1) - (10.6) to allow C_D and C_L to vary during entry.

The altitude variable Z is a poor choice since it is dependent on a (now) non-constant C_D. Instead, we can replace it with a similarly defined variable W :

$$W = \frac{\rho C_{L_*} S}{2m}\sqrt{\frac{r}{\beta}} = \frac{L_*}{f(\lambda)}Z \qquad\qquad \textbf{10.21}$$

With this substitution, the four equations of motion that change become:

$$\frac{dW}{ds} = -\overline{\beta r}W\tan\gamma \qquad\qquad \textbf{10.22}$$

$$\frac{du}{ds} = -\frac{2Wuf(\lambda)\sqrt{\beta r}}{L_*\cos\gamma}\left(1 + \frac{L_*\lambda\cos\sigma\tan\gamma}{f(\lambda)} + \frac{L_*\sin\gamma}{2Wf(\lambda)\sqrt{\beta r}}\right) \qquad\qquad \textbf{10.23}$$

$$\frac{d\gamma}{ds} = \frac{W\lambda\sqrt{\beta r}}{\cos\gamma}\cos\sigma + \left(1 - \frac{\cos^2\gamma}{u}\right) \qquad\qquad \textbf{10.24}$$

$$\frac{d\psi}{ds} = \frac{W\lambda\sqrt{\beta r}}{\cos^2\gamma}\sin\sigma - \cos\psi\tan\phi \qquad\qquad \textbf{10.25}$$

These four equations, together with Eqs. (10.3) and (10.4), can be numerically integrated for variable lift problems with relative ease.

Vinh, <u>et al.</u> showed the equations can be simplified somewhat by introducing another variable change:

$$v = \frac{u}{\cos^2 \gamma} = \frac{{}^R V^2}{gr} \qquad \text{10.26}$$

v is related to the kinetic energy and we will refer to it as a kinetic energy parameter. With this change, the full set of six differential equations of motion becomes:

$$\frac{dW}{ds} = -\overline{\beta r} W \tan \gamma \qquad \text{10.27}$$

$$\frac{dv}{ds} = -\frac{2Wvf\left(\lambda\right)\sqrt{\beta r}}{L_* \cos \gamma} - (2-v)\tan \gamma \qquad \text{10.28}$$

$$\frac{d\gamma}{ds} = \frac{W\lambda\sqrt{\beta r}}{\cos \gamma}\cos \sigma + \left(1 - \frac{1}{v}\right) \qquad \text{10.29}$$

$$\frac{d\theta}{ds} = \frac{\cos \psi}{\cos \phi} \qquad \text{10.30}$$

$$\frac{d\phi}{ds} = \sin \psi \qquad \text{10.31}$$

$$\frac{d\psi}{ds} = \frac{W\lambda\sqrt{\beta r}}{\cos^2 \gamma}\sin \sigma - \cos \psi \tan \phi \qquad \text{10.32}$$

These six equations are "universal" in two respects. First, they are valid during entry as well as in orbit (since they are just a form those in Chapter 9). Second, they are "almost" free of vehicle characteristics. The lift and drag profiles are required to integrate, but not detailed knowledge of the mass, surface area, etc. These are the equations we will use to examine entry with variable lift.

10.3   Parabolic Drag Polar

The drag polar in Eq. (10.9) has been left in its general form to this point. In particular, we have not assigned a value to the exponent n. Regan and Anandakrishnan show it's in the approximate range $1.9958 \le n \le 2.5$ for conical shapes at hypersonic speeds (47:242). Vinh, et al. show $n \approx 1.5$ for thin-winged hypersonic vehicles (58:309). For simplicity (and because it represents many useful situations), we will assume the drag polar is parabolic with $n = 2$ for our purposes.

For a parabolic drag polar, the induced drag parameter becomes simply

$$K = \frac{C_{D_0}}{C_{L*}^{2}}$$

10.33

and the drag polar (Eq. (10.9)) reduces to:

$$C_D = C_{D_0}\left[1 + \lambda^2\right]$$

10.34

Equation (10.16) gives another expression for C_D, where $f(\lambda)$ is now given by:

$$f(\lambda) = \frac{1 + \lambda^2}{2}$$

10.35

296

Comparing these two expressions for C_D reveals:

$$C_{D*} = 2C_{D_0} \qquad \qquad \textbf{10.36}$$

This tells us drag is evenly split between the zero-lift drag and induced drag when the vehicle is at the maximum lift-to-drag ratio L_*. As a check, the induced drag at this point ($\lambda = \lambda_* = 1$) is given by

$$\left(KC_L^{\,2}\right)_* = \left(\frac{C_{D_0}}{C_{L*}^{\,2}}\right)\left(C_{L*}\lambda_*\right)^2 = C_{D_0} \qquad \qquad \textbf{10.37}$$

which is exactly half of the total drag.

Equation (10.10) can be evaluated using these expressions for a parabolic drag polar. After simplifying, the lift-to-drag ratio is:

$$\frac{C_L}{C_D} = \frac{2C_{L*}\lambda}{C_{D*}\left(1+\lambda^2\right)} = 2L_*\left(\frac{\lambda}{1+\lambda^2}\right) \qquad \qquad \textbf{10.38}$$

Figure 10-1 summarizes the key aspects of this type of drag polar. For a symmetrical vehicle capable of negative lift, Figure 10-2 shows the obvious extension to the polar.

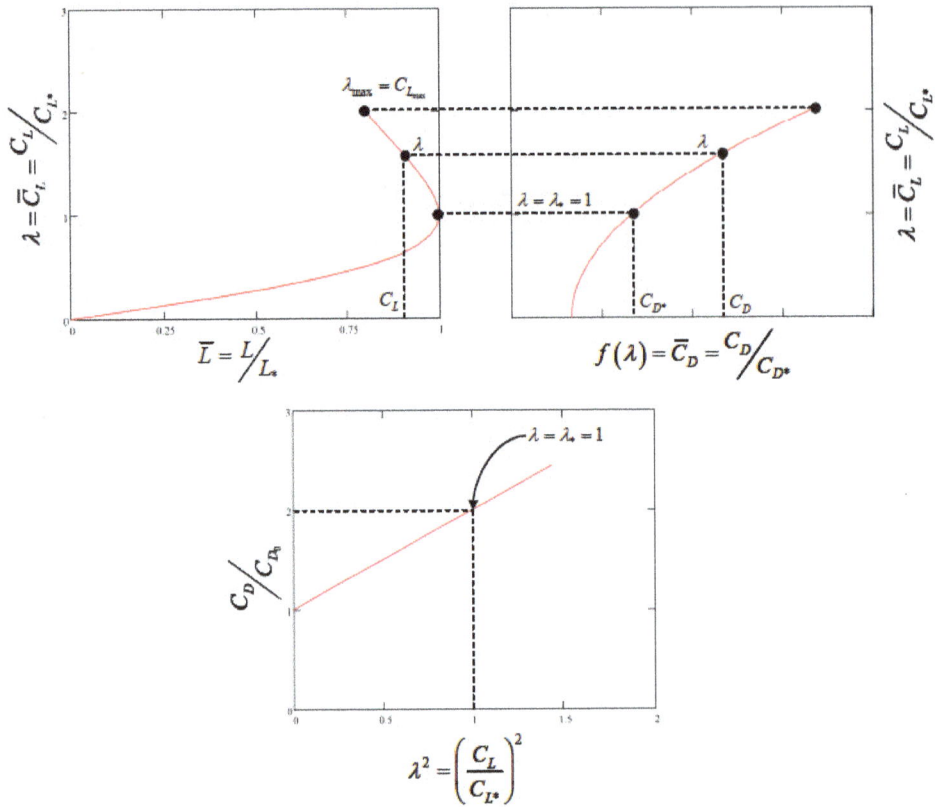

Figure 10-1: Parabolic Drag Polar Relationships

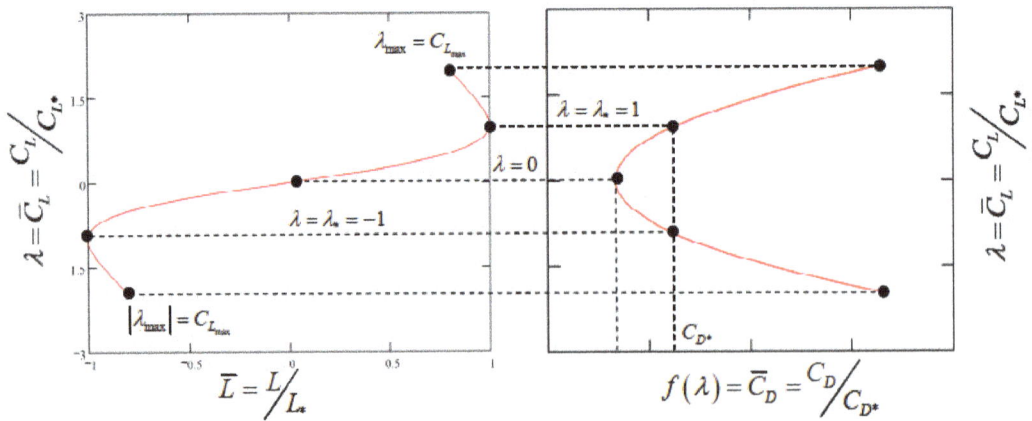

Figure 10-2: Drag Polar Expanded to Include Negative Lift

10.4 Planar Entry

For planar motion about a non-rotating planet, we can define the entry plane to be equatorial without loss of generality. For equatorial entry with $\sigma = 0°$, Eqs. (10.30) - (10.32) can be solved to give $\phi = \psi = 0$ and $d\theta = ds$ throughout the entry. (We can arbitrarily assume θ and s are zero at the same time and let $\theta = s$ if we so desire.) Equations (10.27) - (10.29) are decoupled and describe the motion:

$$\frac{dW}{d\theta} = -\overline{\beta r}W \tan \gamma \qquad \text{10.39}$$

$$\frac{dv}{d\theta} = -\frac{2Wvf(\lambda)\sqrt{\beta r}}{L_* \cos \gamma} - (2-v)\tan \gamma \qquad \text{10.40}$$

$$\frac{d\gamma}{d\theta} = \frac{W\lambda\sqrt{\beta r}}{\cos \gamma} + \left(1 - \frac{1}{v}\right) \qquad \text{10.41}$$

The values spanned by W, v, and γ are the "state space" of solutions to Eqs. (10.39) - (10.41). Following the lead of Vinh, et al., we can visualize the state space in a cylindrical coordinate system, with $h = \frac{1}{W}$ for the "height," v for the "radius," and γ for the angular measurement (58:313-314). This is shown in Figure 10-3.

If the solution to Eqs. (10.39) - (10.41) can be related by a yet-to-be-determined relationship $\mathcal{F}$ between W, v, and γ such that

$$\mathcal{F}(W, v, \gamma) = 0 \qquad \text{10.42}$$

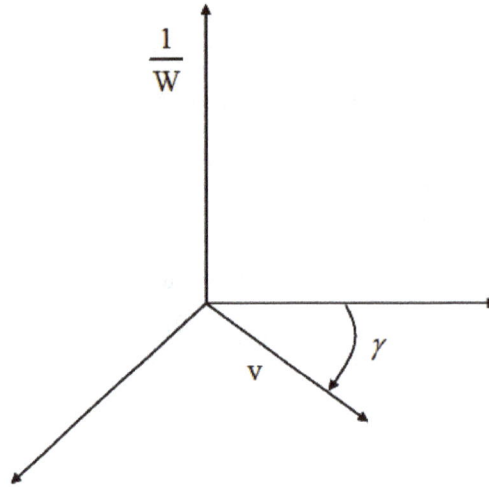

Figure 10-3: Cylindrical Representation of the (W,v,γ) State Space

is true at any θ, then the aggregate of all of the solutions to Eq. (10.42) forms a two-dimensional surface in state space. (We cannot, however, assume $\mathcal{F}$ is an easy or nice, closed-form relationship!) If a second (different) constraint must also be satisfied, the solution (if it exists) is on the intersection of two surfaces. A few of these relationships (and their two-dimensional surfaces) are relatively easy to examine.

10.4.1 Entry at a Constant Flight-Path Angle

If the lift is varied to maintain a constant flight-path angle γ_e, then a relationship between the state variables in the form of Eq. (10.42) is simply:

$$\mathcal{F}(W,v,\gamma) = \gamma - \gamma_e = 0 \qquad \textbf{10.43}$$

The entire entry trajectory is contained in the $\left(\frac{1}{W}, v\right)$ plane as shown in Figure 10-4. The "shape" shown within the plane is notional, but the plane's orientation is exact since it is set at the fixed angle γ_e. As a result, the projection of the entire entry onto the (v, γ) plane is simply a line along v. The actual shape in the $\left(\frac{1}{W}, v\right)$ plane, as well as how it maps from cylindrical coordinates to "physical" space, can be established by evaluating Eqs. (10.39) - (10.41) in more detail.

The left-hand-side of Eq. (10.41) becomes zero and can be solved for the lift profile required to maintain γ_e during entry:

$$\lambda = \frac{\cos \gamma_e}{\sqrt{\beta r}} \left(\frac{1-v}{Wv} \right) \qquad \text{10.44}$$

At any particular point in the entry, the values of W and v "set" the required value of lift λ required to maintain γ_e. This "lift profile," given by Eq. (10.44), would continue throughout the entry until a physical limit such as $\lambda > \lambda_{max}$ or

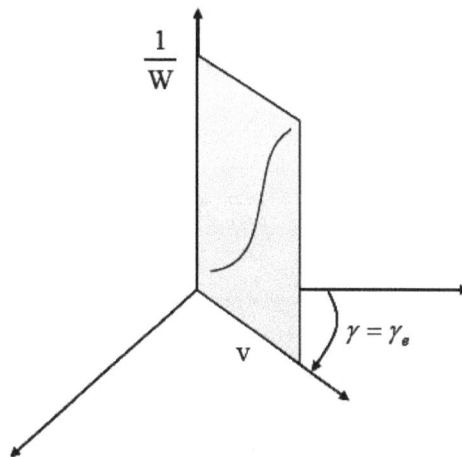

Figure 10-4: Entry at Constant Flight-Path Angle

$\lambda < \lambda_{\min}$ is reached. At that point, the vehicle can no longer maintain the same flight-path angle and the dynamics would change.

Equation (10.39) is separable and can be easily integrated to give the altitude W at any point during the entry

$$W = W(\theta) = W_e \exp\left[-\overline{\beta r} \tan \gamma_e (\theta - \theta_e)\right] \qquad \textbf{10.45}$$

where W_e is the initial altitude. This is the equation for a logarithmic spiral. Simply put, in the physical entry plane of the vehicle, the non-dimensional altitude falls exponentially with "range" θ.

Equation (10.40) can be divided by (10.39) to relate W and v. Assuming a parabolic drag polar and incorporating Eq. (10.44), the relationship becomes:

$$\frac{dv}{dW} = \frac{v}{L_* \sqrt{\beta r} \sin \gamma_e}\left[1 + \frac{\cos^2 \gamma_e}{\beta r}\left(\frac{1-v}{Wv}\right)\right] + \frac{2-v}{W\beta r} \qquad \textbf{10.46}$$

Rather than running screaming to a computer to solve this rather hideous differential equation, we can attempt to simplify it. For most atmospheres of interest, $\dfrac{1}{W\beta r} \ll 1$ once the vehicle is sufficiently deep into the atmosphere.

Keeping only the largest term on the right-hand-side leaves us with a rather simple equation:

$$\frac{dv}{dW} = \frac{v}{L_* \sqrt{\beta r} \sin \gamma_e} \qquad \textbf{10.47}$$

This is easily integrated to give

$$\frac{v}{v_e} = \exp\left(\frac{W - W_e}{L_* \sqrt{\beta r} \sin \gamma_e}\right) \qquad \textbf{10.48}$$

where the e subscript again denotes entry conditions. This relationship gives the shape of the trajectory in the $\left(\frac{1}{W}, v\right)$ plane shown in Figure 10-4. Substituting for W using Eq. (10.45), gives us an expression for v as a function of θ:

$$v = v(\theta) = v_e \exp\left(\frac{W_e\left\{\exp\left[-\overline{\beta r}\tan\gamma_e\left(\theta - \theta_e\right)\right]-1\right\}}{L_*\sqrt{\overline{\beta r}}\sin\gamma_e}\right) \qquad \textbf{10.49}$$

Here, we see the non-dimensional kinetic energy falls off as an exponential to an exponential as the range θ increases.

Alternatively, we could approach finding $v(\theta)$ directly from Eq. (10.40). Using Eqs. (10.44) and (10.45) and assuming a parabolic drag polar, we get:

$$\frac{dv}{d\theta} + f(\theta)v + g(\theta)\frac{(1-v)^2}{v} + (2-v)\tan\gamma_e = 0 \qquad \textbf{10.50}$$

In this equation,

$$f(\theta) = \left(\frac{W_e\sqrt{\overline{\beta r}}}{L_*\cos\gamma_e}\right)\exp\left[-\overline{\beta r}\tan\gamma_e\left(\theta - \theta_e\right)\right] \qquad \textbf{10.51}$$

and

$$g(\theta) = \frac{\cos\gamma_e}{W_e L_*\sqrt{\overline{\beta r}}}\exp\left[\overline{\beta r}\tan\gamma_e\left(\theta - \theta_e\right)\right] \qquad \textbf{10.52}$$

are used as shorthand for functions of θ only. Equation (10.50) could be integrated (maybe even analytically with a lot of pain and suffering) to find v as a function of θ (i.e., $v(\theta)$). For our purposes, we will simply assume the solution $v(\theta)$ has been found. (It can be shown numerically that Eq. (10.49) is a very good approximation for realistic entries so it can be used as the solution.)

As a final note, Eq. (10.43) is not unique in describing the relationship between W, v, and γ. Equation (10.48) can also be arranged to give:

$$\frac{v}{v_e} - \exp\left(\frac{W - W_e}{L_* \sqrt{\beta r} \sin \gamma_e}\right) = 0 = \mathcal{F}(W, v, \gamma) \qquad \textbf{10.53}$$

The single constraint ($\gamma = $ constant) resulted in the $\mathcal{F}(W, v, \gamma)$ defining motion within the $\left(\frac{1}{W}, v\right)$ plane. While not quite as obvious, the totality of solutions satisfying Eq. (10.53) also lie within the $\left(\frac{1}{W}, v\right)$ plane.

10.4.2 Entry at a Constant Sink Rate

The sink rate of an entry vehicle is the vertical component of its velocity:

$$^{R}V_s = {}^{R}V \sin \gamma \qquad \textbf{10.54}$$

Putting this into our current dimensionless variables, we can write

$$v \sin^2 \gamma = \frac{{}^{R}V_s^2}{gr} \qquad \textbf{10.55}$$

For all of the planetary atmospheres of interest, $gr \approx$ constant during the entry (because of the "thin" atmosphere). Thus, for a constant sink rate, Eq. (10.55) becomes

$$v \sin^2 \gamma = k_1 \qquad \textbf{10.56}$$

where k_1 is simply a new constant. We can then write

$$\mathcal{F}(W, v, \gamma) = v \sin^2 \gamma - k_1 = 0 \qquad \textbf{10.57}$$

as a relationship between the states. This constraint will limit the motion in a two-dimensional "sheet." (As we'll see, it is not a plane in cylindrical coordinates.)

The projection of the entry path onto the (v, γ) plane is shown in Figure 10-5 as the curve labeled $v \sin^2 \gamma = k_1$. As the kinetic energy drops (and γ becomes more negative), the projection approaches the circle with a radius k_1. For a large sink rate, the radius is large. For a small sink rate, the radius is small.

Like in the last section, we can solve for the lift profile for this type of entry. Differentiating Eq. (10.56) gives us:

$$\frac{dv}{d\theta} \sin^2 \gamma + \frac{d\gamma}{d\theta} 2v \sin \gamma \cos \gamma = 0 \qquad \textbf{10.58}$$

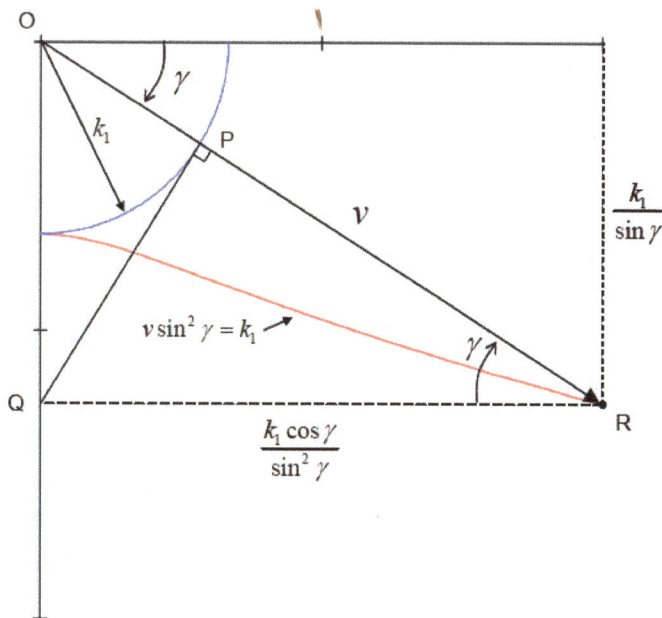

Figure 10-5: Entry at Constant Sink Rate as Seen in (v, γ) Plane

Equations (10.40) and (10.41) can be used to replace the derivatives. After assuming a parabolic drag polar and simplifying, we are left with a quadratic equation to solve for λ anywhere along the entry:

$$\lambda^2\left(\sin\gamma\right)+\lambda\left(-2L_*\cos\gamma\right)+\sin\gamma+\frac{2L_*}{Wv\sqrt{\beta r}}\left[1+v\left(\frac{\sin^2\gamma}{2}-1\right)\right]=0 \qquad \textbf{10.59}$$

This has both positive and negative roots (giving positive and negative lift, respectively) when:

$$-4\sin\gamma\left\{\sin\gamma+\frac{2L_*}{Wv\sqrt{\beta r}}\left[1+v\left(\frac{\sin^2\gamma}{2}-1\right)\right]\right\}>0 \qquad \textbf{10.60}$$

(The choice of the root may depend on the capabilities of the vehicle and will impacts other aspects of the entry, such as the deceleration experienced.) Assuming $\gamma<0$ throughout entry, this can be rearranged to give:

$$W<\frac{L_*\left[-2\sin^2\gamma+k_1\left(2-\sin^2\gamma\right)\right]}{k_1\sqrt{\beta r}\sin\gamma} \qquad \textbf{10.61}$$

When this condition on altitude is reversed, only negative lift is possible. At some point in the entry, a "control limit" may be reached. A control limit occurs when λ falls outside its allowed bounds ($\lambda_{min}\leq\lambda\leq\lambda_{max}$) or no longer has any real values. The roots become imaginary when:

$$4L_*^2\cos^2\gamma-4\sin\gamma\left\{\sin\gamma+\frac{2L_*}{Wv\sqrt{\beta r}}\left[1+v\left(\frac{\sin^2\gamma}{2}-1\right)\right]\right\}<0 \qquad \textbf{10.62}$$

Vinh, <u>et al.</u> arrived at an expression similar to Eq. (5.59) after assuming a flat-planet entry. We can match their result by making the same flat-planet assumption. Start by letting the dimensionless horizontal distance, altitude, and kinetic energy be defined by:

$$x = \overline{\beta} r \theta \qquad \textbf{10.63}$$

$$z = \frac{\sqrt{\overline{\beta} r}}{W} \qquad \textbf{10.64}$$

$$u = \overline{\beta} r v = \frac{\beta^R V^2}{g} \qquad \textbf{10.65}$$

Substituting these definitions into Eq. (10.59) and simplifying gives:

$$\lambda^2 \left(\sin \gamma\right) + \lambda \left(-2 L_* \cos \gamma\right) + \sin \gamma + \frac{2 L_* z}{u}\left[1 + \frac{u}{\overline{\beta} r}\left(\frac{\sin^2 \gamma}{2} - 1\right)\right] = 0 \qquad \textbf{10.66}$$

In this scale $u \ll \overline{\beta} r$, so we can simplify the result to just:

$$\lambda^2 \left(\sin \gamma\right) + \lambda \left(-2 L_* \cos \gamma\right) + \sin \gamma + \frac{2 L_* z}{u} = 0 \qquad \textbf{10.67}$$

Or, letting k_2 be a positive constant such that $k_2^2 = \overline{\beta} r k_1$, Eq. (10.67) becomes:

$$\lambda^2 \left(\sin \gamma\right) + \lambda \left(-2 L_* \cos \gamma\right) + \sin \gamma + \frac{2 L_* z \sin^2 \gamma}{k_2^2} = 0 \qquad \textbf{10.68}$$

Physically, k_2 is the absolute value of the non-dimensional sink rate. Equation (10.68) is the expression derived by Vinh, et al. for a constant sink rate entry over a flat planet (58:337; 62:1620-1621). With Eq. (10.68), two solutions are possible when:

$$z > \frac{-k_2^2}{2 L_* \sin \gamma} \qquad \textbf{10.69}$$

(One solution is for positive lift and the other is for negative lift.) When this condition on altitude is reversed, negative lift is the only possible solution. For the flat-planet approximation, λ becomes imaginary when:

$$z < \left(\frac{k_2^2 \left(L_*^2 \cos^2 \gamma - \sin^2 \gamma \right)}{2 L_* \sin^3 \gamma} \right)$$ 10.70

(Again, we have assumed $\gamma < 0$ in the derivation of Eqs. (10.69) and (10.70).)

Unlike entries with a constant flight-path angle, the equations of motion for these entries are not readily solved analytically. Finding W, v, and γ as functions of θ (or each other) is probably best left to numerical methods and won't be presented here. For the interested reader, Vinh, et al. provide a detailed look at the relationship between γ and λ (58:337-343; 62:1620-1622). Even without finding W, v, and γ, we can say with certainty that the motion is confined to a "sheet" whose projection on the (v, γ) plane is the curve shown in Figure 10-5 (as long as the sink rate remains constant).

10.4.3 Entry at a Constant Velocity

If the lift is varied to maintain a constant velocity ^RV_e, then

$$v = v_e = \frac{^RV_e^2}{gr} \approx \text{constant}$$ 10.71

for a "thin" atmosphere. Therefore, a relationship between the state variables can be written as:

$$\mathcal{F}(W, v, \gamma) = v - v_e = 0$$ 10.72

In cylindrical state space, the entire entry trajectory is contained on a cylinder of radius v_e centered on (and oriented along) the $\frac{1}{W}$ axis as shown in Figure 10-6. (Only the valid half of the cylinder, for $-90° < \gamma < 90°$, is shown.) The motion on the surface of the cylinder cannot be easily found analytically, so it won't be presented here. However, finding the lift profile is straightforward.

Substituting $v = v_e$ into the differential equation for v (Eq. (10.40)) and assuming a parabolic drag polar gives us the required lift profile to maintain a constant velocity:

$$\lambda = \pm\sqrt{(v_e - 2)\left(\frac{L_* \sin\gamma}{Wv_e\sqrt{\beta r}}\right) - 1} \qquad \textbf{10.73}$$

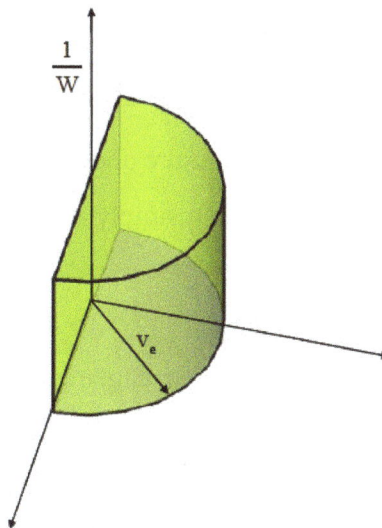

Figure 10-6: Entry at Constant Velocity

The choice of the sign, once again, affects other aspects of the entry such as the deceleration. Once

$$(v_e - 2)\left(\frac{L_* \sin \gamma}{Wv_e \sqrt{\beta r}}\right) - 1 < 0 \qquad \textbf{10.74}$$

the roots are imaginary and the velocity can no longer be held at the same value. Similarly, when λ falls outside the allowed boundaries $\lambda_{\min} \le \lambda \le \lambda_{\max}$, the velocity must change.

10.4.4 Entry at a Constant Dynamic Pressure

The aerodynamic forces acting on a vehicle as it enters the atmosphere is proportional to the dynamic pressure $\frac{1}{2}\rho^R V^2$ (Ref. 22). In non-dimensional terms, we can write the dynamic pressure as:

$$\frac{\frac{1}{2}\rho^R V^2}{\left(\dfrac{mg}{SC_{L*}}\right)} = \sqrt{\beta r}\,Wv \qquad \textbf{10.75}$$

For an entry with constant dynamic pressure and a thin atmosphere (so that $g \approx$ constant), Eq. (10.75) can be written simply as

$$Wv = k_3 \qquad \textbf{10.76}$$

where k_3 is a constant. With a simple rearrangement, we can write this as

$$\mathcal{F}(W, v, \gamma) = \frac{v}{\left(\dfrac{1}{W}\right)} - k_3 = 0 \qquad \textbf{10.77}$$

In our cylindrical state-space, this is the equation for a circular cone just touching the origin and centered on the $1/W$ axis. This is shown in Figure 10-7. (Only half of the cone is shown since $-90° < \gamma < 90°$.)

The motion on the cone surface (e.g., $W(\gamma)$ and $v(\gamma)$) is too complicated to bother finding analytically, but the lift profile can be found by differentiating Eq. (10.76)

$$\frac{dv}{d\theta} = \frac{-k_3}{W^2}\frac{dW}{d\theta}$$

10.78

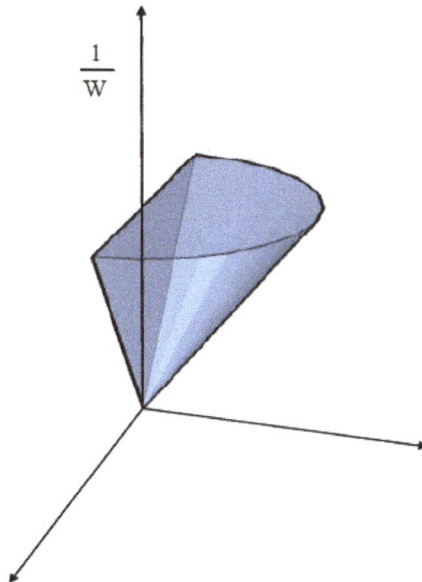

Figure 10-7: Entry at Constant Dynamic Pressure

and substituting Eqs. (10.39) and (10.40) for the derivatives. If a parabolic drag polar is assumed, the profile is:

$$\lambda = \pm \sqrt{\frac{-L_* \sin\gamma}{W\sqrt{\beta r}}\left[\overline{\beta r} + \frac{2W}{k_3} - 1\right] - 1}$$

10.79

As long as the value for λ remains real and $\lambda_{min} \le \lambda \le \lambda_{max}$, a vehicle entering with this lift profile will maintain a constant dynamic pressure.

10.4.5 Entry at a Constant Heat Flux

In Chapter 7, we saw that the average heat flux and stagnation heat flux were proportional to $\rho^R V^3$ and $\rho^{1/2} {}^R V^3$, respectively. When looking at an entry with a constant heat flux, we can simply write

$$\rho^a {}^R V^3 = k_4$$

10.80

to cover both cases. For average (wall) heat flux $a = 1$ and for stagnation heat flux $a = \frac{1}{2}$. Rewriting in terms of the current state variables, this relationship becomes:

$$W^{2a} v^3 r^{-(2a+3)} = k_5$$

10.81

For a thin atmosphere $r \approx$ constant, our constraint equation can be written:

$$\mathcal{F}(W, v, \gamma) = \frac{v^3}{\left(\frac{1}{W}\right)^{2a}} - k_6 = 0$$

10.82

For $a = 1$ (constant average heat flux), Eq. (10.82) is the equation for the surface of a "semicubic paraboloid," created by rotating a semicubic equation about the $1/W$ axis. For $a = \frac{1}{2}$ (constant stagnation heat flux), the surface is that of a "cubic paraboloid," created by rotating a cubic equation about the $1/W$ axis. The intersection of each of these paraboloids with the $\left(\frac{1}{W}, v \right)$ plane (for the same value of k_6) is shown in Figure 10-8. A three-dimensional rendering of both constraint surfaces is shown in Figure 10-9.

Using a process similar to how we found the lift profile for entry with a constant dynamic pressure, Eq. (10.82) can be differentiated to give:

$$3v^2 \frac{dv}{d\theta} = \frac{-2ak_6}{W^{2a+1}} \frac{dW}{d\theta} \qquad \textbf{10.83}$$

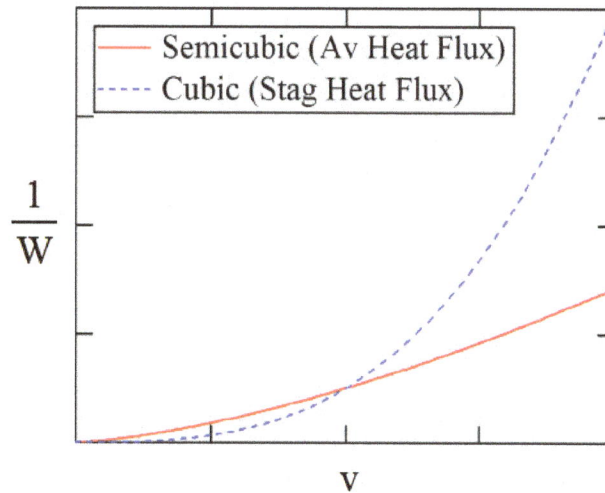

Figure 10-8: Semicubic and Cubic Equations for Constant Heat Flux as Seen in the $\left(\frac{1}{W}, v \right)$ ***Plane***

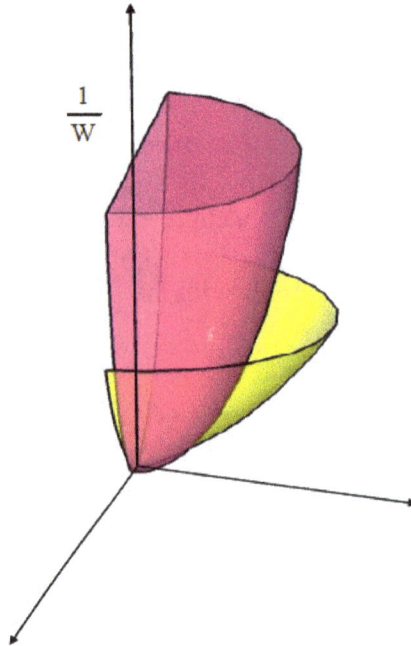

Figure 10-9: Entry at Constant Heat Flux

Equations (10.39) and (10.40) can be used to replace the derivatives and a parabolic drag polar assumed. Once done, a solution for $\lambda(W,\gamma)$ or $\lambda(v,\gamma)$ can be obtained with minimal effort. The solution is left as an exercise for the reader as is the solution to the motion on the paraboloids.

10.4.6 Other Planar Entries

The five planar entries already discussed are just a few of the interesting planar trajectories. They were singled out because various aspects (like λ or the "plane of motion") can be found analytically with relative ease. Each had relatively simple (and constant) constraints (e.g., constant flight-path angle or constant sink rate). It is possible, however, for constraints to change during an entry. We have already seen two situations that might force a change: reaching a limit on the allowable range of λ or failing to find a real solution for λ. Another situation is a *planned* change in one or more constraint.

Certain entry vehicles, including the US and the (then) Soviet space shuttles, have multiple entry "legs." The vehicles switch between different constraints on the motion in discrete "segments" (perhaps triggered by altitude or speed). For example, early in the entry the control system may maintain the maximum drag (subject to heating and deceleration limits) in order to slow the vehicle while high in the atmosphere. Later, the control might shift to maintaining a fixed heat flux. Finally, when the vehicle is low enough, a constant sink rate might be commanded. To study these types of vehicles, each leg can be investigated independently and sequentially, beginning at entry.

If a vehicle's entry is designed as an "optimal" trajectory (e.g., maximum range or minimum total heating), the control can become even more complex. A full treatment and solution requires the application of modern optimal control theory. Optimal control is beyond the scope of this text and won't be covered. References 9 and 10, as well as many other books, can provide a good introduction to the theory. Some examples of optimal entry trajectories, while not necessarily planar, are discussed in References 11, 13, 33, 50, 54, 55, 60, and 61.

10.5 Non-Planar Entry

For entry with lift and a non-zero bank angle, all six differential equations are required. After assuming a parabolic drag profile, these become:

$$\frac{dW}{ds} = -\overline{\beta r}W \tan \gamma \qquad\qquad \textbf{10.84}$$

$$\frac{dv}{ds} = -\frac{Wv\sqrt{\beta r}\left(1+\lambda^2\right)}{L_* \cos\gamma} - (2-v)\tan\gamma \qquad \textbf{10.85}$$

$$\frac{d\gamma}{ds} = \frac{W\lambda\sqrt{\beta r}}{\cos\gamma}\cos\sigma + \left(1-\frac{1}{v}\right) \qquad \textbf{10.86}$$

$$\frac{d\theta}{ds} = \frac{\cos\psi}{\cos\phi} \qquad \textbf{10.87}$$

$$\frac{d\phi}{ds} = \sin\psi \qquad \textbf{10.88}$$

$$\frac{d\psi}{ds} = \frac{W\lambda\sqrt{\beta r}}{\cos^2\gamma}\sin\sigma - \cos\psi\tan\phi \qquad \textbf{10.89}$$

In general, solving these equations will require numerical methods. In three-dimensional motion, both lift λ and bank σ can be adjusted to meet restrictions (constraints) on the trajectory. (In control theory, λ and σ would be called the "control variables.") The next few sections introduce a few examples of non-planar motion and the controls used to meet their requirements.

10.5.1 Equilibrium Glide

In Section 5.2, the concept of a quasi-equilibrium glide was introduced, albeit for planar entries. In that section, we used the small angle approximation to replace:

$$\sin\gamma \approx \gamma \qquad \textbf{10.90}$$

$$\cos \gamma \approx 1 \approx \text{constant} \qquad \textbf{10.91}$$

If we take this one step further and restrict the flight-path angle to a small, nearly constant value, we have something even closer to equilibrium glide and Eq. (10.90) becomes:

$$\sin \gamma \approx \gamma_g \qquad \textbf{10.92}$$

where γ_g is the "gliding" flight-path angle (which may or may not be equal to the entry flight-path angle).

With this assumption, four of the differential equations simplify:

$$\frac{dW}{ds} = -\overline{\beta r} W \gamma_g \qquad \textbf{10.93}$$

$$\frac{dv}{ds} = -\frac{W v \sqrt{\overline{\beta r}}\left(1 + \lambda^2\right)}{L_*} - (2 - v)\gamma_g \qquad \textbf{10.94}$$

$$\frac{d\gamma}{ds} = W \lambda \sqrt{\overline{\beta r}} \cos \sigma + \left(1 - \frac{1}{v}\right) = 0 \qquad \textbf{10.95}$$

$$\frac{d\psi}{ds} = W \lambda \sqrt{\overline{\beta r}} \sin \sigma - \cos \psi \tan \phi \qquad \textbf{10.96}$$

Equation (10.93) can be integrated if the flight-path angle changes slowly enough to be considered constant:

$$W = W_0 \exp\left[-\overline{\beta r}\gamma_g \left(s - s_0\right)\right] \qquad \textbf{10.97}$$

The "0" subscript denotes the conditions at the beginning of the (near) equilibrium glide. The flight-path angle equation (Eq. (10.95)) reduces to a simple algebraic equation that can be solved:

$$\lambda \cos \sigma = \frac{(1-v)}{Wv\sqrt{\beta r}}$$

10.98

Or, substituting the solution for $W(s)$ into this, we find:

$$\lambda \cos \sigma = \frac{(1-v)}{W_0 v \sqrt{\beta r}} \exp\left[\overline{\beta r}\gamma_g (s - s_0)\right]$$

10.99

The left-hand-side is the radial (local vertical) component of the lift force. These two equations show the radial lift force balances the combined force of gravity and centrifugal force. (If it isn't clear why this is true, look back at Eq. (5.2) and Figure 5-1.)

The four remaining equations of motion are coupled and not easily solved. Eliminating W in favor of the control variables using Eq. (10.98), they can be written:

$$\frac{dv}{ds} = -\frac{(1-v)(1+\lambda^2)}{L_* \lambda \cos \sigma} - (2-v)\gamma_g$$

10.100

$$\frac{d\theta}{ds} = \frac{\cos\psi}{\cos\phi}$$

10.101

$$\frac{d\phi}{ds} = \sin\psi$$

10.102

$$\frac{d\psi}{ds} = \frac{(1-v)}{v}\tan\sigma - \cos\psi\tan\phi$$

10.103

In the region of interest, Eq. (10.100) is dominated by the first term on the right (because $W\sqrt{\beta r} \gg \gamma_g$), so it can be replaced with

$$\frac{dv}{ds} \approx -\frac{(1-v)(1+\lambda^2)}{L_*\lambda \cos \sigma}$$

10.104

to yield a set of equations free from the flight-path angle. (This equation is "exact" when $\gamma_g = 0$.) This set of equations can be numerically integrated to produce the entry trajectory.

For a "nearly true" equilibrium glide $\gamma_g = 0$ and the equation for W (Eq. (10.93) fails to provide a solution for the changing altitude. Instead, it is found with Eq. (10.98). Equations (10.101) - (10.104) must be integrated to obtain the remaining states of the solution ($v(s)$, $\theta(s)$, $\phi(s)$, $\psi(s)$) to completely solve for the trajectory of an equilibrium glide.

10.5.2 Maximum Cross Range at a Fixed Bank Angle (Assumed Equilibrium Glide)

One of the advantages of lifting entry vehicles is their ability to shift the landing/impact point off of their initial entry plane. This "shift" gives the vehicles a cross range capability. Vinh and Gell, in separate publications, built on the work of Eggers to find a "sub-optimum" control law to maximize the cross range of a vehicle entering the atmosphere in an equilibrium glide at near circular speed (23:13.51-13.53; 27:28-37; 57:346-363; 58:347-356). Since their solution allows some analytical insight without delving into modern control theory, it is a good example to present here.

For entry about a non-rotating, spherical planet, we can define the entry point to be

$$\theta_e = 0 \qquad\qquad\qquad \textbf{10.105}$$

$$\phi_e = 0 \qquad\qquad\qquad \textbf{10.106}$$

$$\psi_e = 0 \qquad\qquad\qquad \textbf{10.107}$$

without loss of generality. For equilibrium glide at near circular velocity, we can write:

$$\gamma = \gamma_e = 0 \qquad\qquad\qquad \textbf{10.108}$$

$$v_e \approx 1 \qquad\qquad\qquad \textbf{10.109}$$

Lift and bank are to be selected such that they will maximize the cross range (measured in terms of ϕ).

To maximum distance, we'll assume the velocity is reduced all the way to zero. It is reasonable to assume that the "longer" the distance (measured in terms of s) it takes for the velocity to be reduced to zero, the more distance can be covered by the vehicle. (This is one of the assumptions that make this a sub-optimal solution.) Minimizing the change in velocity means $\dfrac{dv}{ds}$ should have the smallest magnitude possible; therefore, the right-hand-side of Eq. (10.104) should minimized. For a given bank angle and speed,

$$\lambda = 1 \qquad\qquad\qquad \textbf{10.110}$$

minimizes $\left| \dfrac{dv}{ds} \right|$. In-other-words, the optimal (or at least "nearly optimal") glide is at the maximum lift-to-drag ratio. The bank angle, however, must be varied in some optimal manner.

In order to obtain an analytic solution, the problem must be simplified a little further. First, we can look at the situation where the bank angle is held constant. This does not strictly limit the application to a vehicle that can only hold a constant bank angle. Vinh suggests we can assume there is an average value $\bar{\sigma}$ that provides comparable performance to the varying value (58:347). Regardless of the reason, we will assume the bank angle is a yet-to-be-determined constant $\bar{\sigma}$. Second, we can assume changes to the heading ψ and latitude ϕ are small and:

$$\sin \psi = \psi \qquad \textbf{10.111}$$

$$\cos \psi = 1 \qquad \textbf{10.112}$$

$$\sin \phi = \phi \qquad \textbf{10.113}$$

$$\cos \phi = 1 \qquad \textbf{10.114}$$

With these simplifications, the equations of motion become:

$$\frac{dv}{ds} = -\frac{2(1-v)}{L_* \cos \bar{\sigma}} \qquad \textbf{10.115}$$

$$\frac{d\theta}{ds} = 1 \qquad \textbf{10.116}$$

$$\frac{d\phi}{ds} = \psi \qquad \textbf{10.117}$$

$$\frac{d\psi}{ds} = \frac{(1-v)}{v} \tan \bar{\sigma} - \phi \qquad \textbf{10.118}$$

The equation for velocity is separable and easily integrated if the entry speed is *not exactly* circular ($v_e \neq 1$):

$$v = 1 - (1 - v_e) \exp \left[\frac{2(s - s_e)}{L_* \cos \bar{\sigma}} \right]$$

10.119

Equations (10.117) can be differentiated with respect to s and combined with Eq. (10.118) to form a single second-order differential equation:

$$\frac{d^2 \phi}{ds^2} + \phi = \frac{(1 - v)}{v} \tan \bar{\sigma}$$

10.120

For simplicity, Equation (10.119) has not been used to eliminate v in favor of s yet. The complete solution to this differential equation involves first finding the homogeneous solution and then the particular solution.

The homogeneous solution to this differential equation is simple

$$\phi_h (s) = c_1 \cos (s) + c_2 \sin (s)$$

10.121

where c_1 and c_2 are constants that can be evaluated using initial conditions once the entire solution is known. The particular solution is more difficult to find, but is manageable with a few modifications. Let

$$\tau = 1 - v$$

$$= (1 - v_e) \exp \left[\frac{2s}{L_* \cos \bar{\sigma}} \right]$$

$$= \tau_e \exp \left[\frac{2s}{L_* \cos \bar{\sigma}} \right]$$

10.122

be a new independent variable. (Note that $s_e = 0$ has been arbitrarily assumed.)

With this change, the relationships between the differential operators become:

$$\frac{d(\ )}{d\tau} = \left(\frac{L_* \cos\bar{\sigma}}{2\tau}\right)\frac{d(\ )}{ds}$$

10.123

$$\frac{d^2(\ )}{d\tau^2} = -\left(\frac{1}{\tau}\right)\frac{d(\ )}{d\tau} + \left(\frac{L_*^2 \cos^2\bar{\sigma}}{4\tau^2}\right)\frac{d^2(\ )}{ds^2}$$

10.124

Using these to transform Eq. (10.120) gives

$$\tau^2\left(\frac{d^2\phi}{d\tau^2}\right) + \tau\left(\frac{d\phi}{d\tau}\right) + \left(\frac{L_* \cos\bar{\sigma}}{2}\right)^2 \phi = \left[\frac{L_*^2 \sin(2\bar{\sigma})}{8}\right]\left(\frac{\tau}{1-\tau}\right)$$

10.125

where the trigonometry identity

$$\sin(2\bar{\sigma}) = 2\cos\bar{\sigma}\sin\bar{\sigma}$$

10.126

was used to simplify the right-hand-side slightly. By its definition, $\tau < 1$ for all but the very last instant of the entry, so the right-hand-side can be expanded in a binomial series:

$$\tau^2\left(\frac{d^2\phi}{d\tau^2}\right) + \tau\left(\frac{d\phi}{d\tau}\right) + \left(\frac{L_* \cos\bar{\sigma}}{2}\right)^2 \phi = \left[\frac{L_*^2 \sin(2\bar{\sigma})}{8}\right]\sum_{n=1}^{\infty}\tau^n$$

10.127

A particular solution can be built by assuming a power series of the form:

$$\phi_p(\tau) = \sum_{n=1}^{\infty} a_n\tau^n$$

10.128

Substituting this into Eq. (10.127) and equating like powers of τ, gives the coefficients of the power series:

$$a_n = \frac{L_*^2 \sin(2\bar{\sigma})}{8n^2 + 2L_*^2 \cos^2 \bar{\sigma}}$$

10.129

Converting the homogeneous solution to a function of τ and combining with the other pieces of the total solution for ϕ, we get:

$$\phi(\tau) = c_1 \cos\left[\left(\frac{L_* \cos \bar{\sigma}}{2}\right) \ln\left(\frac{\tau}{\tau_e}\right)\right] + c_2 \sin\left[\left(\frac{L_* \cos \bar{\sigma}}{2}\right) \ln\left(\frac{\tau}{\tau_e}\right)\right]$$

$$+ \left[\frac{L_*^2 \sin(2\bar{\sigma})}{2}\right] \sum_{n=1}^{\infty} \frac{\tau^n}{4n^2 + L_*^2 \cos^2 \bar{\sigma}}$$

10.130

At this point, we can use the initial conditions $\phi_e = 0$ and $\psi_e = 0$ to evaluate c_1 and c_2. The latter condition requires a slight amount of work. In terms of τ, Eq. (10.117) becomes

$$\frac{d\phi}{d\tau} = \left(\frac{L_* \cos \bar{\sigma}}{2\tau}\right)\psi$$

10.131

telling us $\left.\dfrac{d\phi}{d\tau}\right|_{\tau=\tau_e} = 0$. With these boundary conditions, we can easily solve for c_1 and c_2:

$$c_1 = -\left[\frac{L_*^2 \sin(2\bar{\sigma})}{2}\right] \sum_{n=1}^{\infty} \frac{\tau_e^n}{4n^2 + L_*^2 \cos^2 \bar{\sigma}}$$

10.132

$$c_2 = -2L_* \sin\bar{\sigma} \sum_{n=1}^{\infty} \frac{n\tau_e^{n-1}}{4n^2 + L_*^2 \cos^2\bar{\sigma}} \qquad \textbf{10.133}$$

(Once again, the identity in Eq. (10.126) was used.) Finally, the solution for ϕ can be written:

$$\phi(\tau) = \left\{ -\left[\frac{L_*^2 \sin(2\bar{\sigma})}{2} \right] \sum_{n=1}^{\infty} \frac{\tau_e^n}{4n^2 + L_*^2 \cos^2\bar{\sigma}} \right\} \cos\left[\left(\frac{L_* \cos\bar{\sigma}}{2} \right) \ln\left(\frac{\tau}{\tau_e} \right) \right]$$

$$+ \left(-2L_* \sin\bar{\sigma} \sum_{n=1}^{\infty} \frac{n\tau_e^{n-1}}{4n^2 + L_*^2 \cos^2\bar{\sigma}} \right) \sin\left[\left(\frac{L_* \cos\bar{\sigma}}{2} \right) \ln\left(\frac{\tau}{\tau_e} \right) \right]$$

$$+ \left[\frac{L_*^2 \sin(2\bar{\sigma})}{2} \right] \sum_{n=1}^{\infty} \frac{\tau^n}{4n^2 + L_*^2 \cos^2\bar{\sigma}} \qquad \textbf{10.134}$$

(The solution for ψ could found by substituting this into Eq. (10.131) if desired.)

For an equatorial entry, maximizing the change in latitude ϕ equates to maximizing the cross range. We need to maximize this "simplified" solution with respect to $\bar{\sigma}$ to find the optimal bank angle. The oscillatory terms are small when the entry is near circular speed ($\tau_e \ll 1$ when $v_e \approx 1$) so they can be ignored. After evaluating at the end of the flight ($\tau_f = 1$ when $v_f = 0$), we are left with:

$$\phi(\tau_f) \approx \left[\frac{L_*^2 \sin(2\bar{\sigma})}{2} \right] \sum_{n=1}^{\infty} \frac{1}{4n^2 + L_*^2 \cos^2\bar{\sigma}} \qquad \textbf{10.135}$$

The initial temptation might be to take the first few terms of this series and use the resulting approximation to maximize ϕ. Given enough effort, that tactic can be "forced" to work, but there is a more elegant approach.

A fraction of the form $\dfrac{1}{a+b}$ can be written as:

$$\frac{1}{a+b} = \frac{1}{a} - \frac{b}{a(a+b)}$$

10.136

Applying this identity to the summation in Eq. (10.135), enables us to write:

$$\sum_{n=1}^{\infty} \frac{1}{4n^2 + L_*^2 \cos^2 \bar{\sigma}} = \sum_{n=1}^{\infty} \left[\frac{1}{4n^2} - \frac{L_*^2 \cos^2 \bar{\sigma}}{4n^2 \left(4n^2 + L_*^2 \cos^2 \bar{\sigma} \right)} \right]$$

$$= \frac{1}{4} \sum_{n=1}^{\infty} \frac{1}{n^2} - \sum_{n=1}^{\infty} \frac{L_*^2 \cos^2 \bar{\sigma}}{4n^2 \left(4n^2 + L_*^2 \cos^2 \bar{\sigma} \right)}$$

10.137

A quick search of a math handbook reveals

$$\sum_{n=1}^{\infty} \frac{1}{n^2} = \frac{\pi^2}{6}$$

10.138

which, in turn, enables us to write Eq. (10.135) as:

$$\phi \left(\tau_f \right) \approx \left[\frac{L_*^2 \sin \left(2\bar{\sigma} \right)}{2} \right] \left[\frac{\pi^2}{24} - \sum_{n=1}^{\infty} \frac{L_*^2 \cos^2 \bar{\sigma}}{4n^2 \left(4n^2 + L_*^2 \cos^2 \bar{\sigma} \right)} \right]$$

10.139

If the maximum lift-to-drag ratio L_* is small, the summation can be ignored. Without the summation, we are left with Eggers' formula for cross range (23:13.53; 57:356; 58:351):

$$\phi \left(\tau_f \right) \approx \frac{L_*^2 \pi^2}{48} \sin \left(2\bar{\sigma} \right)$$

10.140

From this equation, we can deduce that a bank angle of about $\bar{\sigma}_* \approx 45°$ maximizes the cross range. (Note, however, Vinh, et al. show that Eq. (10.140) *overestimates* the actual value of $\phi(\tau_f)$ when compared to the value found numerically integrating Eqs. (10.101) - (10.104) with $\bar{\sigma} = 45°$ (58:352).)

Gell found a relatively "inspired" way to analytically maximize Eq. (10.139). Backing up to the previous form of the equation in Eq. (10.135) and "undoing" the trigonometric identity, he started with:

$$\phi(\tau_f) \approx \left(L_*^2 \sin\bar{\sigma}\cos\bar{\sigma} \right) \sum_{n=1}^{\infty} \frac{1}{4n^2 + L_*^2 \cos^2\bar{\sigma}} \qquad \textbf{10.141}$$

With a little rewriting, this becomes

$$\phi(\tau_f) \approx \left(\frac{L_* \sin\bar{\sigma}}{2} \right) \sum_{n=1}^{\infty} \frac{x}{n^2 + x^2}$$

$$= \left(\frac{L_* \sin\bar{\sigma}}{4} \right) \left\{ \left[\frac{1}{x} - \frac{1}{x} \right] + 2\sum_{n=1}^{\infty} \frac{x}{n^2 + x^2} \right\}$$

$$= \left(\frac{L_* \sin\bar{\sigma}}{4} \right) \left\{ \left[\frac{1}{x} + 2\sum_{n=1}^{\infty} \frac{x}{n^2 + x^2} \right] - \frac{1}{x} \right\} \qquad \textbf{10.142}$$

where $x = \dfrac{L_* \cos\bar{\sigma}}{2}$ has been introduced for simplicity. The term in square brackets is the series expansion for a hyperbolic cotangent:

$$\frac{1}{x} + 2\sum_{n=1}^{\infty} \frac{x}{n^2 + x^2} = \pi \coth(\pi x) \qquad \textbf{10.143}$$

(Recognizing this fact is what I call "inspired" on Gell's part!) With the substitution, the expression for $\phi(\tau_f)$ becomes:

$$\phi(\tau_f) \approx \left(\frac{\pi L_* \sin\bar{\sigma}}{4} \right) \coth \left(\frac{\pi L_* \cos\bar{\sigma}}{2} \right) - \frac{\tan\bar{\sigma}}{2} \qquad \textbf{10.144}$$

This expression can be maximized by taking the derivative (with respect to $\bar{\sigma}$) and setting it equal to zero to find the optimum bank angle $\bar{\sigma}_*$. This yields

$$\left(\frac{\pi L_* \cos\bar{\sigma}_*}{4} \right) \coth \left(\frac{\pi L_* \cos\bar{\sigma}_*}{2} \right) - \left(\frac{\pi^2 L_*^2 \sin^2\bar{\sigma}_*}{8} \right) \left[1 - \coth^2 \left(\frac{\pi L_* \cos\bar{\sigma}_*}{2} \right) \right]$$

$$- \left(\frac{1}{2} \right) \left(1 + \tan^2\bar{\sigma}_* \right) = 0 \qquad \textbf{10.145}$$

where we can restrict ourselves to $0 \leq \bar{\sigma}_* \leq \pi/2$ to avoid any quadrant ambiguities. (The problem is symmetrical, so a negative value of $\bar{\sigma}_*$ would simply result getting the corresponding negative value of $\phi(\tau_f)$.) Equation (10.145) must be solved numerically, but it is straight-forward. Table 10-1 gives a few discrete results.

Table 10-1: Optimum Constant Bank Angles to Maximize Cross Range

Maximum Lift-to-Drag ratio,	Optimum Bank Angle, (radians)	Optimum Bank Angle, (degrees)
0.5	0.7953	45.57
1.0	0.8212	47.05
1.5	0.8553	49.00
2.0	0.8910	51.05
2.5	0.9249	52.99
3.0	0.9556	54.75
3.5	0.9830	56.32
4.0	1.0071	57.70

To compare the improvement in cross range using Gell's optimum $\bar{\sigma}_*$ from Eq. (10.145) and Egger's $\bar{\sigma} = 45°$, we can numerically integrate Eqs. (10.101) - (10.104) using both solutions for σ. Note, however, that some care must be used to insure the integration "debanks" the vehicle to $\sigma = 0°$ when the heading angle ψ reaches $\pm 90°$. This eliminates "back-tracking" and ensures the vehicle continues to maximize the cross range. Figure 10-10 notionally illustrates the difference a debank maneuver makes in the final latitude.

For comparison purposes, a near circular equatorial entry into Earth's atmosphere was simulated by integrating Eqs. (10.101) - (10.104) using the initial conditions $v_e = 0.99$, $\phi_e = 0$, $\theta_e = 0$, and $\psi_e = 0$. The equations were integrated forward until $v_f = 0.001$. Solutions using Gell's $\bar{\sigma}_*$ and Egger's $\bar{\sigma} = 45°$ are shown in Figure 10-11. The figure shows the optimum (constant) bank angle is not $45°$ for all L_*. Above about $L_* \approx 2$, the angles computed with Eq. (10.145) provide a greater cross range. For $L_* \geq 4$, the cross range is effectively the entire planet.

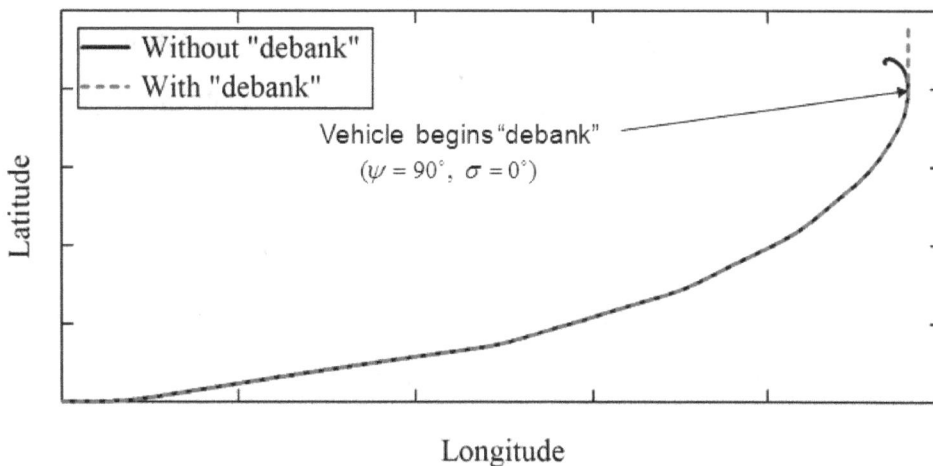

Figure 10-10: "Debanking" to Maximize Cross Range

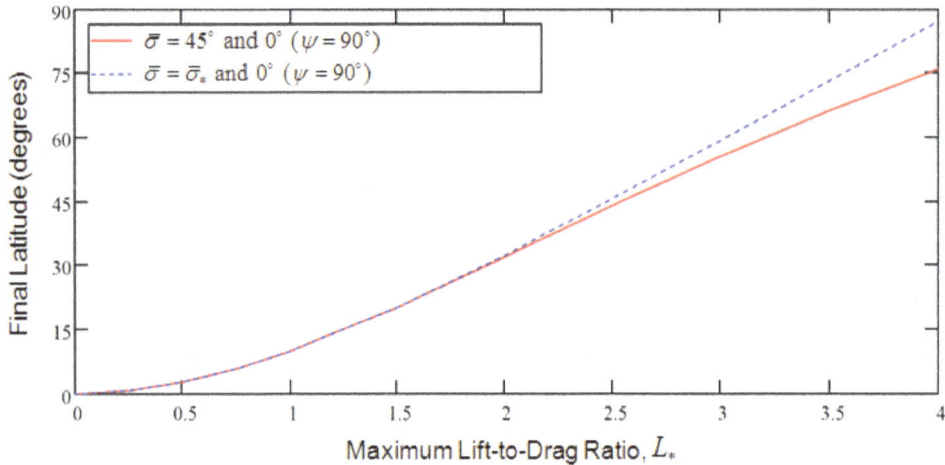

Figure 10-11: Comparison Cross Range using Egger's and Gell's Solutions for Constant Bank Angle

10.5.3 Maximum Cross Range with a Variable Bank Angle (Assumed Equilibrium Glide)

To this point, we've elected to search for a constant bank angle to maximize the cross range. This is a suboptimal solution; however, the actual optimal control for this scenario is to begin with a large bank and decrease it to zero as $\psi \to \pm 90°$ (13:77-81; 27:38; 57:346-351; 58:351). Gell suggests using

$$\sigma = \sigma_0 \tan^{-1}\left(\frac{\cos\phi}{\tan\psi}\right) \qquad \textbf{10.146}$$

since it is a function with the desired behavior. σ_0 is a scaling factor selected to maximize ϕ_f. In his dissertation, he empirically found

$$\sigma_0 = \frac{1}{2}\exp\left(\frac{L_*}{5}\right) \qquad \textbf{10.147}$$

for $0 \leq L_* < 3.5$ (27:37-39). A typical control profile using "Gell's equation"

$$\sigma = \frac{1}{2} e^{\frac{L_*}{5}} \tan^{-1}\left(\frac{\cos\phi}{\tan\psi}\right)$$ **10.148**

for continuous control during a near circular equatorial entry is shown in Figure 10-12. (For reference, $L_* = 2.0$ was used to create the figure.) The resulting cross ranges for different values of L_* are shown in Figure 10-13.

Also shown in Figure 10-13 are the results using the "real" optimal control found by using modern control theory. Chern and Vinh applied that theory to Eqs. (10.101) - (10.104) and found the bank angle profile given by

$$\tan\sigma = \left(\frac{1-v}{v}\right)\left[\frac{\cos\phi\sin\left(\theta_f - \theta\right)}{\cos\left(\theta_f - \theta\right)\sin\psi - \cos\psi\sin\phi\sin\left(\theta_f - \theta\right)}\right]$$ **10.149**

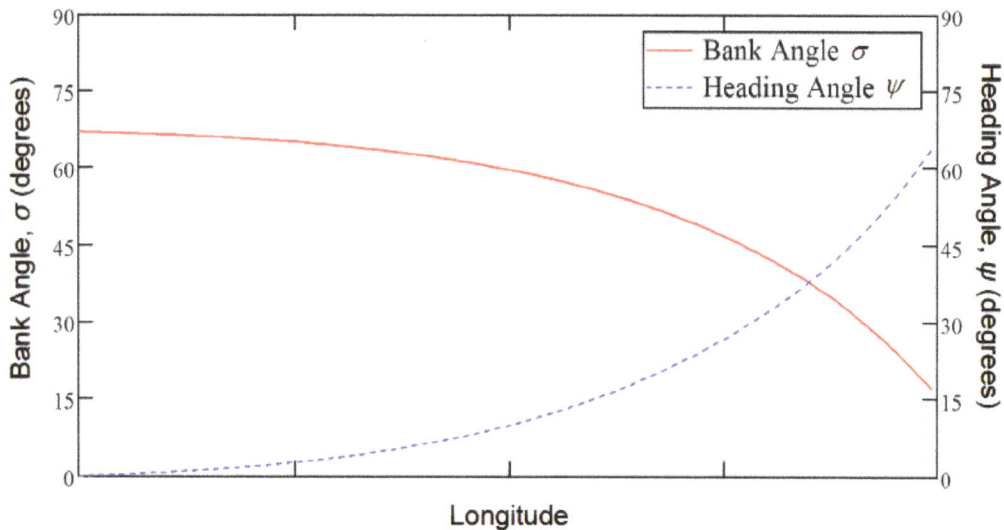

Figure 10-12: Typical Control Profile and Corresponding Heading Angle using Gell's Equation, Eq. (10.148)

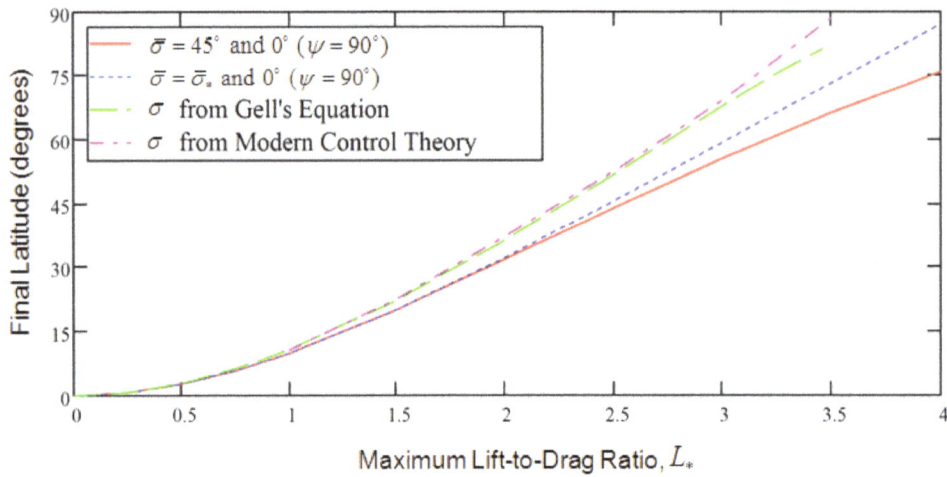

Figure 10-13: Comparison of Cross Range for Various Bank Angle Solutions

maximizes ϕ_f (13:77-81; 57:346-351; 58:351). The subscript "f" denotes final conditions. To use this solution, the final longitude θ_f must be known at the beginning of the maneuver, so its application is more difficult than simply using Gell's equation. But, as Figure 10-13 demonstrates, the solution from modern control theory gives marginally better cross range for $L_* > 3.0$. A typical bank profile found using Eq. (10.149) (and $L_* = 2.0$) is shown in Figure 10-14. Notice its similarity to the one shown in Figure 10-12. (Both of the solutions shown in Figure 10-13 and Figure 10-14 assumed allowable maximum bank angle of $\sigma_{\max} = 85°$.)

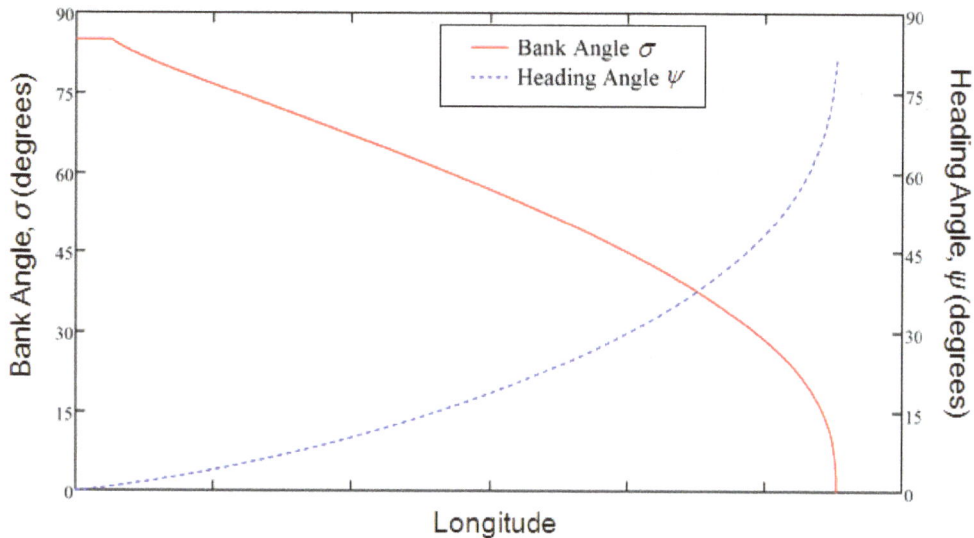

***Figure 10-14: Typical Control Profile and Corresponding Heading Angle
using Modern Control Solution for σ***

10.5.4 Maximum Cross Range with a Variable Bank Angle (Shallow Entry)

All of the optimal bank angle solutions in the previous two sections were found by using equilibrium glide assumptions and then integrating Eqs. (10.101) - (10.104) to find the cross range. It's instructive to look back and use these solutions in the full system of equations (Eqs. (10.84)-(10.89)) in the hope that we've found "near optimal" solutions for the "full dynamics" case. If we wanted to delve into modern control theory, we could actually find the "true" optimal control profile for non-equilibrium flight and then compare the answers. We aren't that industrious, so we'll settle for showing that shallow entries remain "close enough" to equilibrium glide for our bank angle profiles to be "close" to optimal. We'll use results found with Eq. (10.149) convince ourselves.

Using the initial conditions $W_e = 0.0005$, $v_e = 0.99$, $\gamma_e = -4°$, $\phi_e = \theta_e = \psi_e = 0$, a maximum lift-to-drag ratio of $L_* = 2$, a maximum bank of $\sigma_{max} = 85°$, and an Earth-like atmosphere with $\overline{\beta r} = 900$, Eqs. (10.84)-(10.89) were integrated forward until $v_f = 0.001$ to compare with the previous equilibrium glide solutions. Figure 10-15 plots the control history (found with Eq. (10.149)) and corresponding heading angles for this example as a function of longitude. This plot is analogous to Figure 10-14 except it reflects the use of "full dynamics" for this shallow entry instead of assuming equilibrium glide. The basic trends are the same, with the bank angle starting at its maximum value and decreasing to zero as the heading angle approaches ninety degrees.

The results for latitude as a function of longitude are presented in Figure 10-16. Notice that equilibrium glide *overestimates* the longitude (or time-of-flight), but accurately predicts the final latitude. (While this plot is for a solution

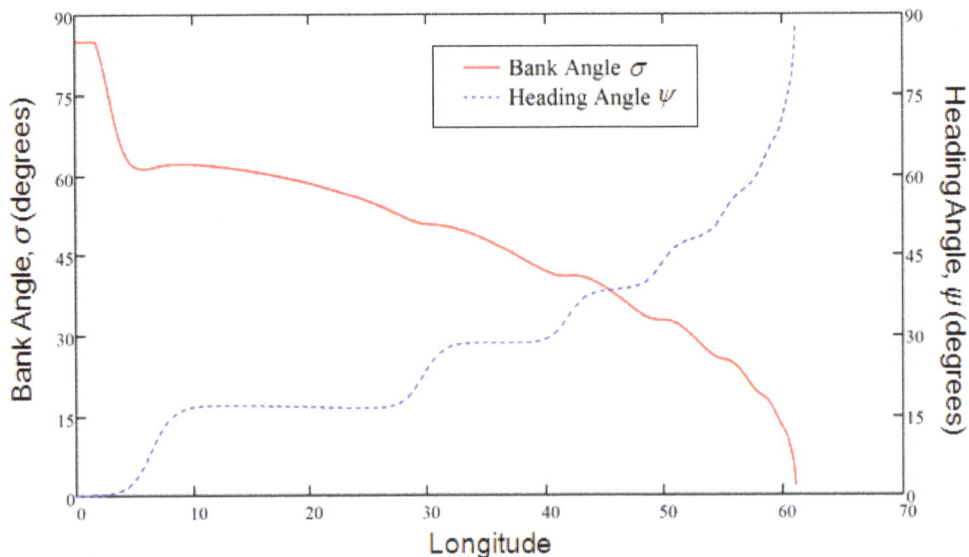

Figure 10-15: Control Profile and Corresponding Heading Angle using Modern Control Solution

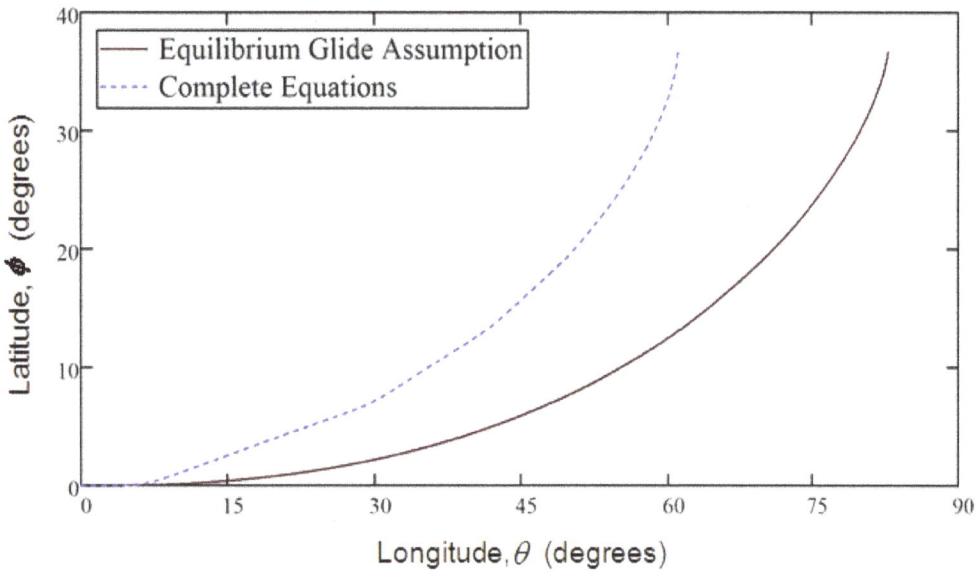

Figure 10-16: Latitude as a Function of Longitude for Flight using Modern Control Solution for σ

using the modern control theory solution, experience with numerous examples shows that this overestimation of θ is typical when comparing equilibrium glide to the "real" solution.) We'll use latitude on the x-axis for the remainder of the comparisons and quietly forget about the longitude discrepancy.

The bank angle solutions are compared in Figure 10-17. (It's difficult to see in the plot, but both solutions start with $\sigma = 85°$.) Given the similarity, we can (rightfully) expect the dynamics solutions to be similar. Solutions for W, γ, v, and ψ are shown in Figure 10-18 - Figure 10-21. (For equilibrium glide, W was found with Eq. (10.98).) Notice how the solutions for equilibrium glide represent (more or less) "averaged" or "smoothed" simplifications of the more general solutions, especially in the first two figures. This fact and the realization (from Figure 10-19) that the flight-path angle remains small gives us confidence we are "close" to equilibrium glide for shallow entries; therefore, the bank angle

335

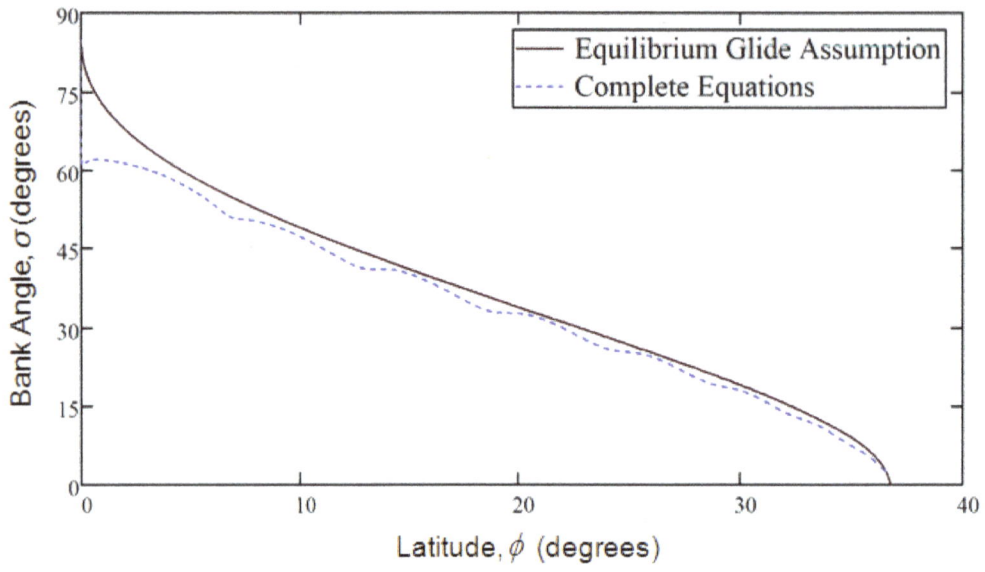

Figure 10-17: *Bank Angle* σ *as a Function of Latitude* ϕ *using Modern Control Solution*

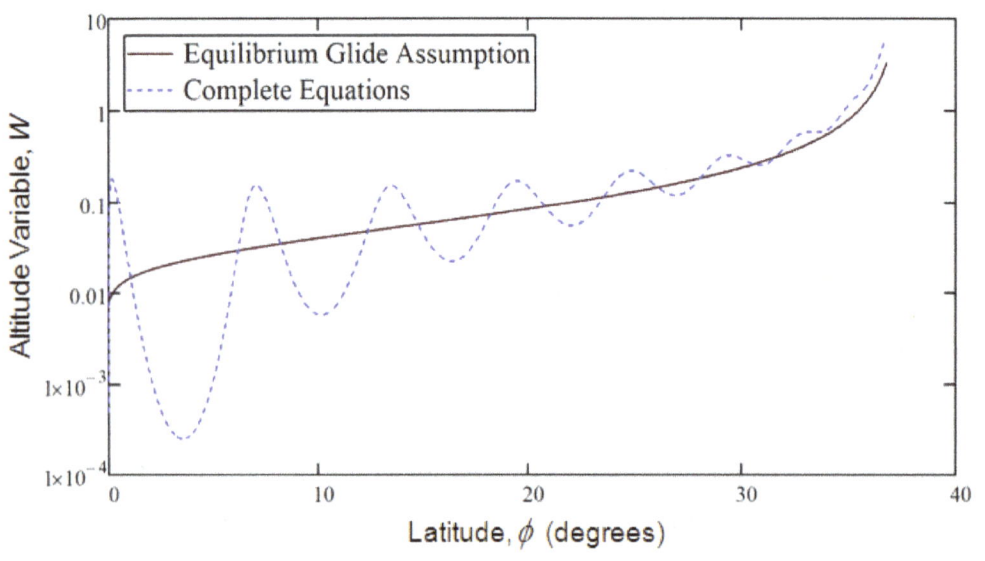

Figure 10-18: *Altitude Variable* W *as a Function of Latitude* ϕ *using Modern Control Solution for* σ

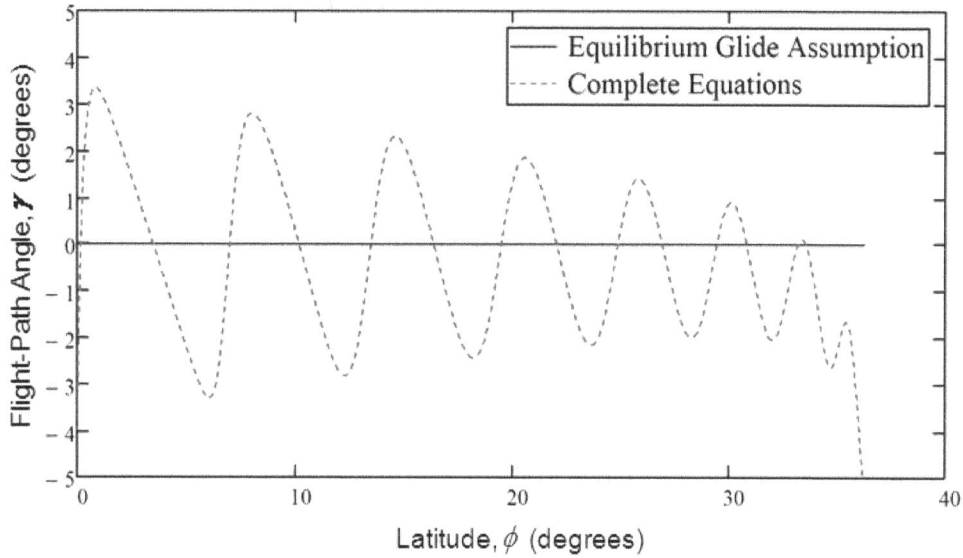

Figure 10-19: Flight-Path Angle γ as a Function of Latitude ϕ using Modern Control Solution for σ

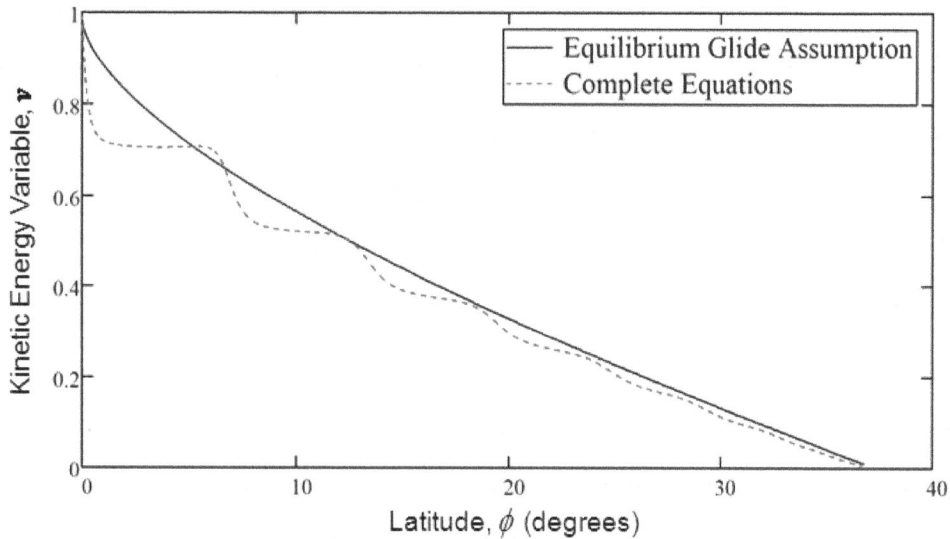

Figure 10-20: Kinetic Energy Variable v as a Function of Latitude ϕ using Modern Control Solution for σ

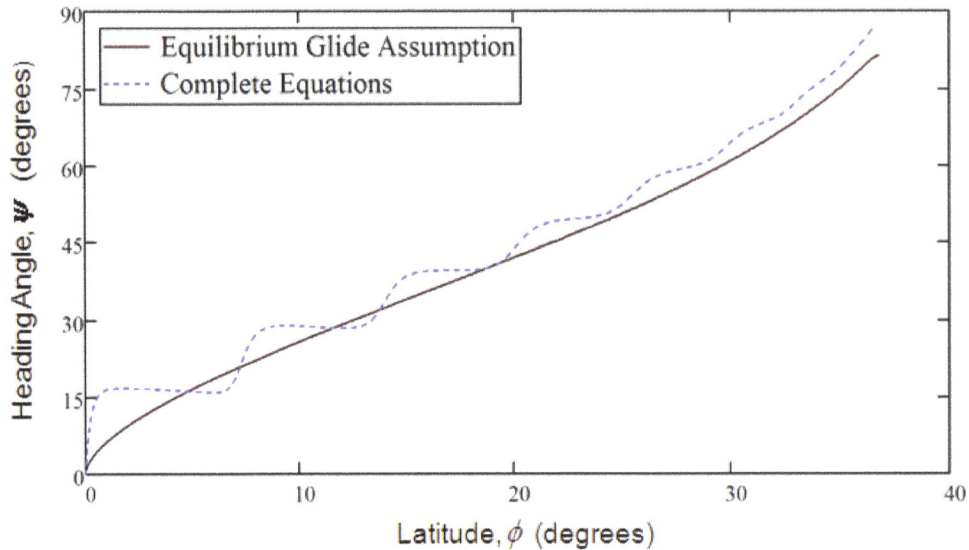

Figure 10-21: Heading Angle ψ as a Function of Latitude ϕ using Modern Control Solution for σ

solutions found in Sections 10.5.2 and 10.5.3 can be applied to shallow entries in general.

If we were to apply each of the solutions we've found to an arbitrary shallow entry, we can expect the results would show similar trends. For example, Figure 10-22 compares the cross range as a function of L_* for the current problem. While not as definitive as when we assumed equilibrium flight (Figure 10-13), the same trends are evident in this figure. In order of increasing "optimality," we have the two constant solutions, $\bar{\sigma} = 45°$ and $\bar{\sigma} = \bar{\sigma}_*$ (from Eq. (10.145)), producing nearly the same results. The latter *generally* gives more cross range. Next, the solutions for σ from Gell's Equation (Eq. (10.148)) and the Modern Control theory (found with Eq. (10.149)) also produce similar results with the latter *generally* giving more cross range at larger values of L_*. A dedicated student might want to explore other examples to verify the trends hold in general, but we won't.

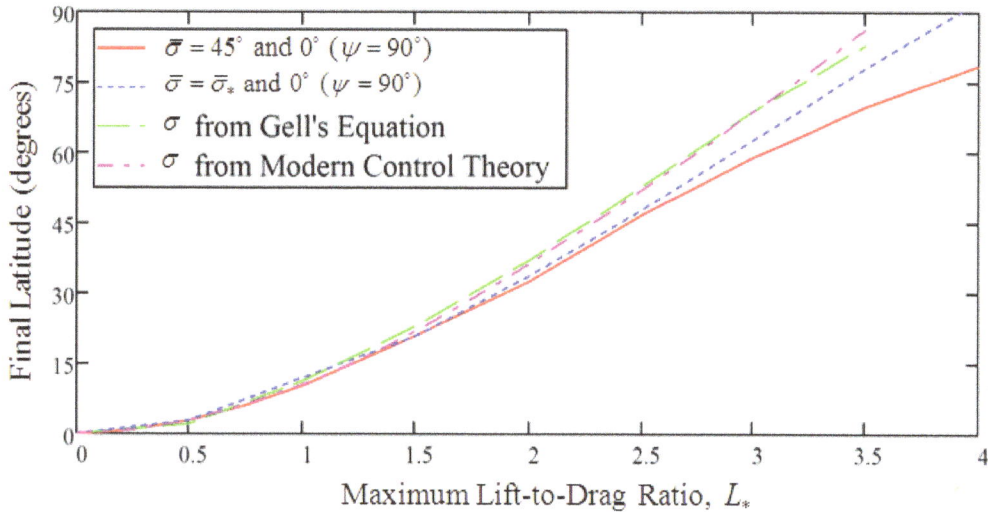

Figure 10-22: Comparison of Cross Range for All Bank Angle Solutions (Shallow Entry Example)

10.5.5 Diveline Guidance

Diveline guidance (DG) is a type of control that provides a trajectory-shaping and trajectory-targeting algorithm. In its use, one or more "divelines" are selected that intersect the Earth (at "targeted points"). When the DG control is initiated, the orientation and magnitude of the vehicle's lift vector $\vec{L}$ is continuously adjusted to guide the vehicle until it is flying "down" the first diveline. (See Figure 10-23.) At some point, it breaks off that approach and the lift vector reorients to maneuver and fly down the next diveline. This continues until the vehicle flies down the final diveline, which intersects the Earth exactly at the desired impact point.

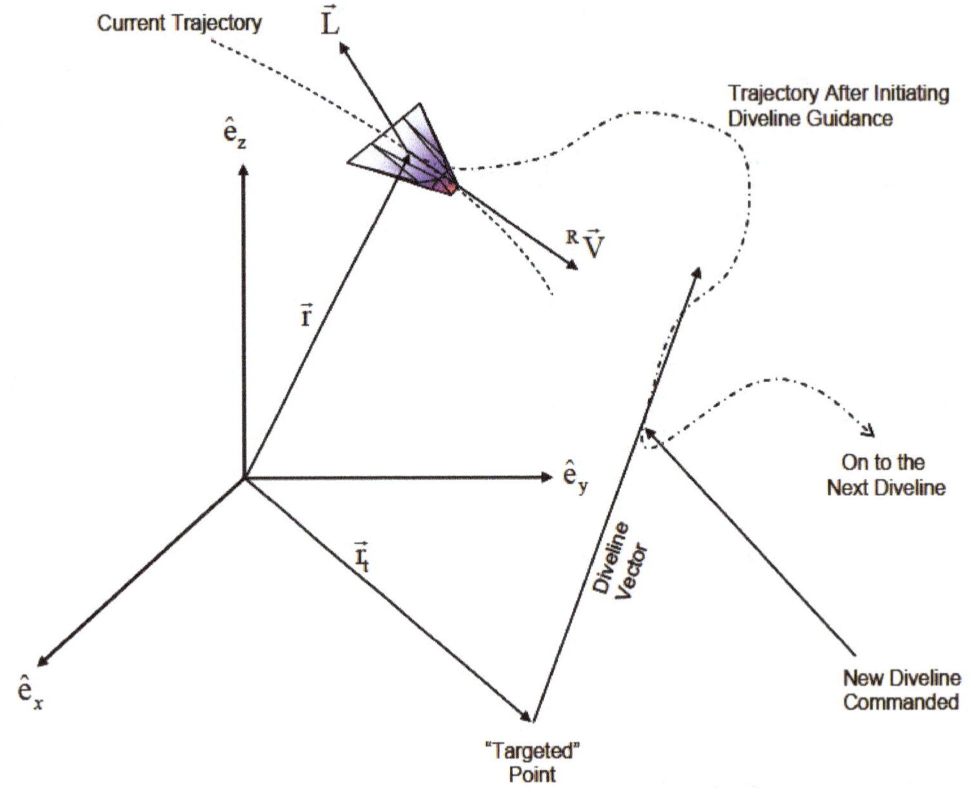

Figure 10-23: Diveline Guidance Geometry

Finding the orientation of $\vec{L}$ is rather simple, so this algorithm is relatively simple to implement (at least in theory). The control is continuously varying, however, and the equations of motion must be numerically integrated. The details for finding the desired $\vec{L}$ will not be presented here, but are rigorously covered in Regan and Anandakrishnan's text (47:252-257). Cameron's and Gracey, et al.'s works also provide a good background on the subject (12:670-678; 28:558-563).

10.5.6 Other Non-Planar Entries

Other useful non-planar trajectories which require finding control profiles (both λ and σ) include a wide assortment of optimized paths (e.g., minimum heating), interceptor avoidance maneuvers (e.g., random changes), and energy management maneuvers (e.g., limiting the speed). Of course, for manned entry (and nuclear warheads), there is always the requirement to maneuver from the *actual* entry trajectory to the *desired* trajectory to insure the vehicle "lands" where the user intends it to land! As the last few sections have no doubt shown, analytically finding control profiles quickly becomes difficult, so we will not delve into these. The interested reader can easily find a plethora of "special cases" to study, including those discussed in Refs. 11, 13, 33, 50, 54, 55, 60, and 61.

10.6 Problems

Material Understanding:

1. Show that Eq. (10.23) for $\dfrac{du}{ds}$ becomes

$$\frac{dv}{ds} = -\frac{2Wvf(\lambda)\sqrt{\beta r}}{L_* \cos \gamma} - (2-v)\tan \gamma$$

   when u is replaced by $v = \dfrac{u}{\cos^2 \gamma}$.

2. Show that k_2 in Vinh's equation for the control law of a planar, constant sink rate entry is a non-dimensional representation of the sink rate. (k_2 is introduced in Eq. (10.68).)

3. Show the more general condition on altitude W given by Eq. (10.60)

$$W < \frac{L_*\left[-2\sin^2\gamma + k_1\left(2 - \sin^2\gamma\right)\right]}{k_1\sqrt{\beta r}\,\sin\gamma}$$

reduces to

$$z > \frac{-k_2^2}{2L_*\,\sin\gamma}$$

by changing to "flat planet" variables.

4. Using a process similar to how we found the lift profile for entry with a constant dynamic pressure, find the solution for $\lambda(W,\gamma)$ for entry at a constant heat flux.

5. In Section 10.5.2, $\left|\dfrac{dv}{ds}\right|$ was minimized with respect to λ to maximize the "time" it took for the velocity to drop to zero. Explain why the other control variable, σ, cannot be used in a similar manner to maximize the cross range.

Computational Insights:

6. In Section 10.4.1, we assumed a vehicle was "sufficiently deep into the atmosphere" to set $\dfrac{1}{W\,\overline{\beta r}} \ll 1$. For entry into Earth's atmosphere ($\overline{\beta r} = 900$) with $\gamma_e = -5°$, $v_e = 0.98$, $L_* = 2$, and $C_{L*} = 2$, graphically compare the solutions to Eqs. (10.46) and (10.48) to show that for entry from $W_e = 1x10^{-4}$, the "sufficiently deep" requirement is rapidly met. Compare the lift profile (C_L/C_D versus entry "time" θ) for the two solutions to show they are consistent.

This page intentionally blank.

Chapter 11

Orbital Contraction

11.1 Introduction

In Chapter 9, a set of equations originally presented by Vinh and Brace were derived as a "unified theory" valid for both orbital motion and entry into a *non-rotating* planetary atmosphere. Those equations can be modified somewhat to examine the motion of a non-lifting satellite perturbed by a *rotating* atmosphere. The process is documented in detail in several works (40:6-18; 58:274-283). While useful, the full set of equations is beyond what we need for a cursory look at the effects of drag on a satellite's orbit. Instead, we will take a somewhat simplified approach and eventually end up with equations consistent with Vinh's but more suitable for "simplified" analyses.

11.2 Relating Inertial and Relative Velocity

Our usual expression for drag

$$\vec{D} = -\frac{\rho C_D S}{2} \, {}^R V \, {}^R \vec{V} \qquad\qquad \textbf{11.1}$$

is based on the velocity *relative to the atmosphere* ${}^R \vec{V}$. Orbital motion, on-the-other-hand, is described using the *inertial velocity* ${}^I \vec{V}$. To examine orbits perturbed by drag, we need an expression relating the two velocities. While the expression is easy to derive, we'll need to expend some effort to put it into a *simple* form written in coordinates we can use.

Using our usual notation, the satellite's velocity relative to the atmosphere ${}^R \vec{V}$ to its inertial velocity ${}^I \vec{V}$ is simply

$$ {}^R \vec{V} = {}^I \vec{V} - \vec{V}_A \qquad\qquad \textbf{11.2}$$

where $\vec{V}_A$ is the inertial velocity of the atmosphere. If it is assumed the atmosphere moves along with the planet, then at a radius r the atmosphere is moving with a speed of

$$V_A = \omega_\oplus r \cos\phi \qquad\qquad \textbf{11.3}$$

in a direction parallel to the equator. ($\omega_\oplus$ is positive when the planet rotates in a positive sense about its north pole.) One way to write this as a vector is to introduce an inertial "nodal" reference frame with one axis pointed along the ascending node $\hat{e}_{node}$, one along $\hat{e}_z$, and one completing the system such that

$\hat{e}_{node} \times \hat{e}_{p-node} = \hat{e}_z$ as shown in Figure 11-1. In this frame:

$$\vec{V}_A = \omega_\oplus r \cos\phi \left[-\sin(\theta - \Omega)\hat{e}_{node} + \cos(\theta - \Omega)\hat{e}_{p-node} \right] \qquad \textbf{11.4}$$

Using the spherical trigonometry identities

$$\cos\phi \cos(\theta - \Omega) = \cos(\omega + \nu) \qquad \textbf{11.5}$$

$$\cos\phi \sin(\theta - \Omega) = \cos i \sin(\omega + \nu) \qquad \textbf{11.6}$$

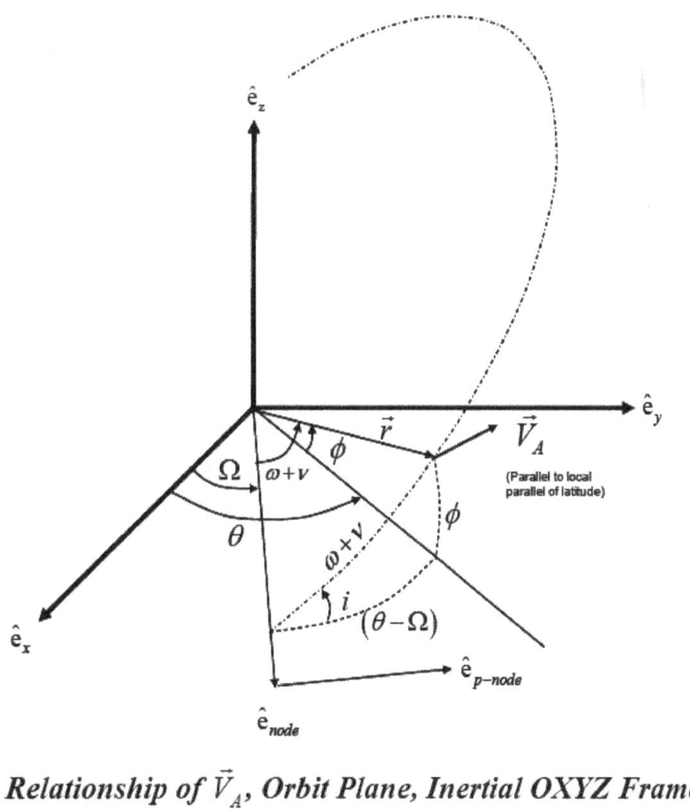

Figure 11-1: Relationship of $\vec{V}_A$, Orbit Plane, Inertial OXYZ Frame, and Nodal Reference Frame

enables us to eliminate the longitude and latitude from Eq. (11.4) and write:

$$\vec{V}_A = \omega_\oplus r \left[-\cos i \sin(\omega + v)\hat{e}_{node} + \cos(\omega + v)\hat{e}_{p-node} \right]$$

11.7

For convenience, the common definition of argument of latitude, $u = \omega + v$, can be introduced:

$$\vec{V}_A = \omega_\oplus r \left[-(\cos i \sin u)\hat{e}_{node} + (\cos u)\hat{e}_{p-node} \right]$$

11.8

Another coordinate frame, aligned with the orbital plane, is convenient to use. As illustrated in Figure 11-2, let $\hat{e}_R$ be in the orbit plane and aligned with the satellite's radius vector. Perpendicular to $\hat{e}_R$ and still in the orbit plane is another vector $\hat{e}_S$. ($\hat{e}_S$ is in the local horizontal plane.) Finally, $\hat{e}_W$ completes the system such that $\hat{e}_W = \hat{e}_R \times \hat{e}_S$. (Note that $\hat{e}_W$ is parallel to the orbit normal.) In the

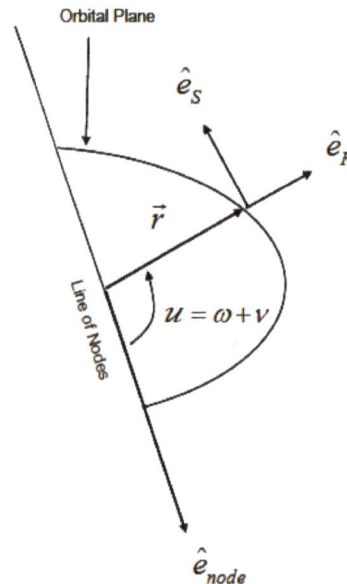

Figure 11-2: The ORSW Reference Frame

convention of Chapter 3, this frame could be labeled as the ORSW frame for easy reference. The ORSW frame is related to the nodal frame by a series of two rotations:

$$
\begin{bmatrix} \hat{e}_R \\ \hat{e}_S \\ \hat{e}_W \end{bmatrix} = R_W(u) R_{node}(i) \begin{bmatrix} \hat{e}_{node} \\ \hat{e}_{p-node} \\ \hat{e}_z \end{bmatrix}
\tag{11.9}
$$

Expanding out the rotations and simplifying, this becomes:

$$
\begin{bmatrix} \hat{e}_R \\ \hat{e}_S \\ \hat{e}_W \end{bmatrix} = \begin{bmatrix} \cos u & \sin u & 0 \\ -\sin u & \cos u & 0 \\ 0 & 0 & 1 \end{bmatrix} \begin{bmatrix} 1 & 0 & 0 \\ 0 & \cos i & \sin i \\ 0 & -\sin i & \cos i \end{bmatrix} \begin{bmatrix} \hat{e}_{node} \\ \hat{e}_{p-node} \\ \hat{e}_z \end{bmatrix}
$$

$$
= \begin{bmatrix} \cos u & \sin u \cos i & \sin u \sin i \\ -\sin u & \cos u \cos i & \cos u \sin i \\ 0 & -\sin i & \cos i \end{bmatrix} \begin{bmatrix} \hat{e}_{node} \\ \hat{e}_{p-node} \\ \hat{e}_z \end{bmatrix}
\tag{11.10}
$$

With the aid of this transform, Eq. (11.8) becomes:

$$
\vec{V}_A = \omega_\oplus r \left[(\cos i) \hat{e}_S - (\sin i \cos u) \hat{e}_W \right]
\tag{11.11}
$$

In this same coordinate frame, the satellite's inertial velocity is simply:

$$
{}^I\vec{V} = \dot{r} \hat{e}_R + r\dot{v} \hat{e}_S
\tag{11.12}
$$

Combining Eqs. (11.2), (11.11), and (11.12) allows us to write an expression for the velocity vector of the satellite relative to the atmosphere

$$
{}^R\vec{V} = \dot{r} \hat{e}_R + (r\dot{v} - \omega_\oplus r \cos i) \hat{e}_S + \omega_\oplus r (\sin i \cos u) \hat{e}_W
\tag{11.13}
$$

and its magnitude:

$$^{R}V^2 = \left|^{R}\vec{V}\right|^2 = {}^{I}V^2\left[1 - \frac{2r^2\dot{v}\omega_\oplus \cos i}{{}^{I}V^2} + \frac{\omega_\oplus^2 r^2}{{}^{I}V^2}\left(\cos^2 i + \sin^2 i \cos^2 u\right)\right] \qquad \textbf{11.14}$$

For most planets, $\omega_\oplus$ is small and terms multiplied by $\omega_\oplus^2$ can be ignored. Thus, we have the approximation:

$$^{R}V^2 = {}^{I}V^2\left(1 - \frac{2r^2\dot{v}\omega_\oplus \cos i}{{}^{I}V^2}\right) + \mathcal{O}\left(\omega_\oplus^2\right)$$

$$\approx {}^{I}V^2\left(1 - \frac{2r^2\dot{v}\omega_\oplus \cos i}{{}^{I}V^2}\right) \qquad \textbf{11.15}$$

(The notation $\mathcal{O}(\ )$ has been used to denote "on the order of $(\ )$" in the equation above.) If the satellite is assumed to be in an elliptical orbit, the radius and time derivatives can be eliminated using well-known two-body solutions

$$r = \frac{a\left(1 - e^2\right)}{1 + e \cos v} \qquad \textbf{11.16}$$

$$\dot{r} = \frac{ane \sin v}{\sqrt{1 - e^2}} \qquad \textbf{11.17}$$

$$\dot{v} = \frac{na^2\sqrt{1 - e^2}}{r^2} \qquad \textbf{11.18}$$

where $n = \sqrt{\dfrac{\mu}{a^3}}$ is the mean motion. (Technically, we must also assume the satellite orbit is simple two-body motion to make these substitutions. Drag makes this assumption *somewhat* false; however, it's a good approximation for short time periods.) Leaving the vector expression intact for the moment, we can use these substitutions to rewrite Eq. (11.15) for the magnitude in terms of orbital elements:

$$^RV^2 \approx {}^IV^2\left(1-\frac{2na^2\omega_\oplus\left(1-e^2\right)^{1/2}\cos i}{{}^IV^2}\right)$$

11.19

Slightly rewriting the small term in the brackets by replacing the mean motion and introducing the vis-viva equation to "partially" replace ^IV yields:

$$\frac{2na^2\omega_\oplus\left(1-e^2\right)^{1/2}\cos i}{{}^IV^2}=\frac{2a\left(1-e\right)^{1/2}\left(1+e\right)^{1/2}r^{1/2}\omega_\oplus\cos i}{{}^IV\left(2a-r\right)^{1/2}}$$

11.20

This term will, eventually, represent the effects of atmospheric rotation on drag. Since that effect is small and the Eq. (11.19) itself is an approximation, a few more approximations won't change the results appreciably. This term is most important near periapsis (where drag effects are greatest), so it is reasonable approximate it by an "average" value based on periapsis conditions. Specifically, we can use the initial values at periapsis: $^IV_{p0}$, r_{p0}, and i_0. (Inclination rarely changes by more than a degree during a satellite's lifetime, so assuming a constant value of $i = i_0$ is easily justified for this situation (35:24).) With these substitutions, Eq. (11.20) becomes:

$$\frac{2na^2\omega_\oplus\left(1-e^2\right)^{1/2}\cos i}{{}^IV^2}\approx\frac{2r_{p0}\omega_\oplus\cos i_0}{{}^IV_{p0}}$$

11.21

Substituting this back into Eq. (11.15) gives a simple expression relating the satellite's relative and inertial velocity magnitudes:

$$^{R}V \approx {}^{I}V\left(1-\frac{2r_{p0}\omega_{\oplus}\cos i_{0}}{{}^{I}V_{p0}}\right)^{1/2}$$

11.22

With one final approximation, this can be rewritten with the help of a binomial series and truncated to two terms:

$$^{R}V \approx {}^{I}V\left(1-\frac{r_{p0}\omega_{\oplus}\cos i_{0}}{{}^{I}V_{p0}}\right) = {}^{I}V\sqrt{F}$$

11.23

where

$$F = \left(1-\frac{r_{p0}\omega_{\oplus}\cos i_{0}}{{}^{I}V_{p0}}\right)^{2}$$

11.24

has been introduced as a notational convenience (and because it matches classical results). While derived somewhat differently, this relationship matches that found by King-Hele in his classic text (35:24-25).

We can now return to the vector equations. Starting with Eq. (11.12) and eliminating r, $\dot{r}$, and $\dot{v}$ with Eqs. (11.16) - (11.18) gives the inertial velocity vector:

$$^{I}\vec{V} = \left(\frac{nae\sin v}{\sqrt{1-e^{2}}}\right)\hat{e}_{R} + \left[\frac{na(1+e\cos v)}{\sqrt{1-e^{2}}}\right]\hat{e}_{S}$$

11.25

For perfect two-body motion, the inertial velocity is along a tangent to the orbit, so Eq. (11.25) can be used to define a tangent direction:

$$\hat{e}_{T} = \frac{{}^{I}\vec{V}}{{}^{I}V}$$

11.26

(Once again, we are assuming the drag has a small effect on the satellite motion, so the orbit "almost" follows two-body dynamics.) With this definition, inertial velocity is simply:

$$^I\vec{V} = \,^IV\hat{e}_T$$

11.27

Define a new ONTW reference frame with one axis along the inertial velocity (orbit tangent) $\hat{e}_T$, another perpendicular to the orbit plane along $\hat{e}_W$ (parallel to the angular momentum vector), and third $\hat{e}_N$ in the orbit plane such that $\hat{e}_N \times \hat{e}_T = \hat{e}_W$. ($\hat{e}_N$ can be thought of as an "outward normal" to the orbit path in the orbit plane.) The ORSW frame and the ONTW frame are related by

$$\begin{bmatrix} \hat{e}_N \\ \hat{e}_T \\ \hat{e}_W \end{bmatrix} = \begin{bmatrix} \dfrac{1+e\cos v}{f} & \dfrac{-e\sin v}{f} & 0 \\ \dfrac{e\sin v}{f} & \dfrac{1+e\cos v}{f} & 0 \\ 0 & 0 & 1 \end{bmatrix} \begin{bmatrix} \hat{e}_R \\ \hat{e}_S \\ \hat{e}_W \end{bmatrix}$$

11.28

where

$$f = \left(1 + 2e\cos v + e^2\right)^{\frac{1}{2}} = \frac{^IV}{na}\left(1 - e^2\right)^{\frac{1}{2}}$$

11.29

has been introduced for convenience (24:166).

Using this rotation (and eliminating r, $\dot{r}$, and $\dot{v}$), the relative velocity vector given in Eq. (11.13) becomes

$$^R\vec{V} = \,^RV_N\hat{e}_N + \,^RV_T\hat{e}_T + \,^RV_W\hat{e}_W$$

11.30

where the components are:

$$^{R}V_{N} = \frac{ae\omega_{\oplus}\left(1-e^{2}\right)\sin\nu\cos i}{f\left(1+e\cos\nu\right)}$$

11.31

$$^{R}V_{T} = \frac{naf}{\left(1-e^{2}\right)^{\frac{1}{2}}}\left[1 - \frac{\omega_{\oplus}\left(1-e^{2}\right)^{\frac{3}{2}}\cos i}{nf^{2}}\right]$$

11.32

$$^{R}V_{W} = \frac{a\omega_{\oplus}\left(1-e^{2}\right)\sin i\cos u}{1+e\cos\nu}$$

11.33

$^{R}V_{N}$ is dominated by the small multiplier $\omega_{\oplus}e$ throughout the orbit. When the satellite is nearest to the planet, $\sin\nu$ makes it even smaller, eventually reaching zero at periapsis (when the drag effect is largest). Thus, $^{R}V_{N}$ will be neglected. $^{R}V_{W}$ is proportional to the small term $\omega_{\oplus}$ and not the (even smaller) $\omega_{\oplus}e\sin\nu$ terms, so it will kept. Our vector becomes:

$$^{R}\vec{V} \approx {}^{R}V_{T}\hat{e}_{T} + {}^{R}V_{W}\hat{e}_{W}$$

11.34

(Note that neglecting $^{R}V_{N}$ relative to $^{R}V_{T}$ also implies the drag component *in the orbital plane* is along the inertial velocity vector and tangent to the orbit.) King-Hele makes the same assumption in his classic formulation as do Vinh, et al. in their text (35:40; 58:279). An excellent derivation which retains $^{R}V_{N}$ can be found in Fitzpatrick's text (24:319-323).

We can rewrite $^{R}V_{T}$ using the definition $f = \dfrac{^{I}V}{na}\left(1-e^{2}\right)^{\frac{1}{2}}$:

$$^{R}V_{T} = \,^{I}V\left[1 - \frac{na^{2}\omega_{\oplus}\left(1-e^{2}\right)^{\frac{1}{2}}\cos i}{^{I}V^{2}}\right] \qquad \textbf{11.35}$$

Comparing this with Eq. (11.21), we can see we have already derived an appropriate approximation:

$$^{R}V_{T} = \,^{I}V\left[1 - \frac{na^{2}\omega_{\oplus}\left(1-e^{2}\right)^{\frac{1}{2}}\cos i}{^{I}V^{2}}\right] \approx \,^{I}V\left(1 - \frac{r_{p0}\,\omega_{\oplus}\cos i_{0}}{^{I}V_{p0}}\right) \qquad \textbf{11.36}$$

(At first glance, it appears Eqs. (11.23) and (11.36) say that $^{R}V = \,^{R}V_{T}$; however, they really only say $^{R}V \approx \,^{R}V_{T}$. Because $^{R}V_{W} \ll \,^{R}V_{T}$, the magnitude of ^{R}V is dominated by $^{R}V_{T}$ and $^{R}V \approx \,^{R}V_{T}$.) Substituting back into Eqs. (11.33) and (11.36) into Eq. (11.34), we can write the relative velocity vector as:

$$^{R}\vec{V} \approx \,^{I}V\left(1 - \frac{r_{p0}\,\omega_{\oplus}\cos i_{0}}{^{I}V_{p0}}\right)\hat{e}_{T} + \left[\frac{a\omega_{\oplus}\left(1-e^{2}\right)\sin i\cos u}{1+e\cos v}\right]\hat{e}_{W} \qquad \textbf{11.37}$$

It could be argued that the $\hat{e}_{W}$ component in Eq. (11.37) should be similarly evaluated near perigee like we have done with the $\hat{e}_{T}$ component; however, there is a somewhat subtle difference in the effect of such an approximation. In the $\hat{e}_{T}$ component, we made an approximation to a term of the form $\left(1-\varepsilon\right)$ where ε is small. The "1" dominates the value and the ε *almost* "lost in the noise" of the calculation. In the $\hat{e}_{W}$ component, we have a small term alone. It is not added to anything "larger," so the term itself dominates the value. Because of this, the $\hat{e}_W$ component is left "exact."

The definition of F can be inserted to simplify Eq. (11.37) (and to match classical results):

$$^R\vec{V} \approx {}^I V \sqrt{F} \hat{e}_T + \left[\frac{a\omega_\oplus \left(1-e^2\right)\sin i \cos u}{1+e\cos\nu} \right] \hat{e}_W \qquad \textbf{11.38}$$

Since $\omega_\oplus$ is small for most planets, Eq. (11.38) tells us most of the velocity relative to the atmosphere is aligned with the inertial velocity vector tangent to the orbit. Some authors choose to ignore the out of plane component altogether, while others retain it (24:319-323; 35:40; 58:274-280). We will retain it for now.

11.3 Drag Acceleration Vector

If Eq. (11.1) is divided by the mass of the satellite, we have an expression for the acceleration caused by drag:

$$\vec{a}_{drag} = \frac{\vec{D}}{m} = -B\rho \, {}^R V \, {}^R\vec{V} \qquad \textbf{11.39}$$

where

$$B = \frac{C_D S}{2m} \qquad \textbf{11.40}$$

is one of many ways of defining the ballistic coefficient (5:423-424; 49:330-333; 56:140; 65:86). (Variations include the reciprocal of this expression and twice this expression!) Substituting for $^R\vec{V}$ and $^R V$ from Eqs. (11.23) and (11.38), this becomes

$$\vec{a}_{drag} = T\hat{e}_T + W\hat{e}_W \qquad \textbf{11.41}$$

where

$$T = -BF\rho^{l}V^{2}$$

11.42

$$W = -B\sqrt{F}\rho^{l}V\left[\frac{a\omega_{\oplus}\left(1-e^{2}\right)\sin i\cos u}{1+e\cos v}\right]$$

11.43

are the components along each axis of the ONTW frame. (If $^{R}V_{N}$ had not been neglected, there would also be a component N in the $\hat{e}_{N}$ direction.)

King-Hele and Fitzpatrick treated T and W as perturbing accelerations to pure two-body orbital dynamics by using Lagrange planetary equations (24:175, 322-323; 35:31-36, 40-42). The advantage in this case is that the drag effects are seen in terms of easy-to-visualize orbital elements. The disadvantage is that the resulting equations are only valid when drag is a minor perturbation to an orbit and not during atmospheric entry.

Vinh, et al. modified the universal theory equations (Eqs. (9.29) - (9.34)) to include the effect of T by modifying the drag coefficient and the effect of W by treating it as "pseudo-lift" (58:279). The advantage is they find a set of equations bridging the gap between orbital motion and planetary entry into a rotating atmosphere. The disadvantage (albeit minor) is that the modification is not entirely intuitive. (Vinh, et al. also formulate a perturbation technique related to the Lagrange planetary equations (LPEs) in order to look at long term effects on orbits (58:280-281).)

We will follow the LPE approach in upcoming sections. But, before proceeding, it is helpful write $\vec{a}_{drag}$ in terms of the eccentric anomaly E instead of true anomaly v. In Chapter 2, several mathematical relationships between v

and E were given. They are repeated below for convenience:

$$\cos E = \frac{e + \cos v}{1 + e \cos v}$$

11.44

$$\sin E = \frac{\left(1 - e^2\right)^{1/2} \sin v}{1 + e \cos v}$$

11.45

$$\tan\left(\frac{v}{2}\right) = \left(\frac{1 + e}{1 - e}\right)^{1/2} \tan\left(\frac{E}{2}\right)$$

11.46

The orbital radius and velocity were also given in terms of E:

$$r = a\left(1 - e \cos E\right)$$

11.47

$$^I V^2 = \left(\frac{\mu}{a}\right)\left(\frac{1 + e \cos E}{1 - e \cos E}\right)$$

11.48

These relationships allow us to rewrite Eqs. (11.42) - (11.43) in terms of eccentric anomaly:

$$T = -BF\rho\left(\frac{\mu}{a}\right)\left(\frac{1 + e \cos E}{1 - e \cos E}\right)$$

11.49

$$W = -\omega_\oplus B\sqrt{F}\rho n a^2 \left(1 + e \cos E\right)^{1/2} \left(1 - e \cos E\right)^{1/2} \sin i \cos u$$

11.50

In Eq. (11.50), the positive roots are always taken because they come from simplifying the product of two positive terms, $^I V = \left(\frac{\mu}{a}\right)^{1/2}\left(\frac{1 + e \cos E}{1 - e \cos E}\right)^{1/2}$ and $\left(1 - e \cos E\right)$. Equations (11.49) and (11.50) are in the form we need to apply the LPEs.

11.4 Lagrange Planetary Equations

The Lagrange planetary equations describe changes in orbital elements when an orbit is perturbed from the nominal two-body solution. The equations are well-documented in many forms and will not be derived here (5:396-407; 20:327; 24:145-175; 42:146-148; 65:78-85, 94-98). The two we need for this chapter are:

$$\frac{da}{dt} = \frac{2}{n}\left(\frac{1+e\cos E}{1-e\cos E}\right)^{1/2} T \tag{11.51}$$

$$\frac{de}{dt} = \frac{1-e^2}{na}\left(\frac{1-e\cos E}{1+e\cos E}\right)^{1/2}\left[\frac{2\cos E}{1-e\cos E}T + \frac{\sin E}{\left(1-e^2\right)^{1/2}}N\right] \tag{11.52}$$

Previous approximations have left us with $N \approx 0$ and we can change the independent variable to eccentric anomaly by dividing Eqs. (11.51) and (11.52) by the identity:

$$\frac{dE}{dt} = \frac{n}{1-e\cos E} \tag{11.53}$$

Making these changes, the differential equations become:

$$\frac{da}{dE} = \frac{2}{n^2}\left(1+e\cos E\right)^{1/2}\left(1-e\cos E\right)^{1/2} T \tag{11.54}$$

$$\frac{de}{dE} = \frac{2\left(1-e^2\right)\cos E}{n^2 a}\left(\frac{1-e\cos E}{1+e\cos E}\right)^{1/2} T \tag{11.55}$$

Finally, T can be replaced with Eq. (11.49):

$$\frac{da}{dE} = -2a^2 BF\rho \frac{(1+e\cos E)^{3/2}}{(1-e\cos E)^{1/2}}$$

11.56

$$\frac{de}{dE} = -2aBF\rho \left(\frac{1+e\cos E}{1-e\cos E}\right)^{1/2} (1-e^2)\cos E$$

11.57

The theory is somewhat simpler to derive if we define

$$x = \beta ae$$

11.58

to replace the eccentricity. Equation (11.56) is unchanged, but Eq. (11.57) is traded for:

$$\frac{dx}{dE} = \beta a \frac{de}{dE} + \beta e \frac{da}{dE} = -2\beta a^2 BF\rho \left(\frac{1+e\cos E}{1-e\cos E}\right)^{1/2} (\cos E + e)$$

11.59

(Strictly speaking, the eccentricity in Eq. (11.59) should be replaced with $e = \dfrac{x}{a\beta}$ to complete the variable change. We will, however, leave it in for the moment.)

To put these in the (nearly) final form, we need to replace the density ρ with an expression in terms of orbital elements. If we assume a strictly exponential atmosphere, we can write

$$\rho(r) = \rho_{r_{p_0}} \exp\left[-\beta(r - r_{p_0})\right]$$

11.60

where $\rho_{r_{p_0}}$ is the density at the initial periapsis radius r_{p_0}. (This definition is consistent with that in earlier chapters. We have merely chosen to use the periapsis radius instead of the planetary radius as the reference point.) Equation

(11.47) can be used to eliminate the radius in favor of the eccentric anomaly and semimajor axis to get:

$$\rho(E) = \rho_{r_{p_0}} \exp\left\{-\beta\left[a(1-e\cos E)-r_{p_0}\right]\right\}$$

11.61

In terms of the initial semimajor axis and eccentricity

$$r_{p_0} = a_0(1-e_0)$$

11.62

so the density can also be expressed as

$$\rho(E) = \rho_{r_{p_0}} \exp\left[\beta(a_0-a_0e_0-a)+\beta ae\cos E\right]$$
$$= \rho_{r_{p_0}} \exp\left[\beta(a_0-a)-x_0\right]\exp(x\cos E)$$

11.63

where $x_0 = \beta a_0 e_0$.

Combining Eq. (11.63) into the two LPEs of concern, we have:

$$\frac{da}{dE} = -2a^2 BF\rho_{r_{p_0}} \exp\left[\beta(a_0-a)-x_0\right]\exp(x\cos E)\frac{(1+e\cos E)^{3/2}}{(1-e\cos E)^{1/2}}$$

11.64

$$\frac{dx}{dE} = -2\beta a^2 BF\rho_{r_{p_0}} \exp\left[\beta(a_0-a)-x_0\right]$$
$$\cdot\exp(x\cos E)\left(\frac{1+e\cos E}{1-e\cos E}\right)^{1/2}(\cos E+e)$$

11.65

Once solved, these two equations will describe the reduction in the semimajor axis and the orbital "circularization" due to atmospheric drag. (Collectively, the two effects will be called "orbital contraction" or "orbit contraction.") Numerically integrating Eqs. (11.64) and (11.65) is definitely possible, but very inefficient due to the long times required to see the small effects of drag on the orbit. Fortunately, the problem can be simplified to study "the trends."

11.5 Average Rates of Orbital Contraction

On the right-hand-side of Eqs. (11.64) and (11.65), the orbital elements a and e change very little in a single orbit (an interval of 2π in E). If "average" values are used for a and e (and, of course, x), then we can average over one orbit to find the average rate of change for a and x. Mathematically, these averages are

$$\left\langle \frac{da}{dE} \right\rangle = \frac{-2a^2 BF \rho_{r_{p_0}} \exp\left[\beta(a_0 - a) - x_0 \right]}{2\pi}$$
$$\cdot \int_0^{2\pi} \exp(x\cos E) \frac{(1 + e\cos E)^{3/2}}{(1 - e\cos E)^{1/2}} dE \qquad \textbf{11.66}$$

$$\left\langle \frac{dx}{dE} \right\rangle = \frac{-2\beta a^2 BF \rho_{r_{p_0}} \exp\left[\beta(a_0 - a) - x_0 \right]}{2\pi}$$
$$\cdot \int_0^{2\pi} \exp(x\cos E) \left(\frac{1 + e\cos E}{1 - e\cos E} \right)^{1/2} (\cos E + e) dE \qquad \textbf{11.67}$$

where the $\langle \ \rangle$ brackets are used to denote "average over an orbit." The integrands can be simplified for integration if they are expanded in a power series in e. For the case when $e < 0.2$, King-Hele suggested an expansion retaining terms to order e^3 (35:43). In Eq. (11.66), the expansion is

$$\frac{(1 + e\cos E)^{3/2}}{(1 - e\cos E)^{1/2}} \approx 1 + 2e\cos E + \frac{3}{4}e^2 \left[1 + \cos(2E) \right]$$
$$+ \frac{1}{4}e^3 \left[3\cos E + \cos(3E) \right] \qquad \textbf{11.68}$$

and in Eq. (11.67) it is:

$$\left(\frac{1+e\cos E}{1-e\cos E}\right)^{\!1/2}(\cos E+e)\approx\cos E+\frac{1}{2}e\left[3+\cos(2E)\right]$$
$$+\frac{1}{8}e^2\left[11\cos E+\cos(3E)\right]$$
$$+\frac{1}{16}e^3\left[7+8\cos(2E)+\cos(4E)\right] \qquad \textbf{11.69}$$

When Eq. (11.68) is substituted into Eq. (11.66), a closed-form answer is obtained

$$\left\langle\frac{da}{dE}\right\rangle\approx-2a^2BF\rho_{r_{p0}}\exp\left[\beta(a_0-a)-x_0\right]$$
$$\bullet\left[I_0+2eI_1+\frac{3}{4}e^2\left(I_0+I_2\right)+\frac{1}{4}e^3\left(3I_1+I_3\right)\right] \qquad \textbf{11.70}$$

where $I_n=I_n(x)$ is shorthand for the modified Bessel functions of the first kind and of order n (35:43-44; 56:606; 58:284). For reference, these Bessel functions are defined as:

$$I_n(x)=\frac{1}{2\pi}\int_0^{2\pi}\cos(nE)\exp(x\cos E)dE \qquad \textbf{11.71}$$

Equation (11.70) gives the average rate of change (with respect to eccentric anomaly) in the semimajor axis. Since the semimajor axis describes the size of an orbit, this could be viewed as one measure of the orbit's "shrinking" due to drag.

Similarly, Eqs. (11.67) and (11.69) give:

$$\left\langle\frac{dx}{dE}\right\rangle\approx-2\beta a^2BF\rho_{r_{p0}}\exp\left[\beta(a_0-a)-x_0\right]$$
$$\bullet\left[I_1+\frac{1}{2}e\left(3I_0+I_2\right)+\frac{1}{8}e^2\left(11I_1+I_3\right)+\frac{1}{16}e^3\left(7I_0+8I_2+I_4\right)\right] \qquad \textbf{11.72}$$

Buried in this equation, through the relationship $x = \beta ae$, is a measure of the "circularization" of the orbit due to drag.

Equations (11.70) and (11.72) have been integrated to find a and x in closed-form by King-Hele, Longuski, Vinh, and others using various formulations and approximations (35:41-73; 40:27-37; 58:284-291). The interested reader can refer to any of their texts for details. As we are only interested in some general trends, these two equations will suit our needs as written.

We can use Eqs. (11.70) and (11.72) to calculate the semimajor axis and eccentricity of an orbit perturbed by drag. Figure 11-3 illustrates the process and typical results for a few orbits are shown in Figure 11-4. The "stair step" appearance is a result of holding the orbital elements constant during each orbit. The stair steps are unnoticeable when the values are plotted for the (typically) large number of orbits in a satellite's lifetime. Several results using this technique will be given in the next section.

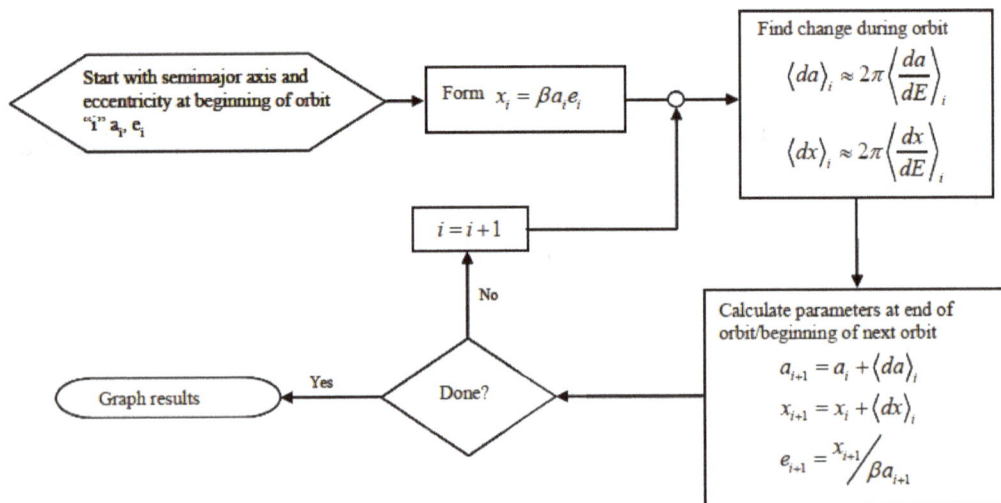

Figure 11-3: Method for Using Averaged Equations

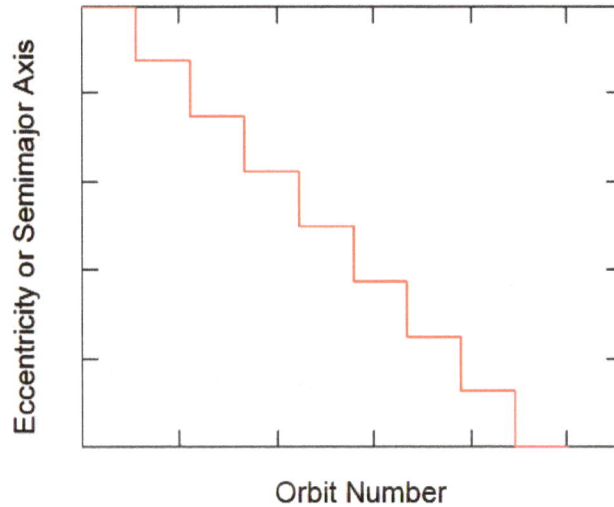

Figure 11-4: Example Results for Averaged Equations

11.6 Observations about Orbital Contraction

Even without delving into complicated solutions, we can already draw some conclusions from the results in Eqs. (11.70) and (11.72):

- Since $I_n(x) \geq 0$ for all x, Eq. (11.70) proves the semimajor axis a decreases with each orbit.
- Equation (11.72) proves $x = \beta ae$ decreases with each orbit for the same reason.
- Both a and x decrease faster when the eccentricity is larger.
- It's not immediately obvious how quickly the eccentricity decreases relative to how quickly the semimajor axis decreases.

Figure 11-5 shows typical orbital contraction profiles computed using the method described in Section 11.5 and shown in Figure 11-3. These graphs visually confirm two of the observations above; a and x decrease with time. The shapes in these plots remain surprisingly constant over a wide range of elliptical orbits with $e \leq 0.2$. Only the scales on the axes change.

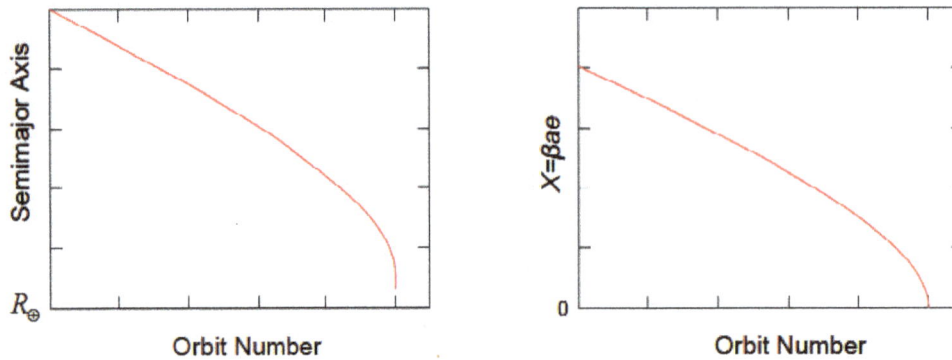

Figure 11-5: Typical Orbital Contraction Profile

Mathematically, we can demonstrate the often stated "fact" that an orbit is circularized by drag before its periapsis begins to significantly drop (56:609; 65:86; 66:85). To make the derivations a little simpler, we start by rewriting Eqs. (11.70) and (11.72) slightly as

$$\left\langle \frac{da}{dE} \right\rangle \approx -Ca^2 e^{-\beta a} \left[I_0 + 2eI_1 + \frac{3}{4}e^2 \left(I_0 + I_2 \right) + \frac{1}{4}e^3 \left(3I_1 + I_3 \right) \right] \qquad \textbf{11.73}$$

$$\left\langle \frac{dx}{dE} \right\rangle \approx -Ca^2 e^{-\beta a}$$
$$\cdot \beta \left[I_1 + \frac{1}{2}e\left(3I_0 + I_2 \right) + \frac{1}{8}e^2 \left(11I_1 + I_3 \right) + \frac{1}{16}e^3 \left(7I_0 + 8I_2 + I_4 \right) \right] \qquad \textbf{11.74}$$

where

$$C = 2BF\rho_{r_{p_0}} \exp\left[\beta a_0 - x_0 \right] \qquad \textbf{11.75}$$

has been used as shorthand for the constants. Retaining terms smaller than order

e is unnecessary to demonstrate trends, so these are further simplified to:

$$\left\langle \frac{da}{dE} \right\rangle \approx -Ca^2 e^{-\beta a}\left(I_0 + 2eI_1 \right) \qquad \textbf{11.76}$$

$$\left\langle \frac{dx}{dE} \right\rangle \approx -Ca^2 e^{-\beta a}\beta\left[I_1 + \frac{1}{2}e\left(3I_0 + I_2\right) \right] \qquad \textbf{11.77}$$

Recall the definitions of periapsis and apoapsis and write them in our current variables:

$$r_p = a(1-e) = a - \frac{x}{\beta} \qquad \textbf{11.78}$$

$$r_a = a(1+e) = a + \frac{x}{\beta} \qquad \textbf{11.79}$$

Differentiating these and assuming we can take averages over an orbit as we did in the previous section, we have:

$$\left\langle \frac{d\left(r_p\right)}{dE} \right\rangle = \left\langle \frac{da}{dE} \right\rangle - \frac{1}{\beta}\left\langle \frac{dx}{dE} \right\rangle \qquad \textbf{11.80}$$

$$\left\langle \frac{d\left(r_a\right)}{dE} \right\rangle = \left\langle \frac{da}{dE} \right\rangle + \frac{1}{\beta}\left\langle \frac{dx}{dE} \right\rangle \qquad \textbf{11.81}$$

After substituting Eqs. (11.76) and (11.77), these become:

$$\left\langle \frac{d\left(r_p\right)}{dE} \right\rangle \approx -Ca^2 e^{-\beta a}\left\{ \left(I_0 - I_1\right) + e\left[2I_1 - \frac{1}{2}\left(3I_0 + I_2\right) \right] \right\} \qquad \textbf{11.82}$$

$$\left\langle \frac{d(r_a)}{dE} \right\rangle \approx -Ca^2 e^{-\beta a} \left\{ (I_0 + I_1) + e \left[2I_1 + \frac{1}{2}(3I_0 + I_2) \right] \right\}$$

11.83

To compare the relative changes, form the ratio of the magnitudes of these rates of changes

$$\frac{\left| \left\langle \frac{d(r_p)}{dE} \right\rangle \right|}{\left| \left\langle \frac{d(r_a)}{dE} \right\rangle \right|} \approx \left| \frac{d(r_p)}{d(r_a)} \right| \approx \frac{\left| \left\{ (I_0 - I_1) + e \left[2I_1 - \frac{1}{2}(3I_0 + I_2) \right] \right\} \right|}{\left| \left\{ (I_0 + I_1) + e \left[2I_1 + \frac{1}{2}(3I_0 + I_2) \right] \right\} \right|}$$

11.84

where we have treated the derivatives as continuous and dropped the $\langle \ \rangle$ brackets on the right-hand-side. Since $I_n(x) \geq 0$, the ratio can be bounded:

$$\left| \frac{d(r_p)}{d(r_a)} \right| \approx \frac{\left| \left\{ (I_0 - I_1) + e \left[2I_1 - \frac{1}{2}(3I_0 + I_2) \right] \right\} \right|}{\left| \left\{ (I_0 + I_1) + e \left[2I_1 + \frac{1}{2}(3I_0 + I_2) \right] \right\} \right|} \leq \frac{|I_0 - I_1|}{|I_0 + I_1|} \leq 1$$

11.85

Equation (11.85) proves the apoapsis radius changes faster than the periapsis radius on a drag-perturbed orbit. This effect circularizes the orbit as the apoapsis lowers to the periapsis radius. Closer examination of Eq. (11.85) also reveals that the larger the eccentricity, the faster the apoapsis drops. Had we carried all of the terms in Eqs. (11.73) and (11.74) through to Eq. (11.85), we would have found the same trends, only with significantly more effort.

Figure 11-6 graphically illustrates the relatively dramatic difference in the rates at which periapsis and apoapsis change for both nearly circular and moderately elliptical orbits. (For reference, these were computed using an Earth-

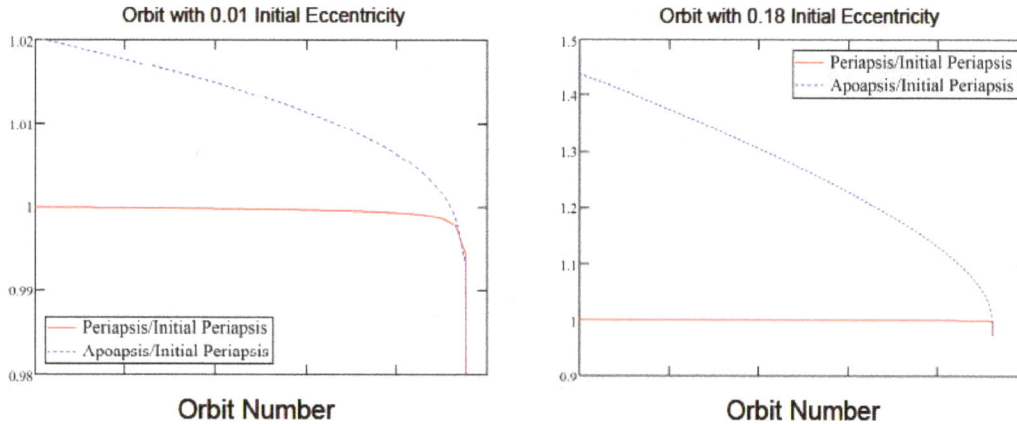

Figure 11-6: Circularization of Two Orbits

like atmosphere and the method described in Section 11.5.) Notice that for both orbits, once the periapsis and apoapsis are equal, they both drop more rapidly. This is the subject of the next few paragraphs.

If we accept the concept that an elliptical orbit tends to circularize when perturbed by drag, then we can treat the "end game" of the orbit as a decay from a circular orbit. Late in the orbit lifetime (or at least once it is circular), $x = 0$, $I_0(x) = 1$, and $I_1(x) = 0$, so Eqs. (11.73) and (11.74) are simply:

$$\frac{da}{dE} = -Ca^2 e^{-\beta a} \qquad\qquad \textbf{11.86}$$

$$\frac{dx}{dE} = 0 \qquad\qquad \textbf{11.87}$$

(Once again the derivatives have been assumed to be continuous and the $\langle\ \rangle$ brackets dropped.) The latter is easily solved, giving $x = 0$ for all values of the eccentric anomaly. This says that a circular orbit remains circular. (More precisely, it says a circular orbit remains circular as long as the drag is only a perturbation to the orbit!) The first equation requires a little more work.

For a circular orbit, the semimajor axis can be replaced in favor of the altitude h and Eq. (11.86) rewritten as

$$\frac{dh}{dE} = -C\left(R_\oplus + h\right)^2 e^{-\beta(R_\oplus + h)}$$

$$= -C_1\left(R_\oplus + h\right)^2 e^{-\beta h} \qquad \textbf{11.88}$$

where $R_\oplus$ is the planetary radius. Since, for most atmospheres of interest $h \ll R_\oplus$, we can make the approximation:

$$\frac{dh}{dE} \approx -C_1 R_\oplus^2 e^{-\beta h} = -C_2 e^{-\beta h} \qquad \textbf{11.89}$$

If the altitude is h_0 when the eccentric anomaly is E_0, this separable differential equation is easily solved to give:

$$h \approx \frac{1}{\beta} \ln\left[e^{\beta h_0} - \beta C_2\left(E - E_0\right)\right]$$

$$= \frac{1}{\beta} \ln\left[e^{\beta h_0} - C_3\left(E - E_0\right)\right] \qquad \textbf{11.90}$$

Equations (11.75), (11.88), and (11.89) can be used to evaluate the constant C_3:

$$C_3 = 2\beta BF\rho_{r_{p0}} e^{\beta h_0} R_\oplus^2$$

$$= 2\beta BF\rho_s R_\oplus^2 \qquad \textbf{11.91}$$

$\rho_s = \rho_{r_{p0}} \exp\left[-\beta\left(R_\oplus - r_{p_0}\right)\right] = \rho_{r_{p0}} \exp\left(\beta h_0\right)$ has been introduced as the density at the planet's surface. With this change, Eq. (11.90) becomes:

$$h \approx \frac{1}{\beta} \ln\left[e^{\beta h_0} - 2\beta BF \rho_s R_\oplus^2 \left(E - E_0 \right) \right]$$

11.92

Finally, to get a *rough* estimate of the time it takes to decay completely from a circular orbit, we can simply replace the eccentric anomaly in favor of time using

$$E - E_0 = \sqrt{\frac{\mu}{a^3}} \Delta t \approx \sqrt{\frac{\mu}{R_\oplus^3}} \Delta t$$

11.93

and solve for when the altitude reaches zero:

$$\Delta t \approx \frac{1}{2\beta BF \rho_s \sqrt{\mu R_\oplus}} \left(e^{\beta h_0} - 1 \right)$$

11.94

As you might expect, Eq. (11.94) shows the lifetime of a low-altitude satellite is inversely proportional to its ballistic coefficient B and the atmospheric density ρ_s. We can't control either of these very easily to extend the life of a satellite. Luckily, however, the equation also shows lifetime is directly proportional to the exponential of its initial altitude h_0. Increasing h_0 by β^{-1} will almost triple Δt.

11.7 Summary

Unlike previous chapters where we looked at trajectories dominated by drag (i.e., "atmospheric entry"), this chapter focused on two-body orbital motion *perturbed* by drag. We simplified the equations and analyzed changes to eccentricity and semimajor axis. These are the dominant effects of drag, but for completeness, we should note it is possible to look at the (lesser) effect of drag on the orbital inclination. (The rotation of the planet, through Eq. (11.50), introduces

an out-of-plane acceleration that causes the inclination to change.) Fitzpatrick
follows a very similar process to that in Sections 11.3 and 11.4 to find
perturbations to all six classical orbital elements (24:319-330).

11.8 Problems

Material Understanding:

1. Show that

$$\frac{2na^2\omega_\oplus\sqrt{1-e^2}\cos i}{{}^IV^2} \approx \frac{2r_{p0}\omega_\oplus\cos i_0}{{}^IV_{p0}}$$

   when $r \approx r_{p0}$, ${}^IV \approx {}^IV_{p0}$, and $i \approx i_0$.

2. Starting with $\vec{a}_{drag} = T\hat{e}_T + W\hat{e}_W$ in terms of true anomaly v (as given in Eqs.
   (11.41) - (11.43)), show the components can be written in terms of eccentric
   anomaly as:

$$T = -BF\rho\left(\frac{\mu}{a}\right)\left(\frac{1+e\cos E}{1-e\cos E}\right)$$

$$W = -\omega_\oplus B\sqrt{F}\rho na^2\left(1+e\cos E\right)^{\frac{1}{2}}\left(1-e\cos E\right)^{\frac{1}{2}}\sin i\cos u$$

3. Not all formulations use the periapsis radius as the reference point when
   writing an expression for the density. Show that density in a strictly
   exponential atmosphere can also be expressed in terms of semimajor axis by

$$\rho(E) = \rho_a\exp\left(\beta ae\cos E\right)$$

   where ρ_a is the density at a radius equal to the semimajor axis.

4. Equation (11.94) gives an estimate for the decay time for a low-altitude satellite. For a non-rotating atmosphere, it becomes

$$\Delta t \approx \frac{1}{2\beta BF\rho_s \sqrt{\mu R_\oplus}}\left(e^{\beta h_0}-1\right)$$

Show that increasing the initial altitude by $\frac{1}{\beta}$ almost triples Δt.

This page intentionally blank.

Chapter 12

Apollo 10 Reentry

12.1 Introduction

In the previous chapters, we've tried to stick to the "quest" to develop universal results and free ourselves from detailed knowledge of the vehicle whenever possible. This enables us to see general trends and narrow down the characteristics a vehicle might require to meet a mission goal. For example, to limit the maximum deceleration, the methods discussed all the way back in Chapter 5 can help us identify the minimum required lift-to-drag ratio, which, in turn, begins to define the vehicle. While this is an excellent approach early in the design (and is a primary goal of this book), there comes a point where we would like to examine a vehicle whose design is "fixed."

The reentry trajectory of *Apollo 10* is quite well-documented, so it makes an excellent example for us to compare methods from this introductory text to actual flight results (Refs. 3, 29, 51). (If you're wondering why I didn't select the more famous *Apollo 11* as the example, it's because a data loss during the mission prevented a detailed postflight analysis (29:2).) We'll assemble the necessary information to simulate the flight and compare the results to "reality." Then, we'll examine a few "perceived" improvements to our simulations, introducing a few ways to analyze results along the way.

12.2 Actual Vehicle and Entry Characteristics

You might expect assembling the necessary entry interface (EI) conditions (γ_e, etc.), control laws (σ, etc.), and vehicle parameters (C_D, etc.) would be relatively easy given the plethora of data that was collected with Apollo. Unfortunately, it requires sorting through multiple reports which are lengthy and sometimes conflicting. We'll start with the easiest information to find and identify any conflicts and estimates as we complete the task.

At entry interface, *Apollo 10*'s telemetry recorded the values in Table 12-1 (51:3). (As we'll discuss later, two different EI times were given in NASA's own postflight analysis report. The values given in the table reflect the time prominently given in the front of the report.) The entries are shown with the same number of significant digits cited in the NASA source document but it's unlikely they are as accurate as they appear. (We'll maintain a similar number of significant figures for consistency.)

Table 12-1: Actual *Apollo 10* States at Entry Interface (51:3)

State	Value
Inertial Velocity, ^IV_e	11.06715 *km/s*
Inertial Flight-Path Angle, $^I\gamma_e$	−6.6198381°
Inertial Azimuth, $^I\beta_e = 90° - {}^I\psi_e$	71.9317°
Longitude, θ_e	174.24384° East
Geodetic Latitude, ϕ_{gd_e}	23.653003° South
Geodetic Altitude, h_{gd_e}	123.55077 *km*

The bank angle σ was varied during entry to target the desired landing point as well as to manage the deceleration. Rather than trying to model the control logic for the capsule, we will simply treat the bank angle as a known function of time. The actual bank angle history for *Apollo 10*'s entry is tabulated in Reference 51 and Figure 12-1 presents it graphically (and assumes instantaneous changes). The drogue chutes deployed 498 seconds after entry so $\sigma = 0°$ has been added to the plot at that point and beyond (51:24).

The magnitude of the lift was not employed as a control parameter during Apollo entries, with one possible exception. Rolling the capsule about the velocity vector (i.e., rotating the bank angle through 360°) could be used to effectively cancel out lift. Such a maneuver could be used to "stay the course" when the capsule's trajectory is already "on target." In the case of *Apollo 10*, however, Figure 12-1 seems to indicate this was not utilized.

Not controlling the lift magnitude doesn't mean the lift-to-drag ratio remains a constant. For real vehicles, it varies with the Mach number. Figure 12-2 shows the actual C_L/C_D values recorded during *Apollo 10*'s entry as well as preflight predictions for it and *Apollo 11* (29:4; 51:36). The data sets are similar

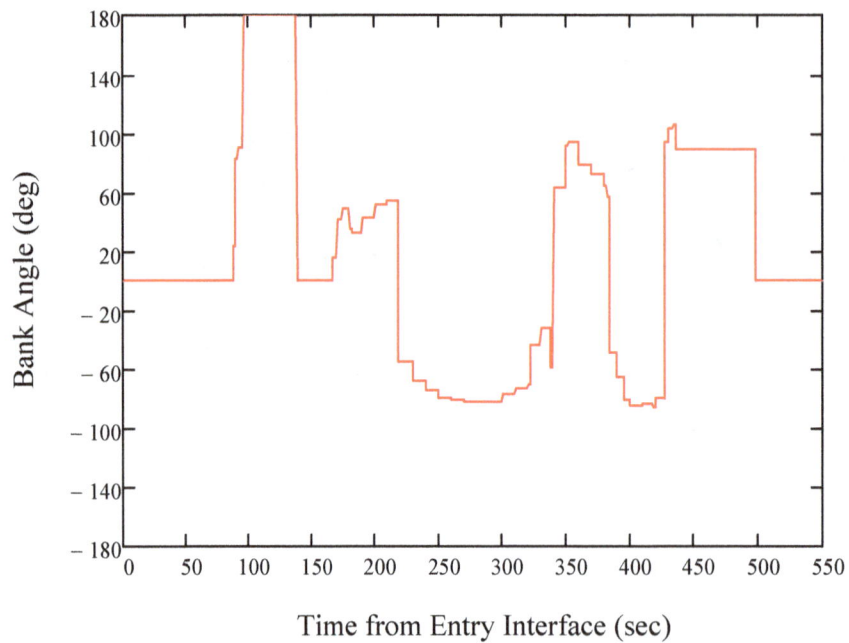

Figure 12-1: Bank Angle History for Apollo 10 Command Module (51:16)

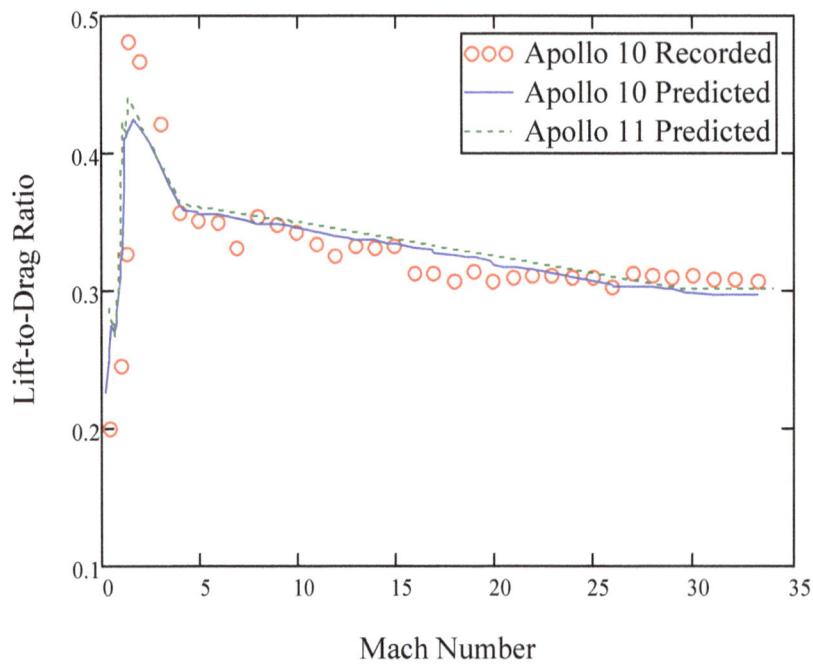

Figure 12-2: Lift-to-Drag Ratio as a Function of Mach Number (29:4; 51:36)

across the range of Mach numbers and all three sets are fairly "flat" (nearly constant) above Mach 25, with the recorded (actual fight) values remaining flat down to about Mach 15.

NASA's postflight analysis for *Apollo 10* did not document separate values for C_L and C_D (51). Without these values, we won't be able to integrate the equations of motion. Fortunately, the preflight estimates for *Apollo 11* are documented (29:4). These two capsules were virtually identical aerodynamically, so we will use the *Apollo 11* values shown in Table 12-2.

Finally, we need a couple of constants for the capsule. The reference area S upon which C_L and C_D are based is the maximum cross section for capsules (21:2-3). Using the design diameter of 154 *in*, we can calculate $S = 12.017\ m^2$ (4:84). The estimated pre-entry mass for *Apollo 10* was $m=5498.22\ kg$ (51:11).

Table 12-2: Aerodynamic Coefficients for the *Apollo 11* Command Module (29:4)

Mach Number	C_L	C_D	C_L/C_D
0.4	0.24465	0.853	0.286811
0.7	0.26325	0.98542	0.267145
0.9	0.32074	1.10652	0.289864
1.1	0.49373	1.1697	0.4221
1.2	0.47853	1.156	0.413953
1.35	0.56282	1.2788	0.440116
1.65	0.55002	1.2657	0.434558
2	0.53247	1.2721	0.418576
2.4	0.5074	1.2412	0.408798
3	0.47883	1.2167	0.393548
4	0.44147	1.2148	0.36341
10	0.42856	1.2246	0.349959
≥29.5	0.38773	1.2891	0.300776

12.3 Setting up the Simulation

12.3.1 Equations of Motion

We have almost all of the information necessary to integrate the dimensional equations of motion we found in Section 3.4 (Eqs. (3.35)-(3.37) and (3.65)-(3.67)). Only a few easily obtained tidbits remain, such as determining the alignment of the inertial and Earth-fixed coordinate frames at the time of entry and setting a value for Earth's rotation rate. For our immediate purposes, we won't need these, however.

We can simplify the dimensional equations of motion by first ignoring the Earth's rotation

$$\dot{r} = {}^{R}V \sin \gamma \qquad\qquad \textbf{12.1}$$

$$\dot{\theta} = \frac{{}^{R}V \cos \gamma \cos \psi}{r \cos \phi} \qquad\qquad \textbf{12.2}$$

$$\dot{\phi} = \frac{{}^{R}V \cos \gamma \sin \psi}{r} \qquad\qquad \textbf{12.3}$$

$$ {}^{R}\dot{V} = -\frac{D}{m} - g \sin \gamma \qquad\qquad \textbf{12.4}$$

$$ {}^{R}V\dot{\gamma} = \frac{L}{m}\cos\sigma - g\cos\gamma + \frac{{}^{R}V^{2}}{r}\cos\gamma \qquad\qquad \textbf{12.5}$$

$$ {}^{R}V\dot{\psi} = \frac{L\sin\sigma}{m\cos\gamma} - \frac{{}^{R}V^{2}}{r}\cos\gamma\cos\psi\tan\phi \qquad\qquad \textbf{12.6}$$

and then making our usual assumptions about gravity, density, lift, and drag:

$$g = g(r) = g_s \left(\frac{R_\oplus}{r} \right)^2$$ **12.7**

$$\rho = \rho_s e^{-\beta(r - R_\oplus)}$$ **12.8**

$$L = \frac{\rho C_L S_R}{2} V^2$$ **12.9**

$$D = \frac{\rho C_D S_R}{2} V^2$$ **12.10**

Implicit in Eq. (12.7) is the fact we've used the Earth's radius as the reference radius ($r_0 = R_\oplus$). The physical constants needed for Eqs. (12.7) and (12.8) are given in Table 12-3. With the proper initial conditions, we can integrate Eqs. (12.1) - (12.6) and compare the results to the actual reentry.

Table 12-3: Physical Properties of the Earth

Radius of the Earth, $R_\oplus$	6378.137 km
Acceleration of gravity at the Earth's surface, g_s	9.81 m/s^2
Atmospheric density at the Earth's surface, ρ_s	1.225 kg/m^3
Inverse scale height, β	0.14 km^{-1}

12.3.2 Conditions at Entry Interface

In addition to the longitude θ, NASA gave the entry position in terms of "geodetic" altitude h_{gd} and latitude ϕ_{gd} instead of the "geocentric" values r and ϕ we've been using throughout this book. Their relationship is shown in Figure 12-3. The radius r is easily found by evaluating the following equation:

$$r^2 = \left(\frac{a_e}{\sqrt{1-e^2 \sin^2 \phi_{gd}}} + h_{gd} \right)^2 \cos^2 \phi_{gd} + \left(\frac{a_e(1-e^2)}{\sqrt{1-e^2 \sin^2 \phi_{gd}}} + h_{gd} \right)^2 \sin^2 \phi_{gd} \qquad \textbf{12.11}$$

(This equation and others relating geodetic and geocentric values are readily found in other references, including References 5 and 68.) For Earth, $a_e = 6378.137\ km$ and $e = 0.08181919$ (5:98; 68:387). Substituting these values and those from Table 12-1, we can calculate the entry radius:

$$r_e = 6498.270\ km$$

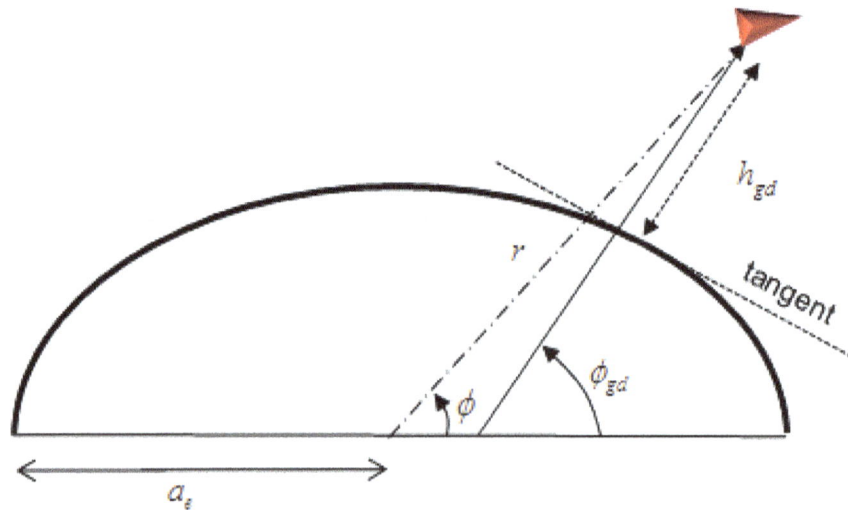

Figure 12-3: Geocentric and Geodetic Latitude (Exaggerated Geometry)

Geodetic and geocentric latitudes are related by the following:

$$\tan\phi = \frac{\left(\dfrac{a_e\left(1-e^2\right)}{\sqrt{1-e^2\sin^2\phi_{gd}}} + h_{gd}\right)}{\left(\dfrac{a_e}{\sqrt{1-e^2\sin^2\phi_{gd}}} + h_{gd}\right)}\tan\phi_{gd} \qquad \textbf{12.12}$$

When evaluated at EI, we get a geocentric latitude of:

$$\phi_e = 23.51457° \text{ South}$$

Combined with the entry longitude value already given in Table 12-1, these values define the position vector at entry.

Consistent with ignoring Earth's rotation in the equations of motion, the values for the inertial entry velocity, flight-path angle, and heading angle in Table 12-1 can be assumed to be Earth-relative:

$$^{R}V_e = {}^{I}V_e = 11.06715 \; km/s$$
$$\gamma_e = {}^{I}\gamma_e = -6.6198381°$$
$$\psi_e = {}^{I}\psi_e = 90° - {}^{I}\beta_e = 18.0683°$$

These three values define the magnitude and direction of the capsule's velocity at entry (i.e., the velocity vector).

Table 12-4 summarizes the entry states (initial conditions) for integrating Eqs. (12.1) – (12.6).

Table 12-4: *Apollo 10* States at Entry Interface

State	Value
Radius, r_e	6498.270 *km*
Flight-Path Angle, γ_e	$-6.6198381°$
Heading Angle, ψ_e	$18.0683°$
Velocity, $^R V_e$	11.06715 *km/s*
Longitude, θ_e	$174.24384°$ East
Latitude (Geocentric), ϕ_e	$23.51457°$ South

12.3.3 Aerodynamic Coefficients During Entry

As we saw in Figure 12-2, the lift-to-drag ratio varies with Mach number. On-the-other-hand, the figure also shows the ratio is fairly constant to down to at least Mach 15 for *Apollo 10*. In fact, an "averaged" fit to the entire graph might be a horizontal line at, more or less, the Mach 15 value. A reasonable first approach is to assume constant values for C_L and C_D. Averaging the top two entries in Table 12-2 gives us

$$C_L = 0.40815$$
$$C_D = 1.2569$$

to use as constant values.

When we (finally) plot our results, we'll see that these constants give excellent results. If they did not, one alternative is to calculate the Mach number (based on the speed and altitude) as we integrate and continuously adjust C_L and C_D by interpolating Table 12-2. (We'll cover this in Section 12.5.2 when we examine improvements to our simulation.) Another option is to simply select C_L and C_D values that give the best results while still being realistic for the problem at hand.

12.4 Comparing the Simulation to Reality

The previous sections established the equations of motion (Eqs. (12.1) - (12.6)) and corresponding initial conditions (Table 12-4) for the *Apollo 10* entry. Figure 12-1 provides the bank angle "program" and the remaining constants in the equations (mass, reference area, C_L, and C_D) have been established. At this point, Eqs. (12.1) - (12.6) can be numerically integrated quickly (and quite easily) using a wide range of software programs.[1]

After integrating to find the trajectory, we need to convert the radius r and (geocentric) latitude ϕ values we find to geodetic altitude h_{gd} and latitude ϕ_{gd} to compare with the actual trajectory published by NASA. (It is easier to convert our results than the other way around. We know our solution completely and can simply apply the conversion before graphing the results.) The conversion is as simple as simultaneously solving Eqs. (12.11) and (12.12) for h_{gd} and ϕ_{gd} for each point on our trajectory. Figure 12-4 shows the small difference between geocentric altitude

$$h = r - R_{\oplus}$$
12.13

and the geodetic altitude for this problem. Figure 12-5 illustrates the similarly small difference in latitude values.

NASA's postflight analysis report presents most of the information in the form of graphs, many of which cannot be read to more than a few significant figures (51). Except where specific values are called out in the text of the report, any comparisons we make are only as accurate as our ability to read the graphs.

---

[1] Results in this chapter were found by integrating Eqs. (12.1) - (12.6) with PTC's *Mathcad*, Version 14.

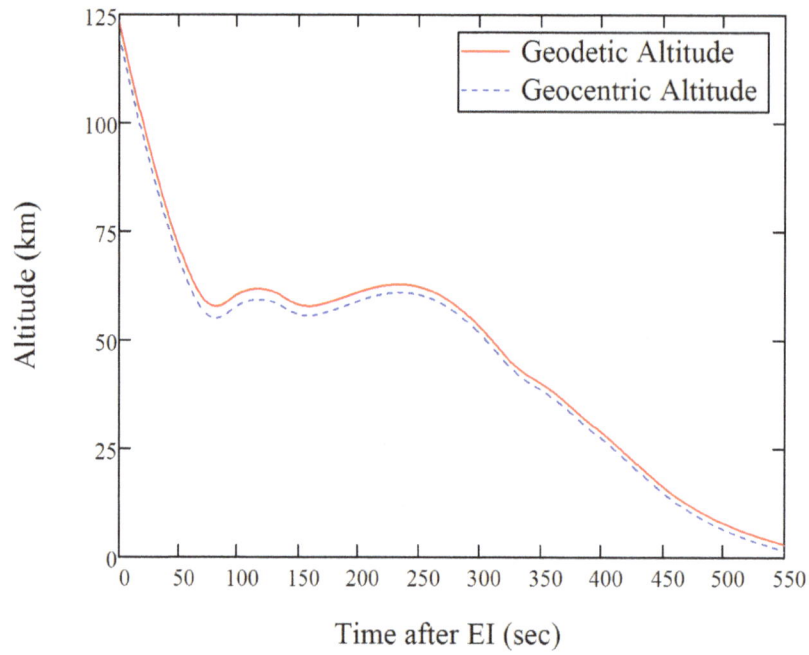

Figure 12-4: Simulation Results for Geodetic and Geocentric Altitude

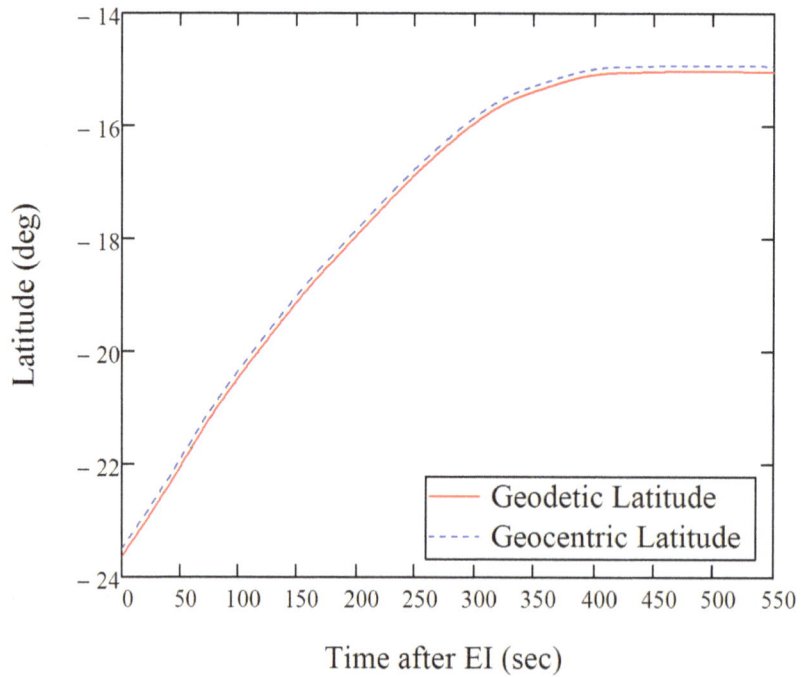

Figure 12-5: Simulation Results for Geodetic and Geocentric Latitude

Further, the "true" values derived from telemetry are plotted as discreet points in the report. For simplicity, values from NASA's graphs have been manually digitized and reconstructed here as continuous curves. Some error is inevitable in the process but is negligible for our purposes. Having made the appropriate disclaimers about accuracy, we can begin to compare simulation results and the actual values recorded by NASA.

Figure 12-6 and Figure 12-7 compare the geodetic altitude and inertial velocity, respectively. (Recall, by ignoring Earth's rotation, we've assumed $^{R}V = {}^{I}V$ throughout the entry.) In lieu of the flight-path angle, NASA computed the altitude (or sink) rate $\dot{h}$. A comparable value from the simulation is easily computed:

$$\dot{h} = {}^{R}V \sin \gamma \qquad \qquad \textbf{12.14}$$

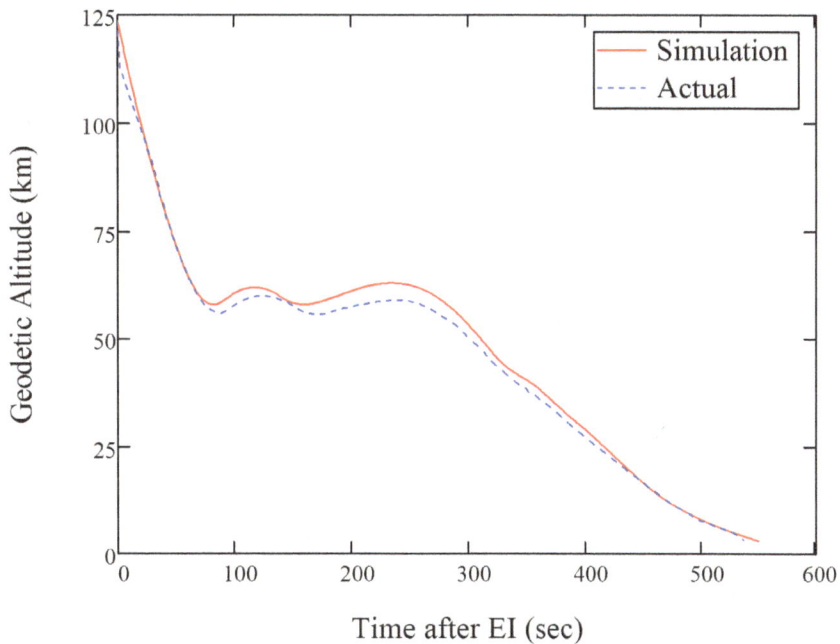

Figure 12-6: Geodetic Altitude Comparison for Apollo 10 Reentry

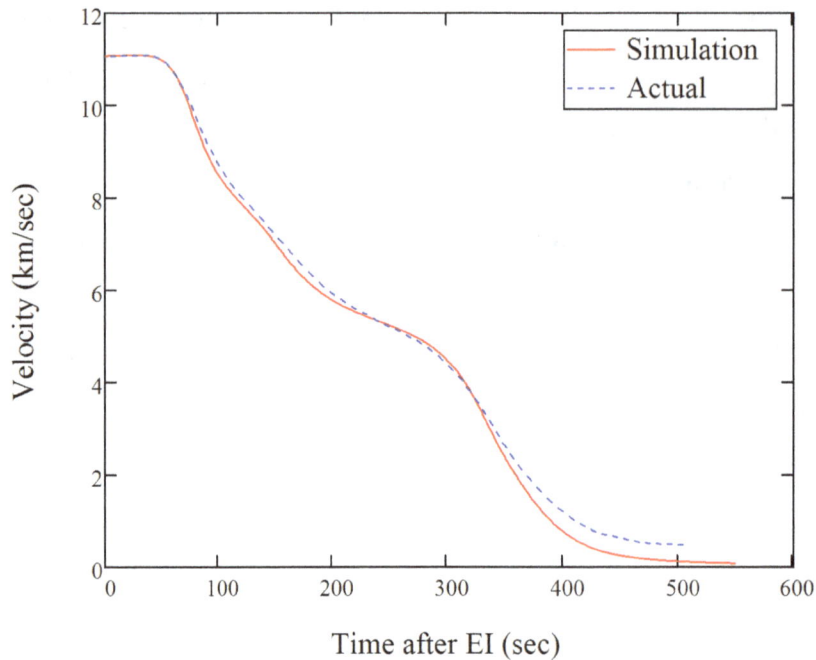

Figure 12-7: Inertial Velocity Comparison for Apollo 10 Reentry

Figure 12-8 compares the altitude rates. Finally, the deceleration profiles are compared in Figure 12-9, where

$$a_{decel} \approx -{}^R\dot{V} = \frac{\rho C_D S}{2m}{}^R V^2 + g\sin\gamma \qquad \textbf{12.15}$$

has been used to calculate an *approximate* deceleration from our simulation.

Visually, the simulated and actual trajectories are very similar, following the same trends and matching closely in magnitudes. A few specific values are called out in NASA's postflight analysis report ; however, the report contains conflicting information. It cites two different times (differing by 1.2 seconds) for the point of entry interface (51:3, 15). Events throughout it are referenced relative to the EI or to the mission clock (starting at launch). Sometimes the same event is

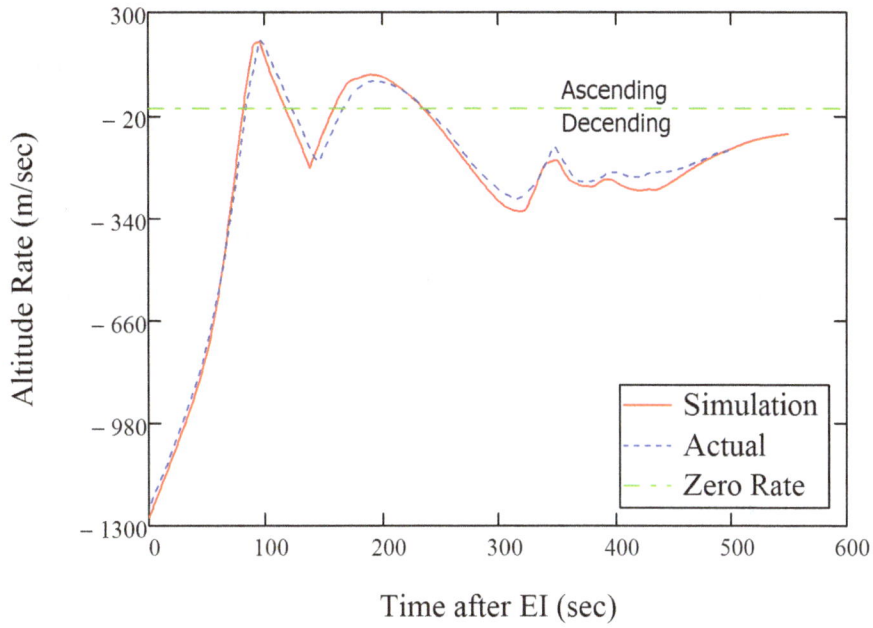

Figure 12-8: Altitude Rate Comparison for Apollo 10 Reentry

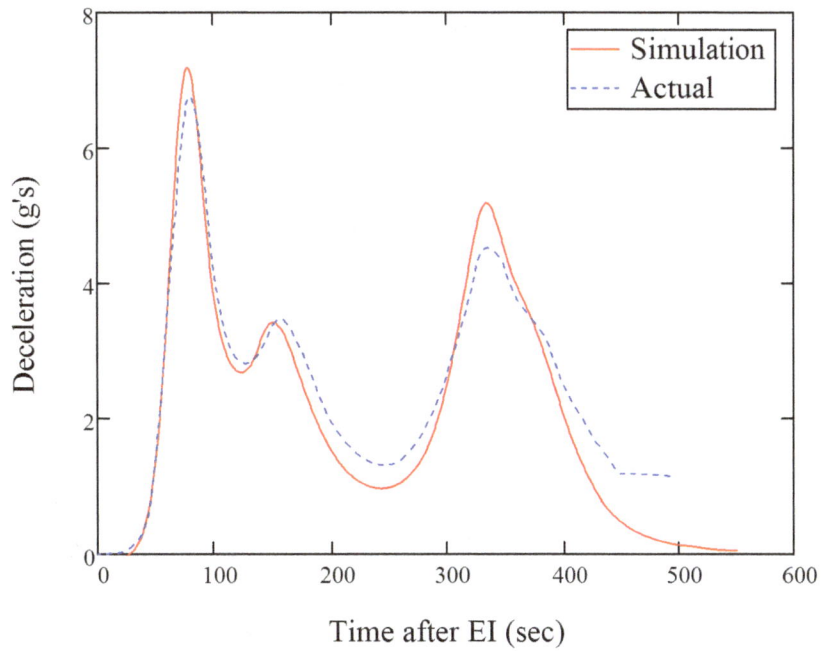

Figure 12-9: Deceleration Comparison for Apollo 10 Reentry

referenced inconsistently to both, so we must accept that either time could be correct and make the appropriate allowances when comparing our simulation results. (To further complicate the comparison, at least two numerical entries in NASA's report have an unexplained 1.4 second discrepancy!)

Table 12-5 compares the geocentric radius at four event times called out in the report. When different times are cited, the times and simulation results are shown as a range. Figure 12-6 and Table 12-5 show the simulation closely matches the actual altitude throughout most of the flight.

Table 12-6 compares the inertial velocity at seven event times cited in the postflight analysis report. Figure 12-7 and Table 12-6 both show the simulation agrees with reality quite well, with the biggest differences occurring near the end of the flight.

A few key deceleration events can also be compared. Specifically, the largest, second largest, and first minimum deceleration values are called out in the postflight report. Table 12-7 compares the telemetry-derived values with our simulation results. Again, the graphic (Figure 12-9) and table confirm fairly good agreement, with the simulation providing slightly "optimistic" lows and "pessimistic" highs.

Table 12-5: Geocentric Radius at Specific Event Times (51:10, 17-18)

Time after EI (sec)	Event	Geocentric Radius (km)	
		Actual	Simulation
80.8 - 81.4	Max. deceleration	6431.159	6433.133 - 6433.129
128.8 - 129.4	First min. deceleration	6435.110	6436.734 - 6436.672
136.8	Final guidance phase	6434.462	6435.636
436.8 - 438.0	Guidance termination	6395.657	6396.021 - 6395.721

Table 12-6: Inertial Velocity at Specific Event Times (51:10, 15, 23)

Time after EI (sec)	Event	Inertial Velocity (km/sec)	
		Actual	Simulation
27.4	First encounters 0.05g	11.0929	11.0886
28.0	Encounters 0.043g	11.0926	11.0886
30.0	Zero bank angle set	11.0935	11.0882
76.8 - 78.0	Initiate "Huntest" guidance	10.0048	9.81044 - 9.72584
128.0	Time of interest	7.89310	7.68931
136.8	Final guidance phase	7.62997	7.44351
436.8 - 438.0	Guidance termination	0.70308	0.320477 - 0.312815

Table 12-7: Deceleration Comparison at Specific Events (51:12)

Deceleration Event	Magnitude (g's)	
	Actual	Simulation
Largest	6.76	7.19
Second largest	4.60	5.21
First minimum	2.80	2.70

A final comparison we can make is the touchdown point. *Apollo 10* landed at 15.07° South (geodetic) latitude and 164.65° West longitude (51:25). Our simulation does not model the trajectory after the drogues deploy, but we can approximate the touchdown point by assuming it is directly below where they deploy. At this time (EI+498 seconds), the simulation results are 15.06° South (geodetic) latitude and 163.67° West longitude. The difference of roughly 110 *km* is certainly acceptable given the numerous simplifications we've used.

12.5 Where to Improve the Simulation?

Even though the results just presented are "good enough" for most purposes (including estimating maximum deceleration, range, and flight time), the engineers among us will always want to push for the maximum accuracy. Several relatively "obvious" enhancements are:

- Include the Earth's rotation
- Include effect of Mach number on lift and drag
- Improve the atmosphere model
- Improve the gravity model

However, it's not as simple as just throwing additional terms into the numerical integration. Assumptions and approximations that went into deriving Eqs. (12.1) - (12.6) tend to "balance out" and removing one or two assumptions may not improve the overall results. To demonstrate, we'll look at each of these four enhancements individually.

12.5.1 Including Earth's Rotation

Consider adding the rotation of the Earth back into the dynamics. Equations (12.4) – (12.6) would be replaced by

$$^R\dot{V} = -\frac{D}{m} - g\sin\gamma + r\omega_\oplus^2\cos\phi\left(\cos\phi\sin\gamma - \sin\phi\sin\psi\cos\gamma\right) \qquad \textbf{12.16}$$

$$^RV\dot{\gamma} = \frac{L}{m}\cos\sigma - g\cos\gamma + \frac{^RV^2}{r}\cos\gamma + 2\,^RV\omega_\oplus\cos\phi\cos\psi$$
$$+ r\omega_\oplus^2\cos\phi\left(\cos\phi\cos\gamma + \sin\phi\sin\psi\sin\gamma\right) \qquad \textbf{12.17}$$

$$
{}^{R}V\dot{\psi} = \frac{L\sin\sigma}{m\cos\gamma} - \frac{{}^{R}V^{2}}{r}\cos\gamma\cos\psi\tan\phi + 2{}^{R}V\omega_{\oplus}\left(\sin\psi\cos\phi\tan\gamma - \sin\phi\right)
$$
$$
- \frac{r\omega_{\oplus}^{2}}{\cos\gamma}\sin\phi\cos\phi\cos\psi \qquad\qquad \textbf{12.18}
$$

where $\omega_{\oplus}$ is Earth's rotation rate. To use these, we'll need to convert the inertial values we are given to values relative to the rotating Earth. In addition to ${}^{R}V$, we can't forget that γ and ψ in the equations of motion are actually *relative* quantities since they are measured with respect to the relative velocity vector ${}^{R}\overline{V}$. (This "subtlety" was called out in Section 3.5 as a warning for this very example.)

Using a geocentric-equatorial coordinate system aligned with the prime meridian at EI as our reference, we can use the values in Table 12-1 (and the conversions in Eqs. (12.11) and (12.12)) to can calculate the inertial position and velocity vectors. These are:

$$
\vec{r}_{e} = \begin{pmatrix} -5928.60 \\ 597.622 \\ -2592.694 \end{pmatrix} km
$$

$$
{}^{I}\vec{V}_{e} = \begin{pmatrix} -1.23773 \\ -10.3795 \\ 3.63549 \end{pmatrix} km/s
$$

(Once again, these numbers are shown with more significant figures than we can reasonably expect to be valid.) The corresponding relative velocity and inertial angles are easily computed:

$$
\left|{}^{R}\vec{V}_{e}\right| = \left|{}^{I}\vec{V}_{e} - \vec{\omega}_{\oplus}\times\vec{r}_{e}\right| = 10.6589 \ km
$$

$$\gamma_e = 90^0 - \cos^{-1}\left(\frac{{}^R\vec{V}_e \bullet \vec{r}_e}{\left|{}^R\vec{V}\right|\left|\vec{r}\right|}\right) = \text{-}6.87459^\circ$$

$$\psi_e = 90^0 - \cos^{-1}\left(\frac{{}^R\vec{V}_e \bullet \hat{e}_z}{\left|{}^R\vec{V}_e\right|}\right) = 19.9425^\circ$$

We can use these as initial conditions to integrate the equations of motion. After integrating, we must assemble the (relative) position and velocity vectors and then transform them to the inertial frame to compare to our previous simulation and to "reality." The simple (but cumbersome) transforms make good use of the matrix rotations derived in Chapter 3 and will not be detailed here.

Figure 12-10 compares the geodetic altitude for the previous (non-rotating Earth) simulation with this one and *Apollo 10*'s actual trajectory. Graphically, the results appear comparable with this simulation appearing to give slightly better results between about 90 and 400 seconds. To help compare overall "goodness" of the simulations, we can use the root mean square (RMS) error between simulated and actual altitudes:

$$RMS_{h_{gd}} = \sqrt{\frac{\sum\limits_{i=1}^{n}\left[\left(h_{gd}\left(t_i\right)\right)_{simulation} - \left(h_{gd}\left(t_i\right)\right)_{actual}\right]^2}{n}} \qquad \textbf{12.19}$$

(The debate of using $n-1$ or n in the denominator of Eq. (12.19) is pointless; the difference is insignificant if we use enough points!) The RMS is comparable to the standard deviation of the error along the entire trajectory. When compared at one

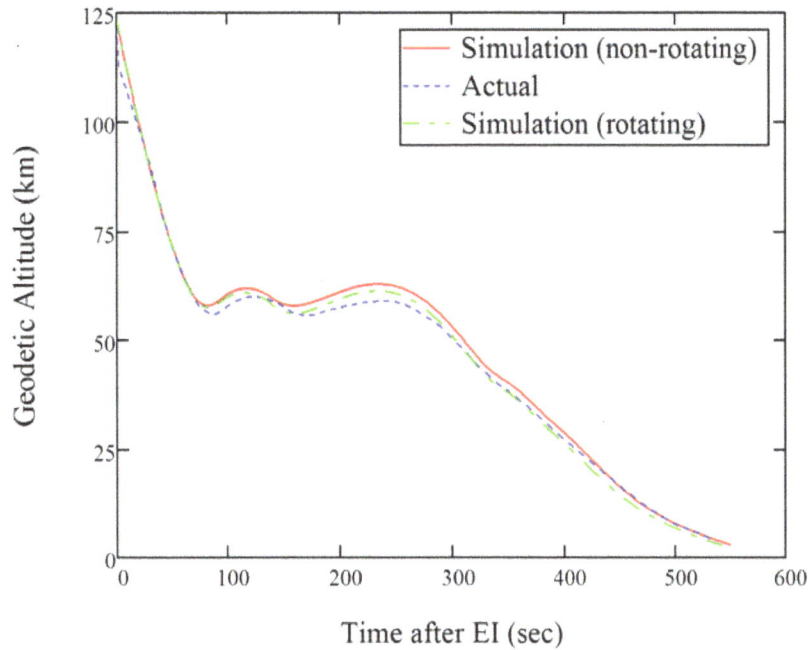

Figure 12-10: Geodetic Altitude for Simulations with and without Rotation

second intervals between entry and drogue deployment, the RMS can be written

as

$$RMS_{h_{gd}} = \sqrt{\frac{\sum_{t=0}^{498}\left[\left(h_{gd}(t)\right)_{simulation} - \left(h_{gd}(t)\right)_{actual}\right]^2}{499}}$$

12.20

or, as

$$RMS_{h_{gd}} = \sqrt{\frac{\sum_{t=0}^{498}\left[\left(h_{gd}\right)_{simulation} - \left(h_{gd}\right)_{actual}\right]^2}{499}}$$

12.21

for simplicity. Using this measure, $RMS_{h_{gd}} = 3.21 \ km$ when Earth's rotation is

included and $RMS_{h_{gd}} = 3.63 \ km$ when isn't. For altitude, it appears we have

marginally improved the simulation accuracy *on average* along the trajectory.

Figure 12-11 provides a similar comparison for the inertial velocity. In this plot, it's more difficult to decide which simulation provides better results. Between 110 and 350 seconds, the first simulation gives better results while this one better matches reality after 350 seconds. (The velocity prediction at drogue deployment is near perfect.) Evaluating the velocity RMS for each simulation

$$RMS_{I_V} = \sqrt{\frac{\sum_{t=0}^{498} \left[\left({}^I V \right)_{simulation} - \left({}^I V \right)_{actual} \right]^2}{499}}$$

12.22

we find $RMS_{I_V} = 187 \ m/s$ with rotation included and $RMS_{I_V} = 241 \ m/s$ without rotation. For inertial velocity, overall accuracy appears to be improved when the Earth's rotation is modeled.

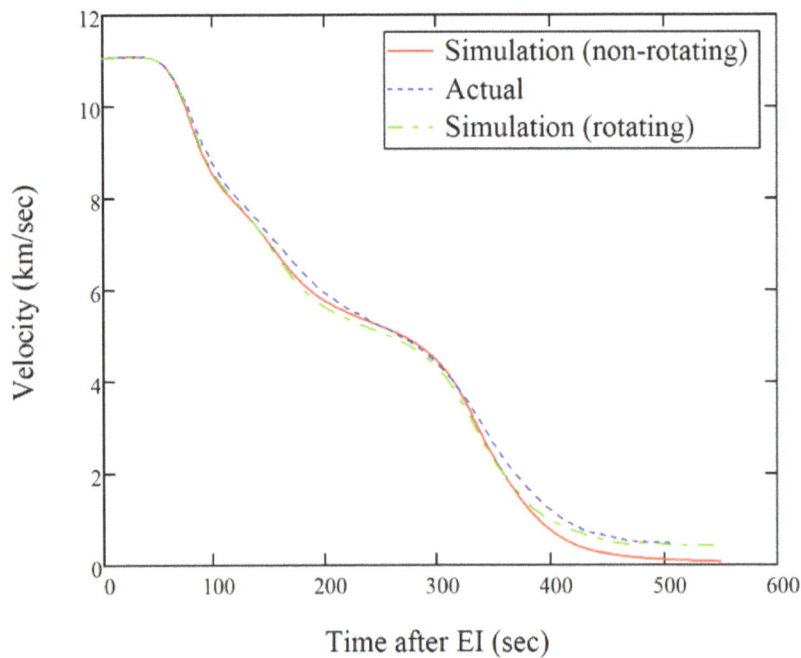

Figure 12-11: Inertial Velocity for Simulations with and without Rotation

Deceleration results are shown in Figure 12-12. As before,

$$a_{decel} \approx -{}^{I}\dot{V}$$

12.23

was used to estimate the deceleration from the simulation results. (The oscillations in the rotating Earth values after about 450 seconds are a numerical artifact caused by round-off error as the velocity approaches a constant.) Both simulations *tend* to overestimate deceleration maximums and underestimate minimums. At two of the three peaks, this simulation produces better estimates but is worse at the third. The RMS

$$RMS_{decel} = \sqrt{\frac{\sum_{t=0}^{498}\left[\left({}^{I}V\right)_{simulation} - \left({}^{I}V\right)_{actual}\right]^2}{499}}$$

12.24

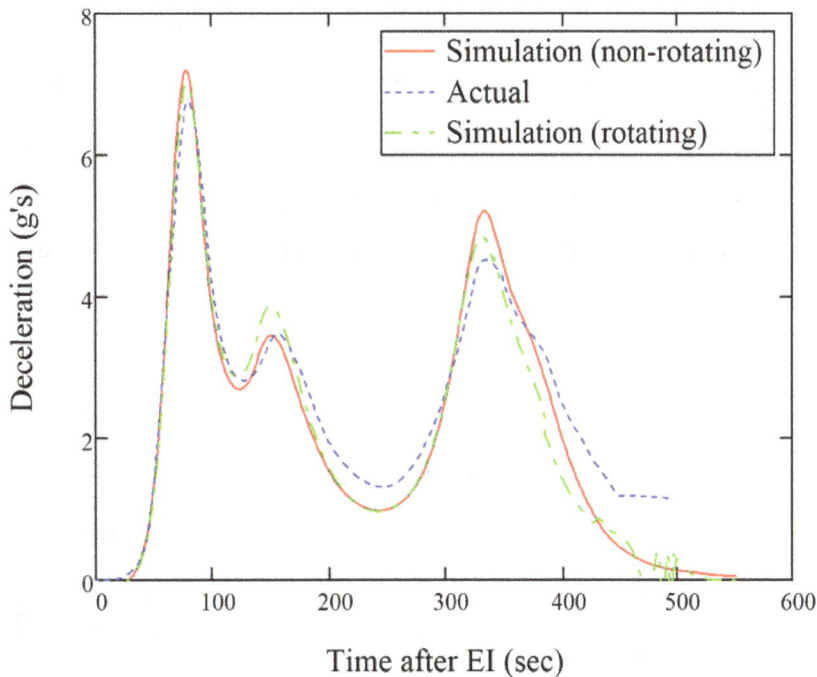

Figure 12-12: Deceleration Comparison for Simulations with and without Rotation

suggests this simulation produces slightly *worse* results overall, with a $RMS_{decel} = 5.40 \ m/s^2$ for it with a $RMS_{decel} = 4.60 \ m/s^2$ when we ignore the Earth's rotation! Most of this difference, however, comes in the last 125 seconds. The RMS values are much closer at $2.94 \ m/s^2$ (with rotation) and $3.21 \ m/s^2$ (without rotation) during the first 375 seconds. Neither simulation is a clear "winner" when it comes to estimating decelerations.

Finally, we can compare the estimates of the touchdown point. Table 12-8 summarizes the results. Both simulations predict the geodetic latitude 0.01° north of the actual location. The non-rotating simulation predicts a longitude 0.98° east of the actual location while the rotating simulation predicts one 0.97° west of the actual location. At first glance, we could conclude these show equal accuracy, but recall we assumed the touchdown point for our simulations was immediately below where the drogues deployed. The capsule was traveling eastward when the drogues deployed and likely continued along that path until the main parachutes deployed (Figure 12-13). As a result, the touchdown point would be more eastward than we've assumed, making the non-rotating simulation results worse and the rotating simulation better. Without knowing the actual latitude and longitude when the drogues deployed, we cannot quantify the results any more accurately.

Table 12-8: Touchdown Point Comparison for *Apollo 10*

	Latitude (geodetic)	Longitude
Actual	15.07° S	164.65° W
Non-rotating simulation	15.06° S	163.67° W
Rotating simulation	15.06° S	165.62° W

Figure 12-13: Apollo 10 Descending on its Main Parachutes

With the possible exception of deceleration predictions, adding Earth's rotation to the simulation does appear to improve the overall accuracy (as measured by the root mean square). However, the differences in RMS for the simulations are only $420\ m$ and $54\ m/s$ for geodetic altitude and inertial velocity, respectively! The dramatic increase in complexity (especially converting between rotating reference frames) may not justify the relatively small improvements in accuracy for many situations.

12.5.2 Including the Effect of Mach Number on Lift and Drag

Before we can incorporate C_L and C_D as functions of Mach number M in our simulation, we must calculate the Mach number from the velocity. For an ideal gas, the speed of sound is given by

$$c = \sqrt{k\mathcal{R}\mathcal{T}}$$ **12.25**

where k is the specific heat ratio, $\mathcal{R}$ is the gas constant per unit weight, and $\mathcal{T}$ is the absolute temperature of the atmosphere (52:38). For Earth,

$$c \approx 20.0468\sqrt{\mathcal{T}} \; m/s$$ **12.26**

when $\mathcal{T}$ is in degrees Kelvin (32:C29). Thus, the Mach number is given by

$$M = \frac{{}^{R}V}{c} \approx \frac{{}^{R}V}{20.0468\sqrt{\mathcal{T}}}$$ **12.27**

at any instant. Our equations (Eqs. (12.1) - (12.6)) don't provide the atmospheric temperature, so we must compute it separately while integrating.

To be completely consistent with our equations of motion, our temperature model should be based on the same assumptions as our atmospheric model. An oft-forgotten assumption in our simple exponential model is that the atmosphere is isothermal (46:15)! The dilemma, of course, is what temperature should be used – an average, sea level, or something else? If we are willing to accept a small inconsistency, we can use a more accurate model of the atmosphere for evaluating Eq. (12.27) and avoid the question all together.

A model of the "1962 Standard Atmosphere" is readily available and can be used with Eq. (12.26) to find the speed of sound at any altitude (32:C28-C29; 46:14-20). Figure 12-14 plots the altitudes of interest. This function of altitude and Eq. (12.27) are used when integrating the equations of motion to determine the Mach number along the reentry trajectory. (As a side note, notice how the speed of sound in Figure 12-14 oscillates around an average value below about 110 km. This gives some "visual" justification for our usual assumption of an isothermal atmosphere.)

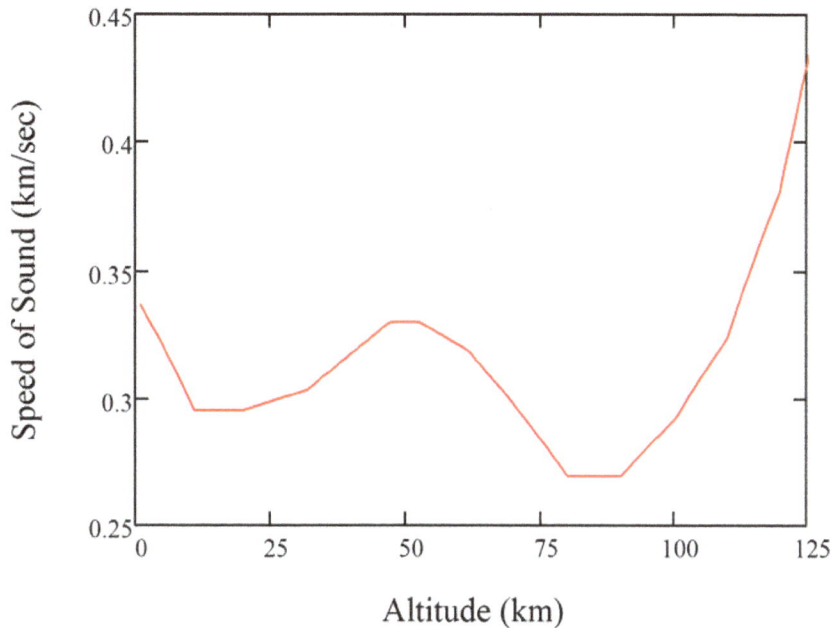

Figure 12-14: Speed of Sound in Earth's Atmosphere

Before the actual reentry, our best estimates for lift and drag coefficients are the *predicted* values in Table 12-2. If we linearly interpolate between values, we can estimate values at any Mach number (Figure 12-15). (These values are actually for *Apollo 11*; as before, we've assumed they are appropriate for *Apollo 10*.) Armed with C_L and C_D as functions of Mach number (and Figure 12-14 and Eq. (12.27) to calculate the Mach number), the equations of motion can be, once again, integrated and compared to the actual trajectory.

Figure 12-16 - Figure 12-18 compare the results using constant and Mach number dependent aerodynamic coefficients to the actual observed trajectory. Unlike in Section 12.5.1, these plots clearly show our "improved" model failed to actually improve the simulation. In fact, the predicted altitude, velocity, and deceleration are *worse* at most points along the trajectory! The root mean square error values summarized in Table 12-9 confirm this observation.

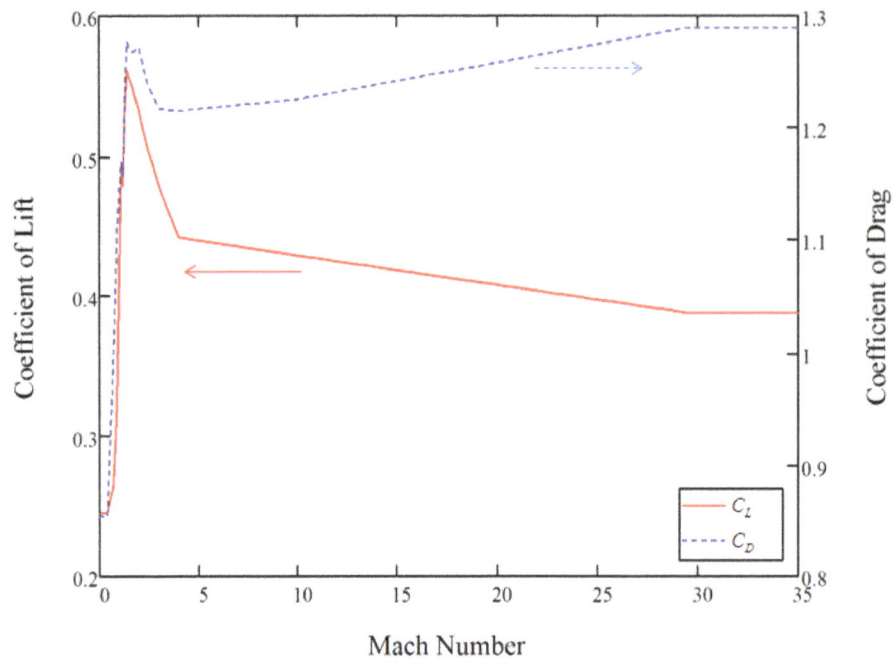

Figure 12-15: Lift and Drag Coefficients as Functions of Mach Number

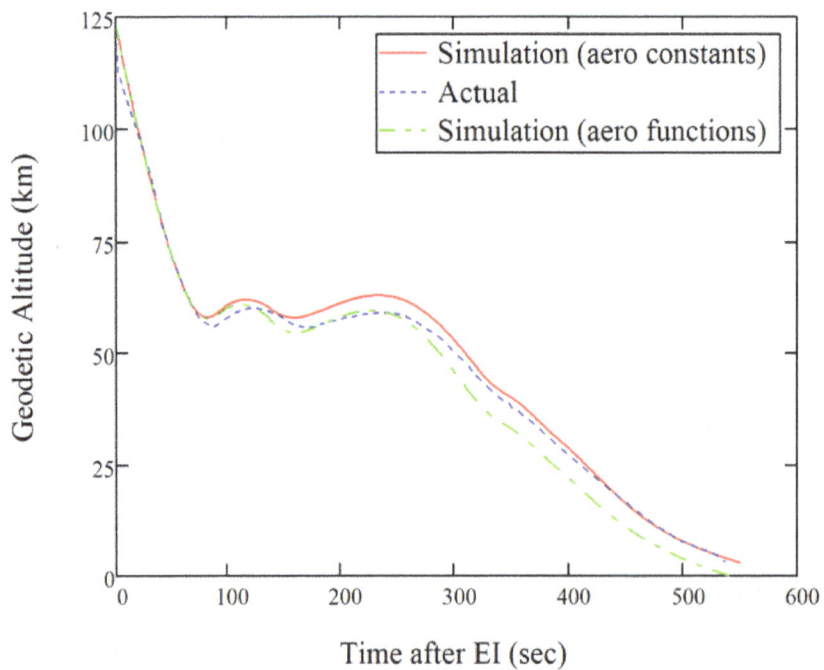

Figure 12-16: Geodetic Altitude with and without Mach-dependent C_L and C_D

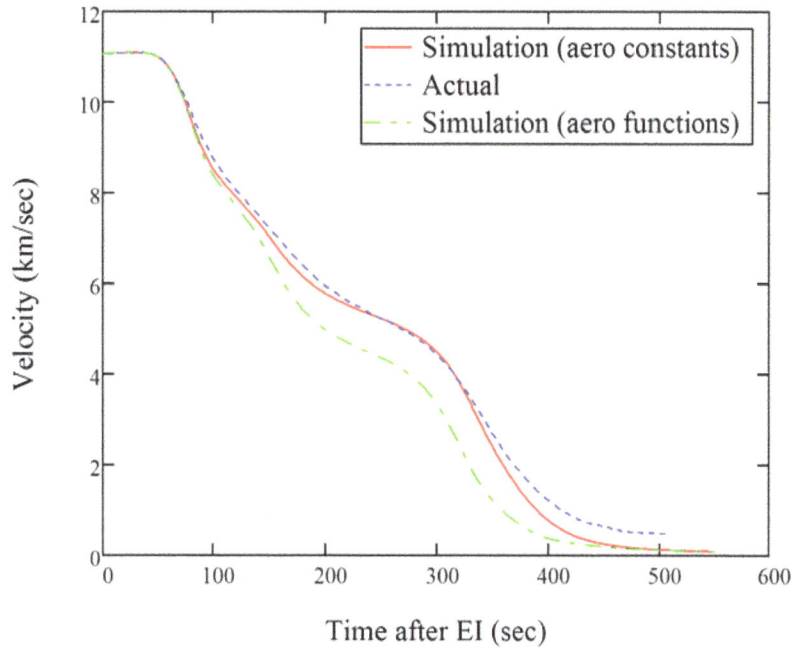

Figure 12-17: Inertial Velocity with and without Mach-dependent C_L *and* C_D

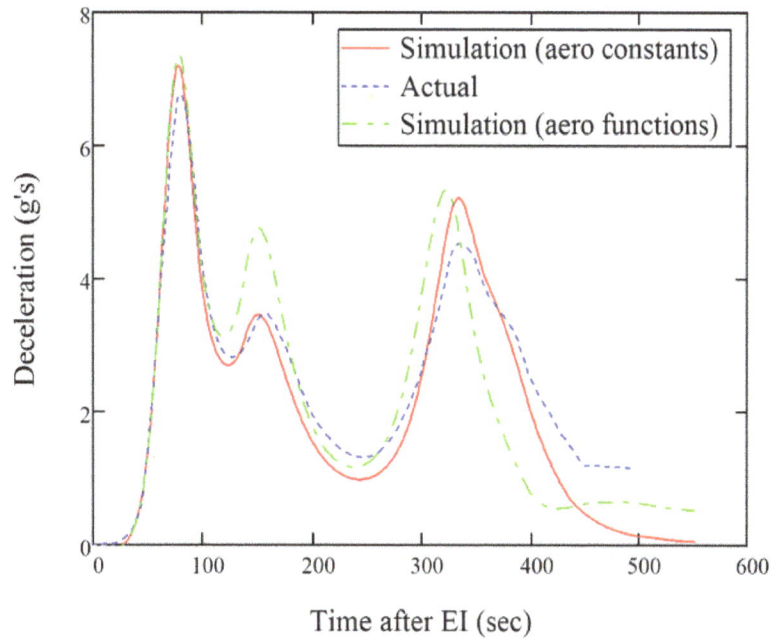

Figure 12-18: Deceleration with and without Mach-dependent C_L *and* C_D

Table 12-9: RMS Errors with and without Mach-dependent C_L and C_D

	C_L and C_D Constant	C_L and C_D Functions
$RMS_{h_{gd}}$	3.63 km	4.48 km
RMS_{I_V}	241 m/s	814 m/s
RMS_{decel}	4.60 m/s^2	8.43 m/s^2

It shouldn't come as much of a surprise that we've failed to improve the simulation accuracy. The functions for C_L and C_D shown in Figure 12-15 are based on preflight estimates of the capsule aerodynamics. Turning back to Figure 12-2, notice the preflight estimates and measured lift-to-drag values are sometimes off by 5%. (*Apollo 11* displayed a similar, if not greater, discrepancy between predicted and observed values (41:21).) It's likely our functions are equally wrong!

Just how much of a difference 5% can make is illustrated in Figure 12-19 and Figure 12-20. The "band" of trajectories shown is generated by changing the constant C_L in our (original) simulation by ±5%. Clearly, unless there's reason to believe the preflight values are extremely accurate, there isn't a need to complicate the simulation. Errors in the aerodynamics are as likely to worsen the predictions as they are to improve them!

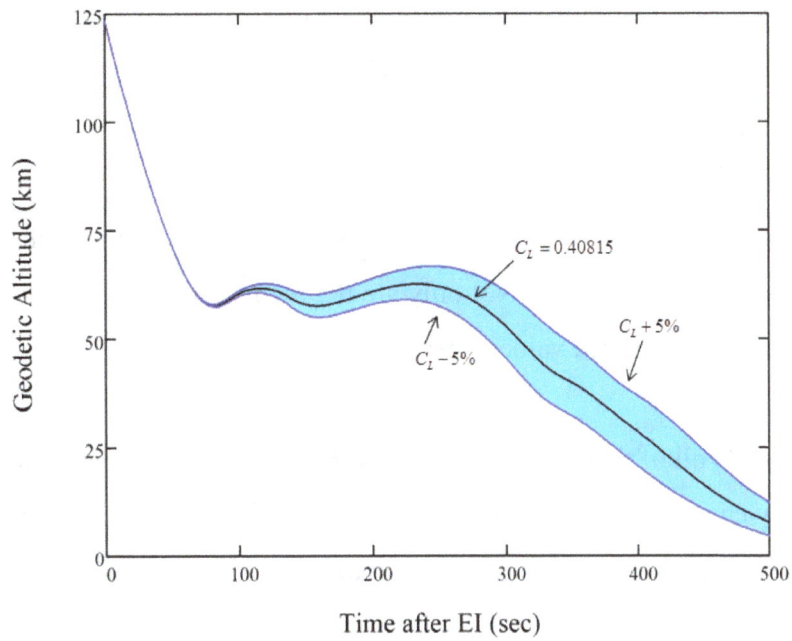

Figure 12-19: Effect of Slight Changes in C_L on Apollo 10 Geodetic Altitude

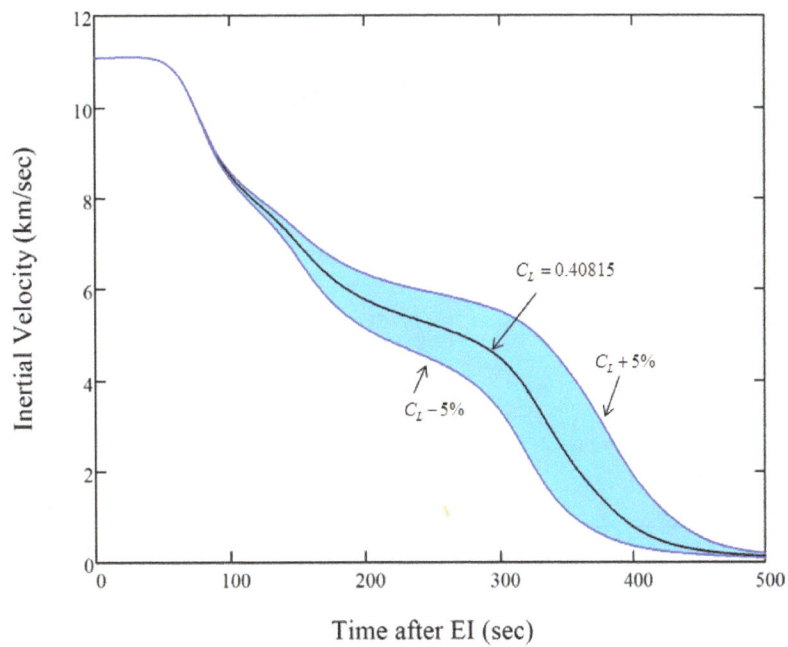

Figure 12-20: Effect of Slight Changes in C_L on Apollo10 Inertial Velocity

NASA didn't use either the preflight estimates or the measured values when they established the "Best Estimated Trajectory" (BET) in the postflight analysis report (51:11, 36). Instead, they "tweaked" the values to match other observed measurements (altitude, velocity, etc.). The "reconstructed" lift-to-drag ratio required to calculate the BET is shown in Figure 12-21 along with the predicted and recorded values. Not only are the reconstructed values *at least 5% different* than the measured values, they are *biased larger* for almost the entire range.

What do all of these graphs tell us? First, aerodynamic uncertainties can dramatically change the predicted trajectory. Second, until you are "certain" of your vehicle properties, you have no reason to believe your simulation is any

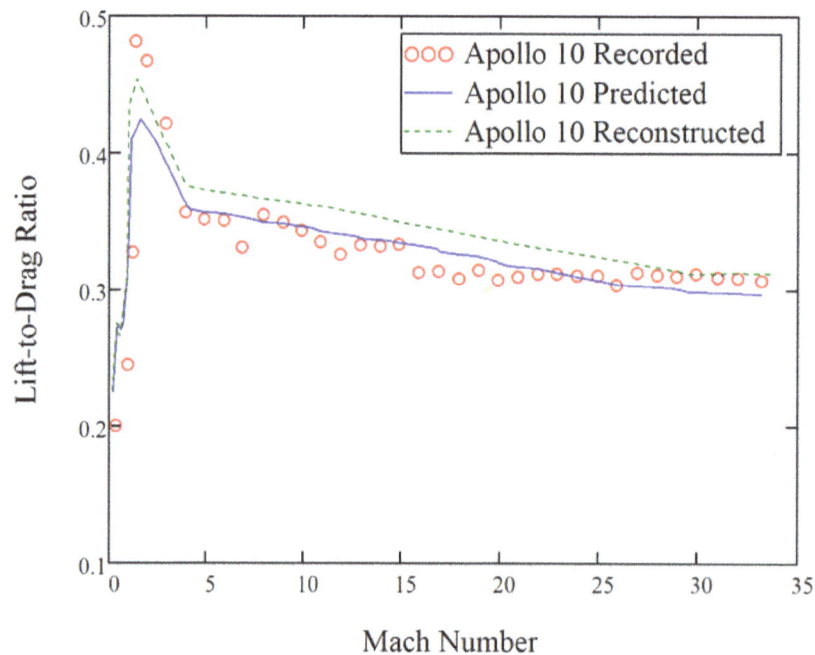

Figure 12-21: Apollo 10 Lift-to-Drag Ratio as a Function of Mach Number (51:36)

more accurate simply because you have modeled it "better." Third, even if you have the best knowledge available (as NASA did), predicting the precise trajectory can be elusive. Finally, "tuning" the (sometimes) uncertain vehicle aerodynamic properties provides a convenient tool for better matching simulations to observed data. The "tuned" aerodynamics can then be used to better predict the *next* reentry.

12.5.3 Improving the Atmospheric Model

Improving our atmospheric model is as straight-forward as replacing the simple exponential in our simulation (Eq. (12.8)) with any other (better) model. (This assumes we are limiting our atmospheric model to predicting the density. If wind predictions, etc. are included, other changes would be needed to incorporate their forces into the equations of motion. Such changes would require us to revisit the original derivations in Chapter 3 and, therefore, won't be considered here.) Once the atmospheric model is replaced, we can integrate the equations of motion and "rate" any improvement in the simulation results.

The 1962 Standard Atmosphere model introduced in the previous section breaks the atmosphere below 150 km into 12 exponential "bands" to better match reality, so it's a reasonable replacement to try (32:C28-C29; 46:14-20). Figure 12-22 and Figure 12-23 compare the two models in the region of interest. The first shows the (almost identical) estimates on a logarithmic scale, while the second plots the "percentage of difference"

$$\% \; diff = 200 \left[\frac{\rho - \rho_{1962}}{\rho + \rho_{1962}} \right] \qquad \textbf{12.28}$$

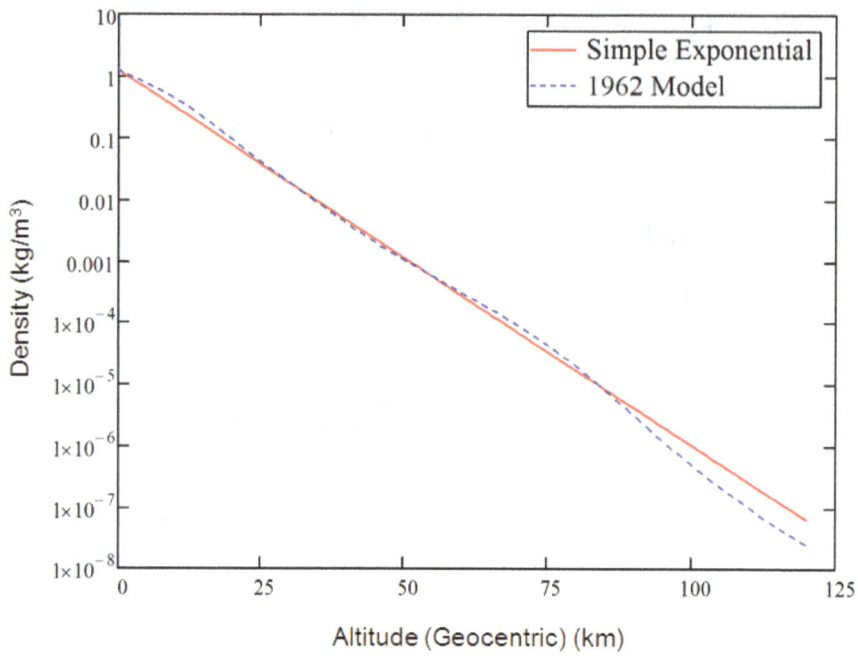

Figure 12-22: Comparison of Atmospheric Models

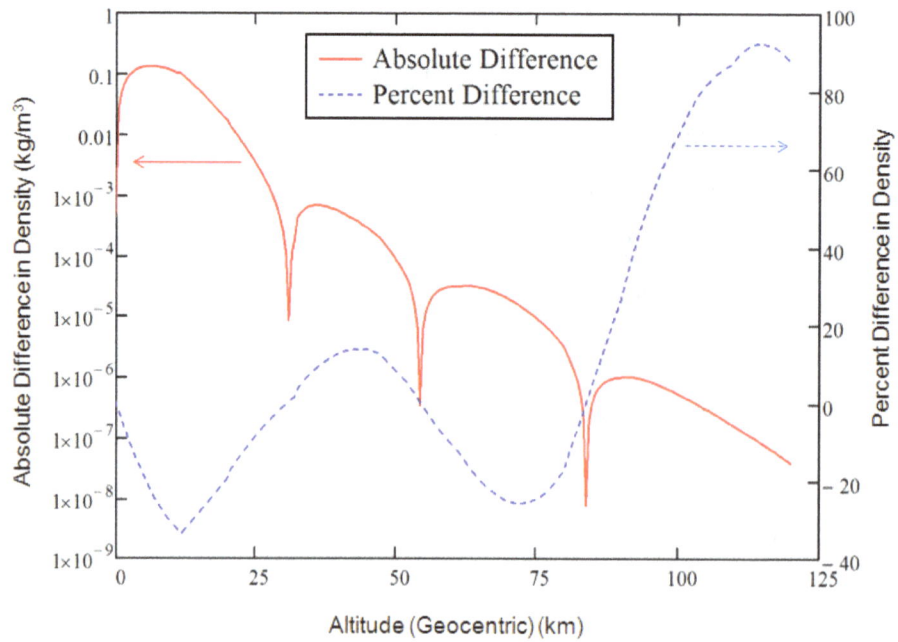

Figure 12-23: Percent Difference Between Atmospheric Density Models

and absolute difference

$$|\Delta\rho| = |\rho - \rho_{1962}|$$ **12.29**

between the models. In these equations, ρ is the density given by Eq. (12.8) and ρ_{1962} is the density given by the model of the 1962 Standard Atmosphere.

We can see good agreement between the two atmospheric models, with the largest percentage differences tending to occur where the absolute difference is the smallest (high altitude). At high altitude, the drag has a lesser effect on the trajectory. At lower altitudes (where drag is much more pronounced), the two models match better (in terms of percentage of difference). If nothing else, this pair of graphs should give you confidence the "simple" atmospheric model we've been using in all of our previous work is actually fairly good (at least below 125 km)! Of course, the real proof is to compare trajectories calculated with our usual (simple exponential) model and the 1962 model.

Figure 12-24 - Figure 12-26 show the nearly identical results for simulations using the two atmospheric models. The altitude and velocity predictions never differ by more than 1.1 km and 100 m/s, respectively. The deceleration predictions are likewise similar. The two atmospheric models produce results within 0.5 g's of each other for the entire trajectory. The maximum difference is at the peak deceleration point (near 80 seconds after EI), with the 1962 Standard Atmosphere matching the actual value better. (This is probably a lucky coincidence.)

The root mean square error values in Table 12-10 help quantify the accuracy of both simulations over the entire reentry. For this particular example, the simple model yields slightly better estimates (when we use the RMS to rate "overall" accuracy). The differences are so small, however, a minor change in an

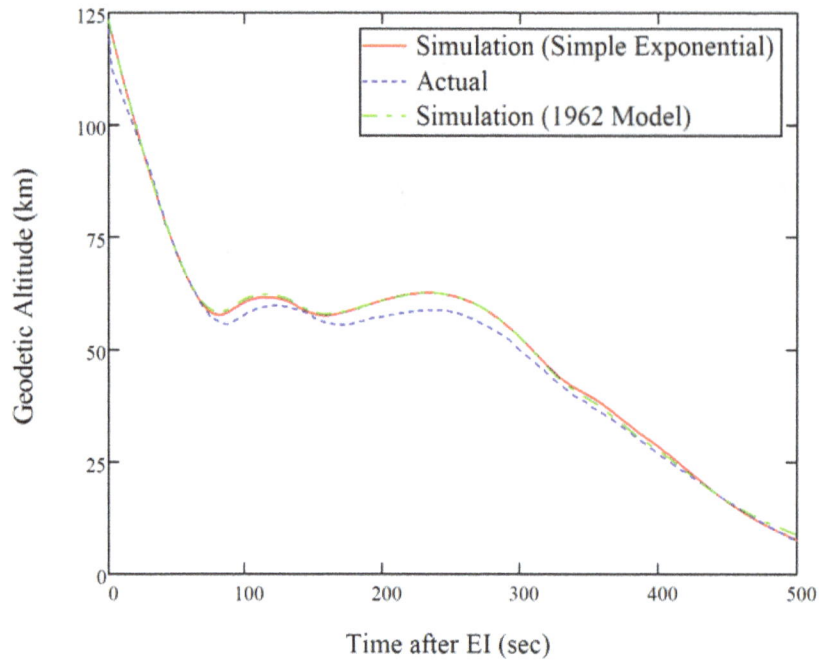

Figure 12-24: Geodetic Altitude with Two Different Atmospheric Models

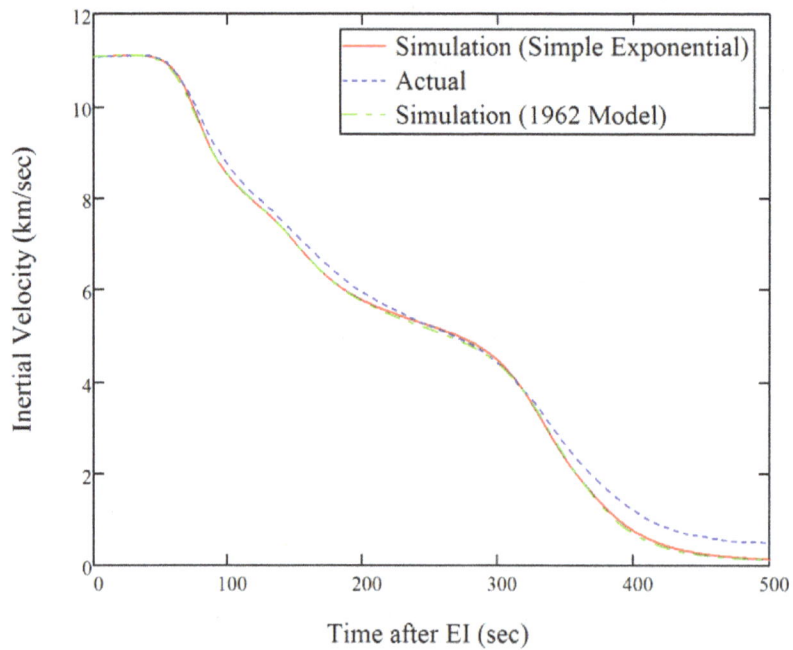

Figure 12-25: Inertial Velocity with Two Different Atmospheric Models

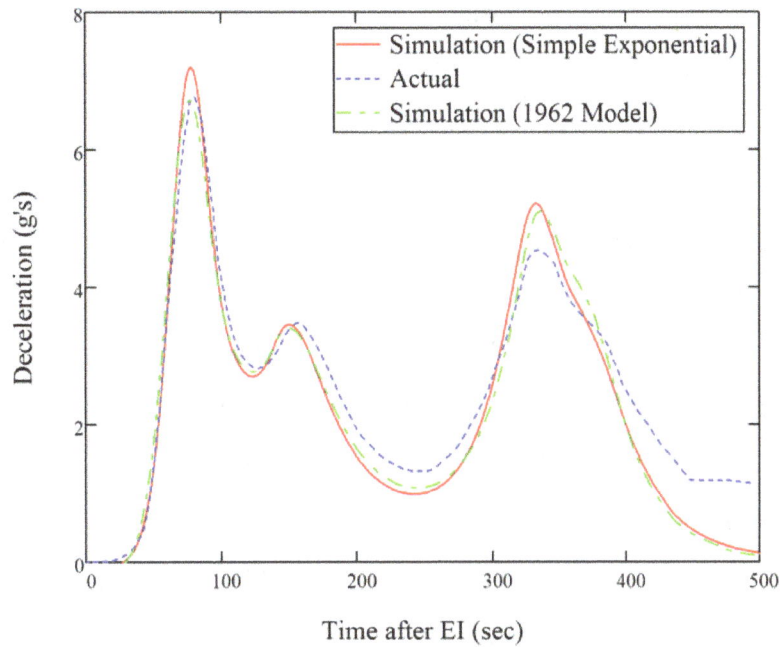

Figure 12-26: Deceleration with Two Different Atmospheric Models

Table 12-10: RMS Errors for Different Atmospheric Models

	Simple Exponential	1962 Standard Atmosphere Model
$RMS_{h_{gd}}$	3.63 *km*	3.64 *km*
RMS_{I_V}	241 m/s	261 m/s
RMS_{decel}	4.60 m/s^2	4.71 m/s^2

aerodynamic coefficient or a slightly different entry profile could easily reverse the results. In short, either model of the atmosphere is just as likely to give "better" results.

By now, it should be obvious there is little to be gained by complicating our equations of motion with the 1962 Standard Atmosphere Model. Assumptions made when we originally derived the equations (point mass, relationship of drag to $^RV^2$, etc.) as well as real-world factors (imperfect aerodynamics, wind gusts, etc.) easily mask potential accuracy gains! Other, more complicated, simulations with better dynamics equations might benefit from a better atmospheric model, but it appears we cannot.

Even with a far superior simulation, there is a bit of "magic" involved with modeling the atmosphere. NASA, with full knowledge of the vehicle and 6-degree-of-freedom modeling, had to "fudge" their best estimated trajectory by using an atmospheric model for 30° North in January (51:11). *Apollo 10* reentered between 16° and 24° *South* in *May*!

12.5.4 Improving the Gravity Model

The largest unmodeled gravity acceleration is due to the Earth's oblateness. When this "J_2" effect is included, the gravity vector remains in the plane containing the radius and the north pole (or, equivalently, in the plane along the current longitude meridian) but is no longer directed at the center of the Earth. Instead, it has components opposite the radius vector $\hat{e}_r$ and perpendicular to the radius (along the $-\hat{e}_\phi$ direction):

$$\vec{g} = -g_r\hat{e}_r - g_\phi\hat{e}_\phi$$

12.30

(The choice to attach the negative signs to the vector equation instead of the components is arbitrary. As shown in Eq. (12.30), we follow the convention used in Chapter 3 when we defined gravity as simply $\vec{g} = -g(r)\hat{e}_r$.) The components in Eq. (12.30) are given by

$$g_r = \frac{\mu}{r^2}\left[1 - \frac{3}{2}J_2\left(\frac{R_\oplus}{r}\right)^2\left(3\sin^2\phi - 1\right)\right] \qquad \textbf{12.31}$$

$$g_\phi = \frac{3\mu J_2}{r^2}\left(\frac{R_\oplus}{r}\right)^2\cos\phi\sin\phi \qquad \textbf{12.32}$$

where μ is the gravitational constant and $J_2 = 1.0827x10^{-3}$ for Earth (46:35, 36; 53:52). (We've shown $\mu = g_\oplus R_\oplus^2$.) Comparing these two, we can see g_r is approximately 1000 times larger than g_ϕ. This large difference is the very reason we have ignored g_ϕ until now.

In terms of the vehicle-pointing system introduced in Section 3.2.3, Eq. (12.30) can be equivalently written:

$$\vec{g} = -g_r\hat{e}_{x_2} - g_\phi\hat{e}_{z_2} \qquad \textbf{12.33}$$

Inserting this directly into Newton's Second Law of Motion (Eq. (3.38)) and solving for the force equations in the same manner as in Chapter 3, we get three new equations of motion:

$$^R\dot{V} = -\frac{D}{m} - g_r\sin\gamma - g_\phi\sin\gamma\cos\gamma \qquad \textbf{12.34}$$

$$^R V\dot{\gamma} = \frac{L}{m}\cos\sigma - g_r\cos\gamma + g_\phi\sin^2\gamma + \frac{^R V^2}{r}\cos\gamma \qquad \textbf{12.35}$$

$$^R V\dot{\psi} = \frac{L\sin\sigma}{m\cos\gamma} - \frac{^R V^2}{r}\cos\gamma\cos\psi\tan\phi - g_\phi\frac{\cos\psi}{\cos\gamma} \qquad \textbf{12.36}$$

These replace Eqs. (12.4)-(12.6) and can be integrated along with Eqs. (12.1)-(12.3) to simulate the trajectory with this "better" gravity model.

The resulting trajectory is even closer to our original solution than when we tried a different atmospheric model; in fact, the two simulation results are virtually indistinguishable when plotted as functions of time. The maximum difference is 280 m, 262 m/s, and 0.075 g's for geodetic altitude, inertial velocity, and deceleration, respectively. The root mean square error values in Table 12-11 help quantify the accuracy of both simulations over the entire reentry.

For this specific example, the "better" gravity model did slightly better at predicting altitude but worse when it came to velocity and deceleration. The differences between the results are so small they are easily overshadowed by other real-world factors including wind effects, seasonal variations in atmospheric density, etc. Clearly, adding the J_2 component of gravity does not improve the overall accuracy of the simulation.

12.6 Summary

The equations of motion we've used (in one form or another) throughout this book can provide solutions accurate enough to study "real" problems. Of the four "obvious" enhancements, none improved all three critical states (altitude, velocity, and deceleration) when the entire trajectory was examined using the

Table 12-11: RMS Errors for Gravity Models

	Spherical	J_2 Included
$RMS_{h_{gd}}$	3.63 km	3.57 km
RMS_{l_V}	241 m/s	253 m/s
RMS_{decel}	4.60 m/s^2	4.69 m/s^2

RMS errors. (See Table 12-12.) Where there did *appear* to be improvements, they were small and could be overshadowed by unmodeled real-world variations in the atmosphere or aerodynamics. (For example, asymmetric ablation of the heat shield can cause the drag coefficient to change *and* could introduce an extra lift force.)

You may wonder how NASA predicted the splash-down points so accurately now that you've seen how sensitive the trajectory is to small changes in the aerodynamics. The answer is simple – NASA didn't have to get it perfect. The active control of the capsule *drove it* to the touchdown point. Their a priori simulations only had cover the range of possible entries to insure the guidance would work. Our simulations are accurate enough to meet that requirement.

Table 12-12: RMS Errors for Different "Improvements" to the Simulations

		Improvement to Dynamics			
	Baseline	Rotation	Aero	Atmosphere	J_2
$RMS_{h_{gd}}$	3.63 *km*	3.21 *km*	4.48 *km*	3.64 *km*	3.57 *km*
RMS_{I_V}	241 m/s	187 m/s	814 m/s	261 m/s	253 m/s
RMS_{decel}	4.60 m/s^2	5.40 m/s^2	8.43 m/s^2	4.71 m/s^2	4.69 m/s^2

12.7 Problems

1. For a non-rotating planet, Newton's Second Law of Motion can be written as

$$m\frac{^R d\left(^R\vec{V}\right)}{dt} = \vec{L} + \vec{D} + m\vec{g}$$ if lift, drag, and gravity are the only forces acting on the

vehicle. (Details can be found in Section 3.4.) Show that substituting gravity in the form

$$\vec{g} = -g_r\hat{e}_r - g_\phi\hat{e}_\phi$$
$$= -g_r\hat{e}_{x_2} - g_\phi\hat{e}_{z_2}$$

and lift and drag vectors in their usual forms into this vector equation gives us the three scalar force equations:

$$^R\dot{V} = -\frac{D}{m} - g_r\sin\gamma - g_\phi\sin\gamma\cos\gamma$$

$$^RV\dot{\gamma} = \frac{L}{m}\cos\sigma - g_r\cos\gamma + g_\phi\sin^2\gamma + \frac{^RV^2}{r}\cos\gamma$$

$$^RV\dot{\psi} = \frac{L\sin\sigma}{m\cos\gamma} - \frac{^RV^2}{r}\cos\gamma\cos\psi\tan\phi - g_\phi\frac{\cos\psi}{\cos\gamma}$$

Hint: Make use of Eq. (3.61) to replace $\dfrac{^R d\left(^R\vec{V}\right)}{dt}$ and equate the vector components on each side of the equation.

Appendix A

Apollo 10 Revisited

Chapter 12 examined *Apollo 10*'s reentry using dimensional equations (Eqs. (12.1) – (12.6)). We can repeat that example to compare the relative accuracy of results using the non-dimensional Universal Equations (from Chapter 9 or Chapter 10). In case you're wondering, the dimensional equations are *far easier* to use in this particular situation (where the vehicle is well-defined and the bank history is given as a function of time) and would normally be the desired approach. However, it's a good opportunity to compare the results.

The Chapter 9 forms of the Universal Equations are the easiest to apply here since we'll be assuming a single (constant) lift-to-drag ratio for the entire flight. These equations are

$$\frac{dZ}{ds} = -\overline{\beta r} Z \tan \gamma \qquad\qquad \textbf{A.1}$$

$$\frac{du}{ds} = -\frac{2Zu\sqrt{\overline{\beta r}}}{\cos \gamma}\left(1 + \frac{C_L}{C_D}\cos\sigma\tan\gamma + \frac{\sin\gamma}{2Z\sqrt{\overline{\beta r}}}\right) \qquad \textbf{A.2}$$

$$\frac{d\theta}{ds} = \frac{\cos\psi}{\cos\phi} \qquad\qquad \textbf{A.3}$$

$$\frac{d\phi}{ds} = \sin\psi \qquad\qquad \textbf{A.4}$$

$$\frac{d\gamma}{ds} = \frac{Z\sqrt{\overline{\beta r}}}{\cos\gamma}\left[\frac{C_L}{C_D}\cos\sigma + \frac{\cos\gamma}{Z\sqrt{\overline{\beta r}}}\left(1 - \frac{\cos^2\gamma}{u}\right)\right] \qquad\qquad \textbf{A.5}$$

$$\frac{d\psi}{ds} = \frac{Z\sqrt{\overline{\beta r}}}{\cos^2\gamma}\left(\frac{C_L}{C_D}\sin\sigma - \frac{\cos^2\gamma}{Z\sqrt{\overline{\beta r}}}\cos\psi\tan\phi\right) \qquad\qquad \textbf{A.6}$$

where

$$u = \frac{{}^R V^2 \cos^2\gamma}{gr} \qquad\qquad \textbf{A.7}$$

$$Z = \frac{\rho C_D S}{2m}\sqrt{\frac{r}{\beta}} \qquad\qquad \textbf{A.8}$$

and $\overline{\beta r} \approx 900$ for Earth. To incorporate the bank profile given by Figure 12-1, we need to integrate

$$\frac{dt}{ds} = \frac{r}{{}^R V \cos\gamma} \qquad\qquad \textbf{A.9}$$

along with Eqs. (A.1) - (A.6) to get time. r and ${}^R V$ are found by simultaneously solving Eqs. (A.7) and (A.8) during the integration.

Using the same initial conditions as those given in Chapter 12 (converted to non-dimensional values where necessary), these equations were integrated. The results were then converted to back to dimensional values and compared to the results of the "baseline" solution in Chapter 12. The results for all six states are shown in Figure A-1 - Figure A-6. As you can see, the results are very close for all of the states. The differences are easily within the accuracy of many of the assumptions. (See Chapter 12 for more details on the impact of errors in the drag coefficients, etc.)

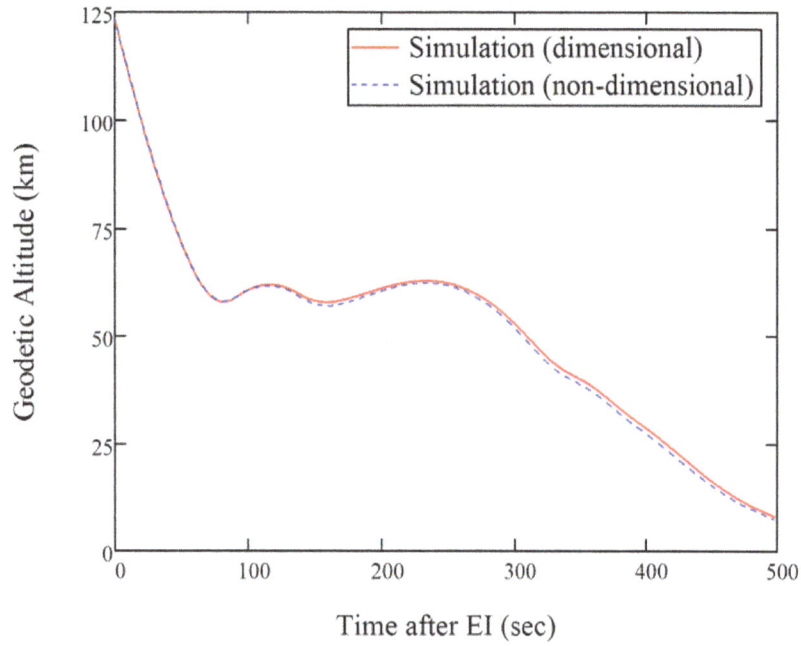

Figure A-1: Geodetic Altitude Comparison for Apollo 10 Reentry

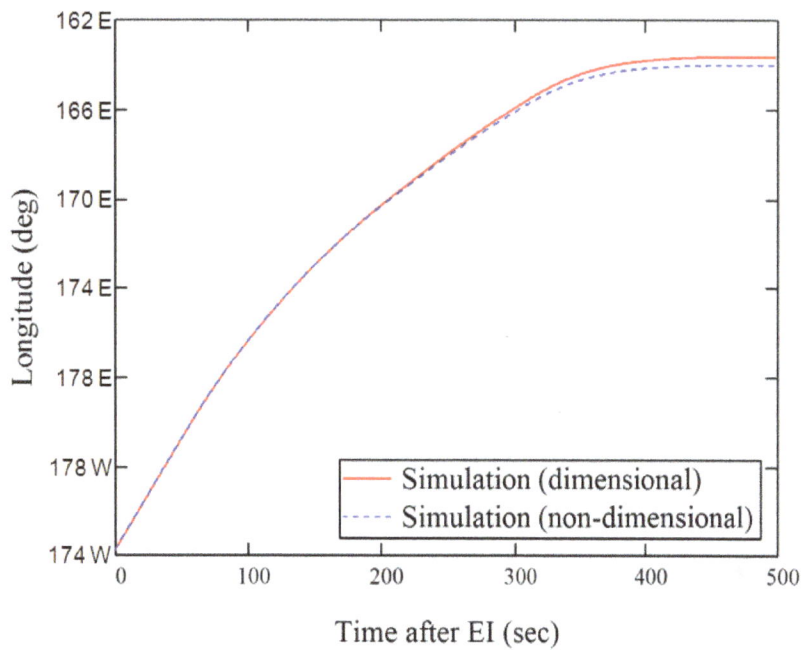

Figure A-2: Longitude Comparison for Apollo 10 Reentry

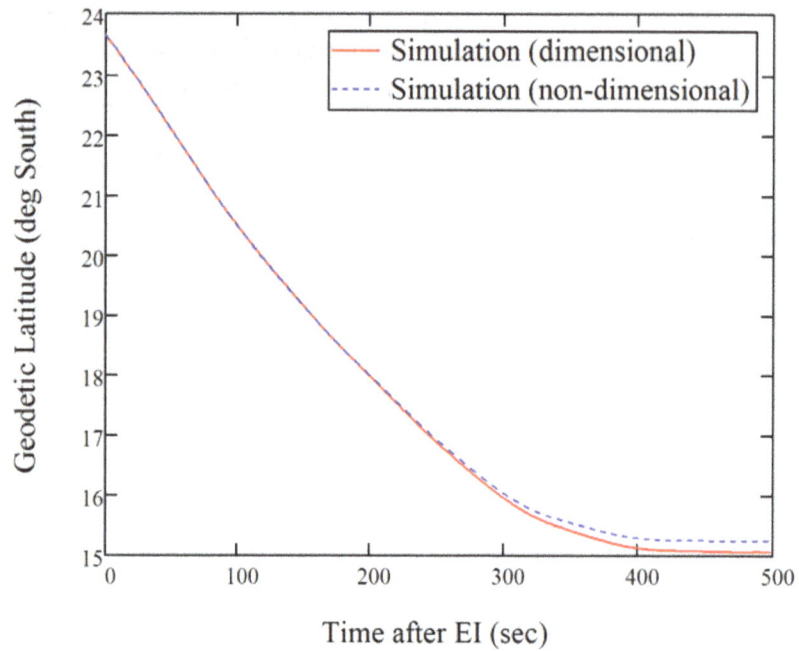

Figure A-3: Latitude Comparison for Apollo 10 Reentry

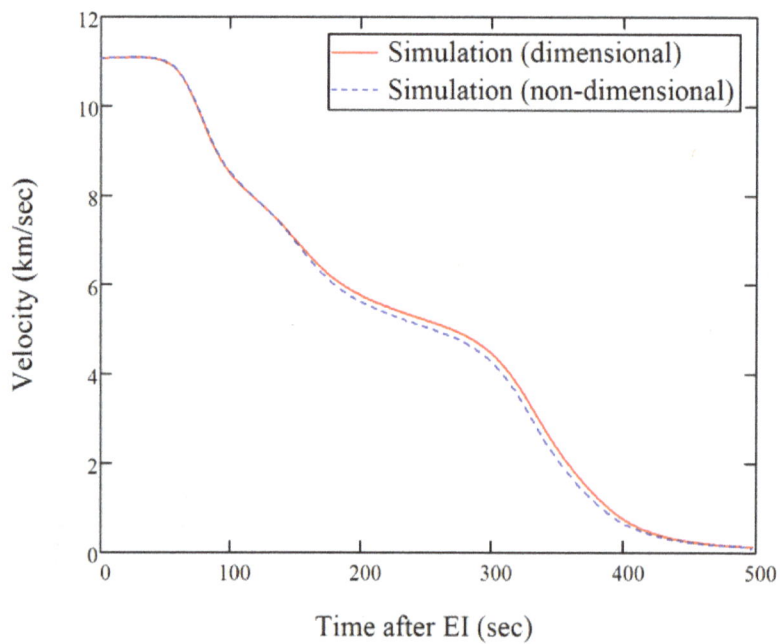

Figure A-4: Velocity Comparison for Apollo 10 Reentry

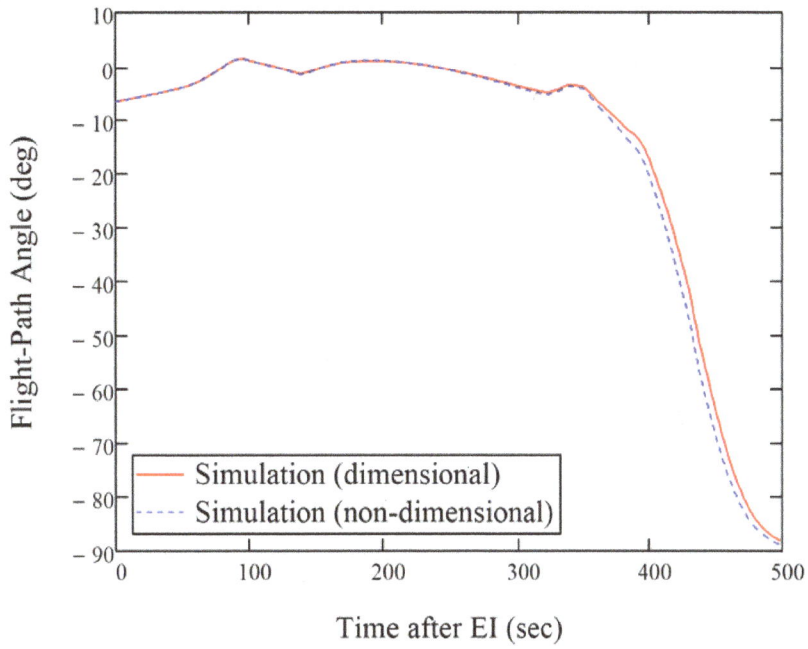

Figure A-5: Flight-Path Angle Comparison for Apollo 10 Reentry

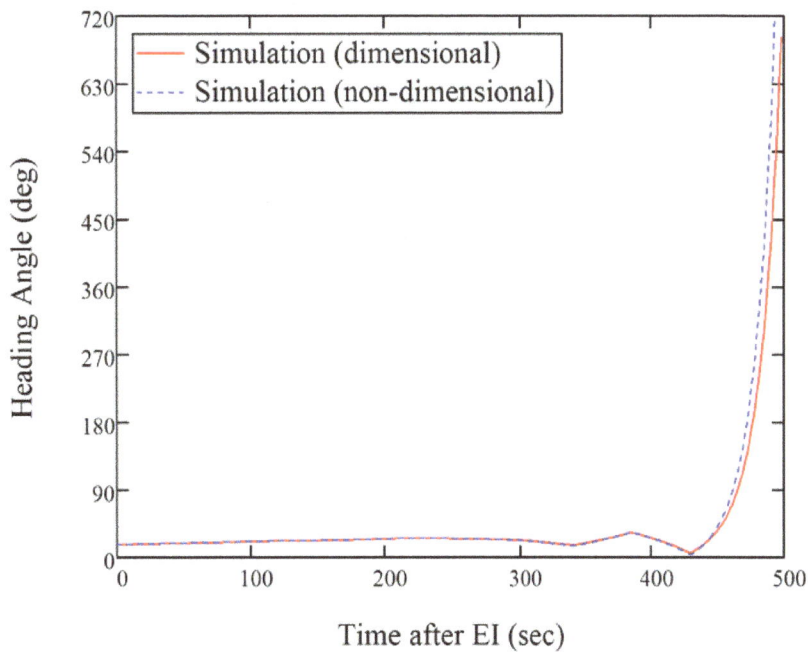

Figure A-6: Heading Angle Comparison for Apollo 10 Reentry

We should have expected similar answers. The Universal Equations are approximations to the dimensional equations, with the primary assumptions being:

$$\frac{1}{\beta r} \ll 1 \qquad\qquad\qquad \text{A.10}$$

$$\beta r \approx \text{constant} = \overline{\beta r} \qquad\qquad\qquad \text{A.11}$$

For Earth, these are both very good approximations.

Appendix B

Capstone Project

B.1 Project Set Up

A good way to tie together all of the material through Chapter 9 is with a capstone project. (Material from Chapter 10 and may be used as well if desired.) This sample project, the *Katcoan Science Probe*, addresses the design of an atmospheric entry mission. The main focus of the project is to practice using the material learned to "solve" and perhaps "optimize" an entry profile. Then, with *a solution* (and there are *many*) in hand, the deceleration and heating profiles are examined.

Creativity is encouraged as is "bending the rules" to meet the mission requirements. For example, some students have added a thrusting maneuver just before hitting the atmosphere the first time to slow the probe down. (Others considered that "cheating" and opted to execute an atmospheric deceleration. Sadly, there was also the one student that relied on a 10,000g "dirt" deceleration, better known as a "splat" maneuver.)

One word of caution is in order; however, this project can eat up tremendous amounts of time. Sometimes it's because students have difficulty finding a trajectory to meet the mission requirements. But, more often than not, it's because they spend so much time "tweaking" their trajectory to get the best performance for the mission. In spite of this, the feedback on this project has been 100% positive. Enjoy.

B.2 The Katcoan Science Probe

The following pages are the project assignment in a format that is "almost" ready to hand out. Feel free to photocopy those pages for use in class. For smaller classes, the project can be worked as a group project to get a "systems engineering" view of the tradeoffs experienced to maximize mission time.

The Katcoan Science Probe

Background: Researchers at the Imperial Science Academy of the planet Katco have been ordered to send a scientific probe to a backwards little planet known to the locals as "Earth."

The Katcoans are trying to design the mission so that the probe will dip into the atmosphere to collect samples. The probe should arrive at Earth traveling no more than 15 km/sec (to avoid detection) and then spend as long as possible in the atmosphere below an altitude of 80 km. At the end of the mission, the probe should be outside the atmosphere so that it can be recovered by a cloaked Katcoan science ship. Your goal is to analyze/design this mission. You must live with the dictated vehicle parameter values given below.

Katcoan technology routinely creates probes with the following parameters:

- $C_D = 0.5$

- A reference area of $S = 100m^2$

- Lift-to-drag ratios that range $0 \leq C_L/C_D \leq 6.6$ without changing C_D or S

- By law, all probes to primitive planets must have a mass of exactly $m = 1000kg$

(Don't ask me how or why -- it's *Katcoan* technology!)

The planned mission will be examined with two things in mind – deceleration and heating. The pertinent questions are summarized below:

- **Deceleration:** What does the deceleration profile look like? What is the maximum and where does it occur?
- **Heating:** What does the non-dimensional stagnation heat flux $\dot{\overline{q}}_s$ look like? What is the maximum and where does it occur?
- **Dynamics:** What does the flight profile look like?

Deliverables:

1. A short "report" to be handed in
   a. Explain your approach to solving the problem
   b. Give your initial conditions (e.g., $^{R}V_e$, γ_e, θ_e, etc.) and assumed values for the vehicle parameters (e.g., C_L/C_D, S, m, etc. where appropriate)
   c. As a minimum, present graphs showing your solution to the trajectory for velocity, altitude, and flight-path angle (as they change with "time").
   d. Present the graphs required for your particular assignment (from the table above).
   e. Discuss any "trends" you think are worth mentioning (such as "any lift in the range $0.5 \leq C_L/C_D \leq 2.0$ will satisfy the mission") and feel free to add extra graphs if they help.

2. You will present your findings to the class during a short (10-15 minutes or so) briefing. You can steal graphs from your report to make most of the briefing. The goal of the briefing is to make sure everyone in the class understands how you solved the problem and what your major findings were.

You will be evaluated on (in no particular order):

1. Correctness in your methods
2. Meeting the constraints (assuming they can all be met)
3. Ease of understanding what you did
4. Completeness of your results (summary, graphs, etc.)
5. Neatness, "readability," and clarity
6. Special consideration will be given for these "separating factors:"
   a. Adding extra (but useful) plots to help explain any special features of your solution
   b. Going beyond the minimum [e.g., showing how selecting a mission for maximum time impacts the maximum deceleration experienced).

This page intentionally blank.

Answers to Selected Problems

Chapter 2

1a. $\vec{r} = \left(-2.483\hat{e}_P + 1.054\hat{e}_Q\right) DU$, $\vec{V} = \left(-0.5331\hat{e}_P - 0.06890\hat{e}_Q\right) DU\!\!\Big/\!\!_{TU}$

b. $\vec{r} = \left(2.5047\hat{e}_x + 1.0018\hat{e}_y\right) DU$,

$\vec{V} = \left(0.4993\hat{e}_x + 0.00050\hat{e}_y + 0.19904\hat{e}_z\right) DU\!\!\Big/\!\!_{TU}$

2a. $r_p = 0.287\ DU$

b. $\gamma_e = -0.982\ rad$, $V_e = 1.10\ DU\!\!\Big/\!\!_{TU}$

3. $a = 2.22\ DU$, $e = 0.100$, $i = 0.524\ rad$, $\Omega = 0\ rad$, $\omega = 0\ rad$, $v = 0\ rad$

4. $\gamma = 0\ rad$

Chapter 3

2.

$$^R\dot{V} = -g\sin\gamma + r\omega_\oplus^2\cos\phi\left(\cos\phi\sin\gamma - \sin\phi\sin\psi\cos\gamma\right)$$

$$^RV\dot{\gamma} = -g\cos\gamma + \frac{^RV^2}{r}\cos\gamma + 2\,^RV\omega_\oplus\cos\phi\cos\psi + r\omega_\oplus^2\cos\phi\left(\cos\phi\cos\gamma + \sin\phi\sin\psi\sin\gamma\right)$$

$$^RV\dot{\psi} = -\frac{^RV^2}{r}\cos\gamma\cos\psi\tan\phi + 2\,^RV\omega_\oplus\left(\sin\psi\cos\phi\tan\gamma - \sin\phi\right) - \frac{r\omega_\oplus^2}{\cos\gamma}\sin\phi\cos\phi\cos\psi$$

Chapter 4

1a. $\dot{r} = {}^R V \sin\gamma,\ {}^R\dot{V} = -g_0\left(\dfrac{r_0}{r}\right)^2 \sin\gamma,\ {}^R V\dot\gamma = \left[\dfrac{{}^R V^2}{r} - g_0\left(\dfrac{r_0}{r}\right)^2\right]\cos\gamma$

b. For $\dot\gamma = 0$, the third equation above becomes ${}^R V = \sqrt{\dfrac{g_0 r_0^2}{r}}$. Comparing this

to the two-body equation for circular velocity, you can see $\mu = g_0 r_0^2$.

2. (Simpler option):

$$c_x = \sin\theta\sin\psi - \cos\psi\sin\phi\cos\theta$$
$$c_y = -\sin\psi\cos\theta - \cos\psi\sin\phi\sin\theta$$
$$c_z = \cos\psi\cos\phi$$

3. Starting with Eq. (4.31):

$$\left(a_{decel}\right)_v = \frac{\rho C_D S\,{}^R V^2}{2m} + g_0\left(\frac{r_0}{r}\right)^2 \sin\gamma$$

$$= \left(\eta\beta\right){}^R V^2 + g_0\left(\frac{r_0}{r}\right)^2 \sin\gamma$$

$$= \left(\eta\beta\right)\left(2g_0 r_0 T\right) + g_0\left(\frac{r_0}{r}\right)^2 \sin\gamma$$

$$= 2\beta r_0 g_0 \eta T + g_0\left(\frac{r_0}{r}\right)^2 \sin\gamma$$

For thin atmospheres $\dfrac{r_0}{r} \approx 1$. Also, $\sin\gamma = \pm\sqrt{1-\cos^2\gamma}$ where the sign is

determined by the sign of γ. Thus, we get

$$\left(a_{decel}\right)_v = 2\beta r_0 g_0 \eta T \pm g_0\sqrt{1-\cos^2\gamma}$$
$$= 2\beta r_0 g_0 \eta T \pm g_0\sqrt{1-\xi^2\gamma}$$

Starting with Eq. (4.32):

$$\left(a_{decel}\right)_L = \frac{-\rho C_L S \,^R V^2}{2m} - \left[\frac{^R V^2}{r} - g_0\left(\frac{r_0}{r}\right)^2\right]\cos\gamma$$

$$= -\left(\frac{\eta\beta}{C_D}\right)\left(C_L \,^R V^2\right) - \left[\frac{^R V^2}{r} - g_0\left(1\right)^2\right]\xi$$

$$= -\eta\beta\left(\frac{C_L}{C_D}\right)^R V^2 - \left(\frac{^R V^2}{r} - g_0\right)\xi$$

$$= -\eta\beta\left(\frac{C_L}{C_D}\right)\left(2g_0 r_0 T\right) - \left[\frac{\left(2g_0 r_0 T\right)}{r} - g_0\right]\xi$$

$$= -2\beta r_0 g_0 \eta T\left(\frac{C_L}{C_D}\right) + g_0\left(1 - 2T\right)\xi$$

Chapter 5

3. For skip entry, η and γ are related by:

$$\eta = \frac{\cos\gamma - \cos\gamma_e}{\left(C_L/C_D\right)} + \eta_e$$

To find a maximum, minimum, or inflection point:

$$\frac{d\eta}{d\gamma} = \frac{-\sin\gamma_*}{\left(C_L/C_D\right)} = 0$$

Solving gives $\gamma_* = 0$. To prove this is a maximum (i.e., a "minimum" in altitude since large η means low "physical" altitude), take the second derivative:

$$\left.\frac{d^2\eta}{d\gamma^2}\right|_{\gamma=\gamma_*} = \left.\frac{-\cos\gamma}{\left(C_L/C_D\right)}\right|_{\gamma=\gamma_*} = \frac{-1}{\left(C_L/C_D\right)}$$

For positive lift, $\left.\dfrac{d^2\eta}{d\gamma^2}\right|_{\gamma=\gamma_*} < 0$, so this point is a maximum for η.

6. $c_0 = 1$, $c_1 = \dfrac{1}{6}$, $c_2 = \dfrac{1}{24}$, $c_3 = \dfrac{47}{4752}$

7. Answers may vary, but this is a typical plot:

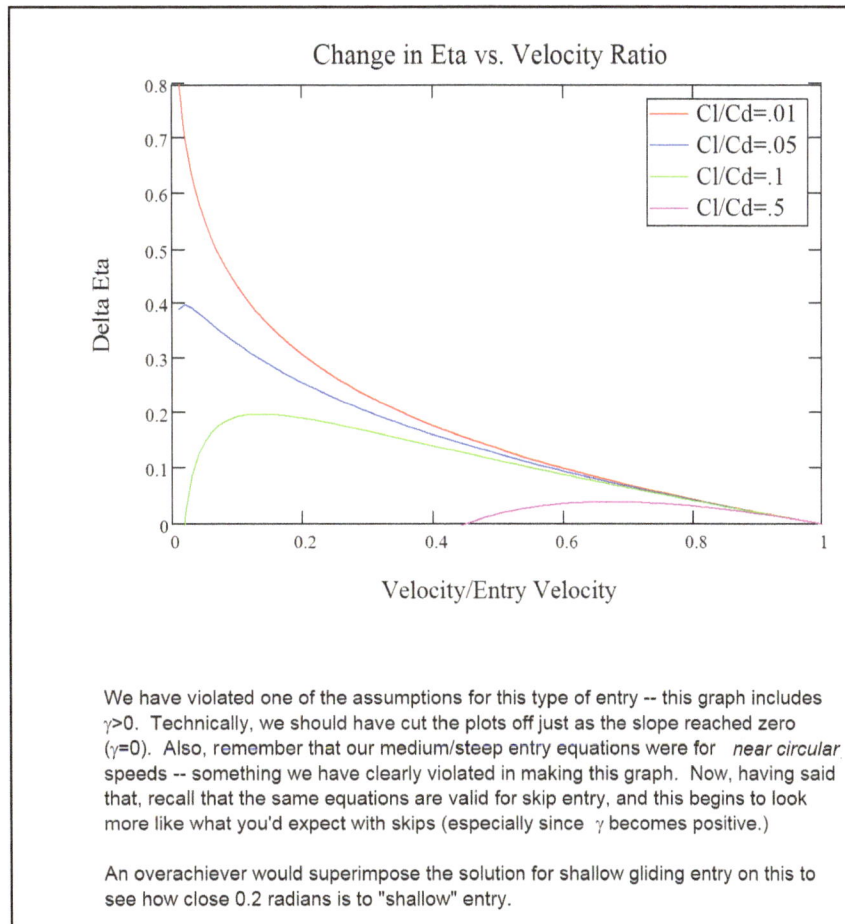

Change in Eta vs. Velocity Ratio

We have violated one of the assumptions for this type of entry -- this graph includes γ>0. Technically, we should have cut the plots off just as the slope reached zero (γ=0). Also, remember that our medium/steep entry equations were for *near circular* speeds -- something we have clearly violated in making this graph. Now, having said that, recall that the same equations are valid for skip entry, and this begins to look more like what you'd expect with skips (especially since γ becomes positive.)

An overachiever would superimpose the solution for shallow gliding entry on this to see how close 0.2 radians is to "shallow" entry.

8. Part a. Consider the two cases shown in the table to the right. The pair of figures below compare the solutions from numerically integrating Eq. (5.78) with the closed-form solutions given by Eq. (5.80).

Case #1	Case #2
$T_e = 0.51$	$T_e = 0.60$
$\gamma_e = -0.15$ radians	$\gamma_e = -0.10$ radians
$C_L/C_D = 0.2$	$C_L/C_D = 0.2$
$\beta r_0 = 910, \ \eta_e = 10^{-4}$	

Part b. The next figure plots the

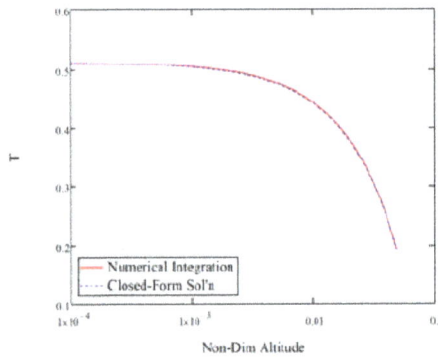

Case 1

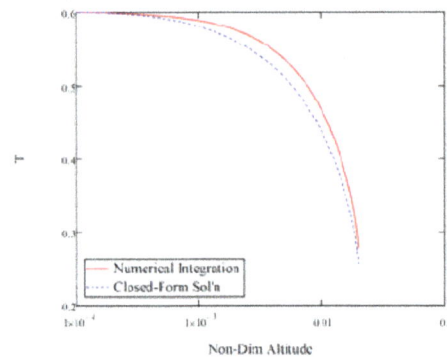

Case 2

"constant" $\dfrac{C_L}{C_D} + \dfrac{1}{\beta r_0 \eta}\left(1 - \dfrac{1}{2T_e}\right)$ for both cases. Notice how the curves asymptotically approach C_L/C_D. The fact that this term is dominated by the lift-to-drag ratio is what enables the assumption of "constant" to work.

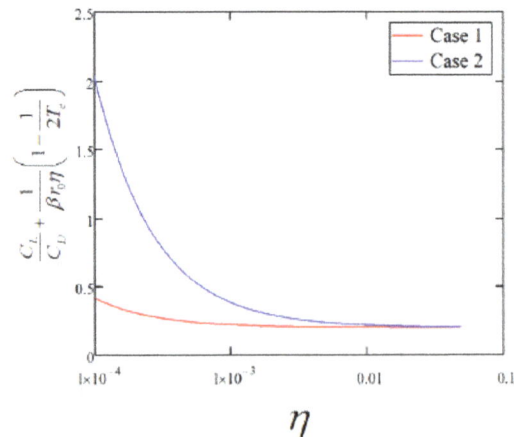

9. A more than sufficient answer that covers the entire range of entry angles is given in the plots below. Two other lift-to-drag ratios are plotted to help illustrate the effect of changing the lift.

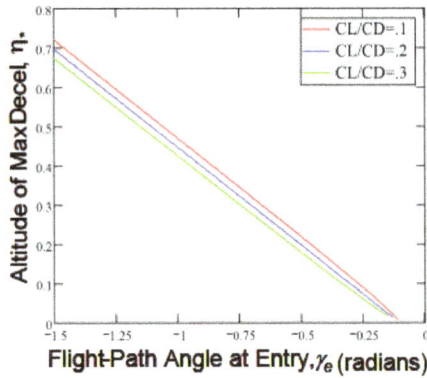

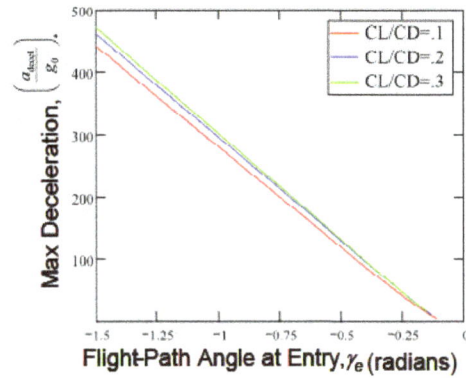

Chapter 6

1. Results may vary slightly (especially with Loh's solutions) depending on the initial guess values and initial altitude selections.

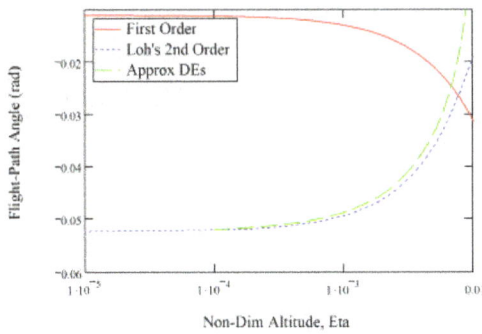

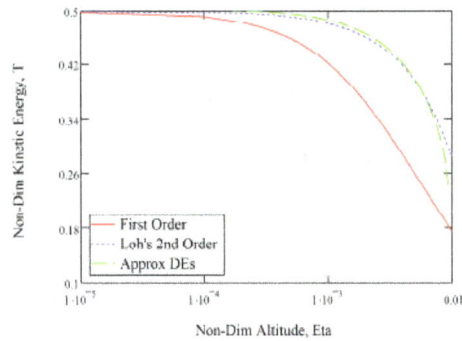

Chapter 7

1.

$$\frac{d}{dt}\left({}^{R}V\right) = -\frac{\rho C_D S^{\,R}V^2}{2m}$$

$$\frac{d}{dt}\left(\sqrt{2g_0 r_0 T}\right) = -\frac{\rho C_D S}{2m}\left(2g_0 r_0 T\right)$$

$$\frac{1}{2}\sqrt{\frac{2g_0 r_0}{T}}\frac{dT}{dt} = -\eta\left(2g_0 r_0 T\beta\right)$$

$$\frac{dT}{dt} = -2\beta\sqrt{2g_0 r_0}\left(\eta T^{3/2}\right)$$

Inverting both sides gives the desired result:

$$\frac{dt}{dT} = -\frac{1}{2\beta\sqrt{2g_0 r_0}}\frac{1}{\eta T^{3/2}}$$

4. Since $\gamma_e < 0$ and $\left|\sin\gamma_e\right| < 1$, we can write the true statement:

$$-\sin\gamma_e < \sqrt{-\sin\gamma_e}$$

Recognizing $3e > \sqrt{6e}$, we can divide the left side by $3e$ and the right side by $\sqrt{6e}$ and still have an inequality:

$$\frac{-\sin\gamma_e}{3e} < \sqrt{\frac{-\sin\gamma_e}{6e}}$$

Finally, if we multiply by $T_e^{3/2}$:

$$\frac{-T_e^{3/2}\sin\gamma_e}{3e} < T_e^{3/2}\sqrt{\frac{-\sin\gamma_e}{6e}}$$

$$\dot{q}_{W*} < \dot{q}_{S*}$$

7. The heat flux graph is shown here. This corresponds to the dynamics graphed in Problem 1 of Chapter 6.

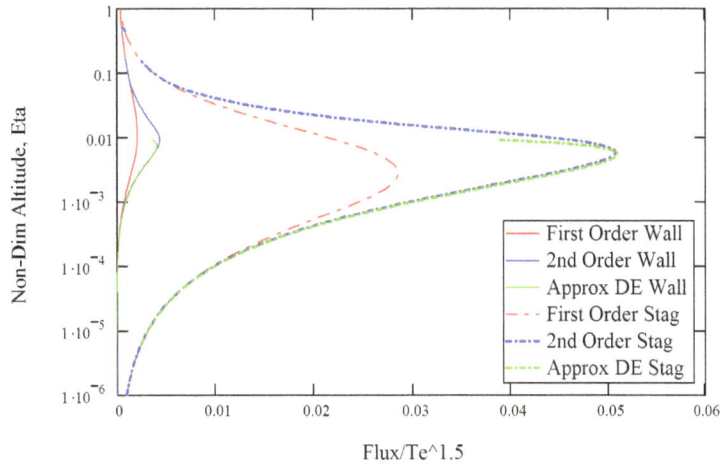

10. An example is shown below for $\gamma_e = -80°$ and $\beta r_0 = 910$.

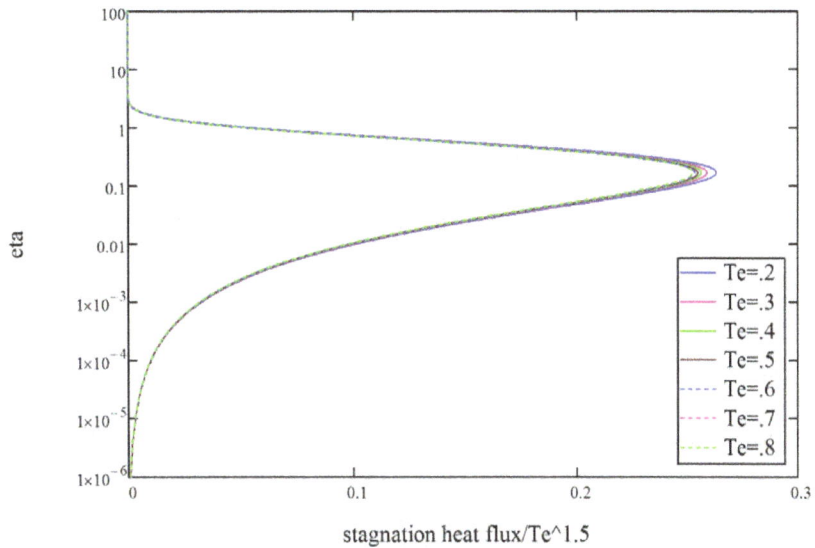

Chapter 8

2. Starting first with the Loh's equation given by Eq. (6.36) and setting $\eta_f = \eta_e$:

$$\cos\gamma_f = \frac{\cos\gamma_e + \dfrac{C_L}{C_D}\eta\left(1-\dfrac{\eta_e}{\eta_f}\right)}{1+\dfrac{1}{\beta r_0}\left(\dfrac{1}{2T}-1\right)\left(1-\dfrac{\eta_e}{\eta_f}\right)}$$

$$\cos\gamma_f = \frac{\cos\gamma_e + 0}{1+\dfrac{1}{\beta r_0}\left(\dfrac{1}{2T_f}-1\right)(0)}$$

$$\cos\gamma_f = \cos\gamma_e$$

which gives the expected result $\gamma_f = -\gamma_e$. Turning to the other of Loh's

equations, given by Eq. (6.34), and setting $T_f = \frac{1}{2}$ and $\gamma_f = -\gamma_e$

$$\ln\left(\frac{T_f}{T_e}\right) = \frac{-2(\gamma_f - \gamma_e)}{\dfrac{C_L}{C_D}+\dfrac{1}{\beta\, r_0\eta_f}\left(1-\dfrac{1}{2T_f}\right)\cos\gamma_f}$$

$$\ln\left(\frac{1}{2T_e}\right) = \frac{4\gamma_e}{\dfrac{C_L}{C_D}+\dfrac{1}{\beta\, r_0\eta_f}(0)\cos\gamma_f}$$

$$\ln\left(\frac{1}{2T_e}\right) = \frac{4\gamma_e}{\left(\dfrac{C_L}{C_D}\right)}$$

and solving:

$$T_e = \frac{1}{2}\exp\left[\frac{-4\gamma_e}{\left(\frac{C_L}{C_D}\right)}\right]$$

Proving the first-order solution reduces to this same expression is trivial and won't be duplicated here.

5a. $T_e = 3.69$, b. $r_{p_{over}} = 0.977r_e$

Chapter 9

5. Given the definition $F_p = \frac{\rho_p SC_D}{2m}\sqrt{\frac{r_p}{\beta}} = \left(\frac{SC_D}{2m}\sqrt{\frac{r_p}{\beta}}\right)\rho_s \exp\left[-\beta\left(r_p - R_\oplus\right)\right]$, we

can form the ratio between the "high" and "low" values:

$$\frac{F_{p_{high}}}{F_{p_{low}}} = \sqrt{\frac{r_{p_{high}}}{r_{p_{low}}}}\exp\left[-\beta\left(r_{p_{high}} - r_{p_{low}}\right)\right]$$

For thin atmospheres, $\dfrac{r_{p_{high}}}{r_{p_{low}}} \approx 1$. Using this fact and the given information on

the range of F_p, we can solve for $\left(r_{p_{high}} - r_{p_{low}}\right)$:

$$\left(r_{p_{high}} - r_{p_{low}}\right) = \frac{-1}{\beta}\ln\left(\frac{F_{p_{high}}}{F_{p_{low}}}\right) = \frac{-1}{0.14km^{-1}}\ln\left(\frac{0.065}{0.295}\right) = 10.8km$$

7. A graph of the overshoot boundary is shown below. A line at

$$\sqrt{\frac{u_e}{\cos^2 \gamma_e}} = {}^R V = \sqrt{2}$$ helps us to read the chart at parabolic entry. The graph

easily shows $Z_{p_{over}} \approx 0.015$ as the point of interest. You must still convert that

to an altitude (for Earth) and compare it to the undershoot value of

$Z_{p_{under}} \approx 0.295$. (Problem #5 from this chapter solves a similar problem for a

strictly ballistic problem. In the problem we find that corridor width to be

10.8 km.) A question you should ask yourself – because it makes a great test

question – is "Can can I expand the corridor even more by using positive lift

to change the undershoot boundary?"

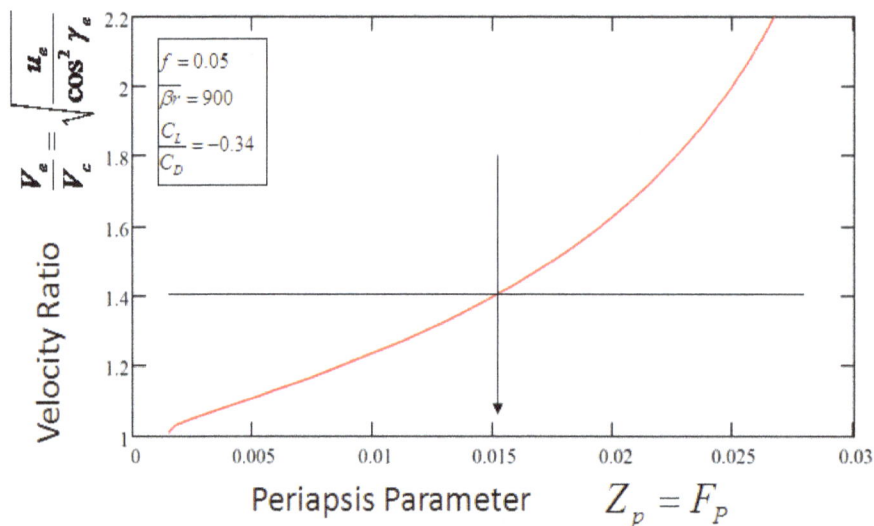

Chapter 10

2. k_2 was defined just above Eq. (10.68) as $k_2^2 = \overline{\beta r} k_1$. Replacing k_1 using Eq. (10.56), we have $k_2^2 = \overline{\beta r}\left(v \sin^2 \gamma\right)$. Using Eq. (10.26) to replace v and simplifying:

$$k_2^2 = \overline{\beta r}\left(\frac{{}^R V^2}{gr}\right)\sin^2 \gamma$$

$$k_2 = \sqrt{\frac{\overline{\beta r}}{gr}}\,{}^R V \sin \gamma$$

For thin atmospheres, $gr \approx$ constant and $\sqrt{\dfrac{gr}{\beta r}}$ has units of velocity and can be used as a scaling factor to "non-dimensionalize" a velocity. Since ${}^R V \sin \gamma$ is the (dimensional) sink rate, dividing it by $\sqrt{\dfrac{gr}{\beta r}}$ creates a non-dimensional form of a sink rate.

6. The first plot below shows the altitude versus velocity curves are, for all intents and purposes, identical. The second shows some difference in the commanded lift, but not enough to significantly alter the trajectory. The final plot helps explain why there is so little difference – the vehicle has dropped to 10% of its initial kinetic energy ($v/v_e \leq 0.1$) by the time the control laws diverge. Recalling that lift is proportional to velocity squared (or proportional to kinetic energy), you can see that the difference, while growing, has less and less of an effect on the motion. (A log scale was used on the x-axis of the final plot simply to "blow up" the region where the results differed.)

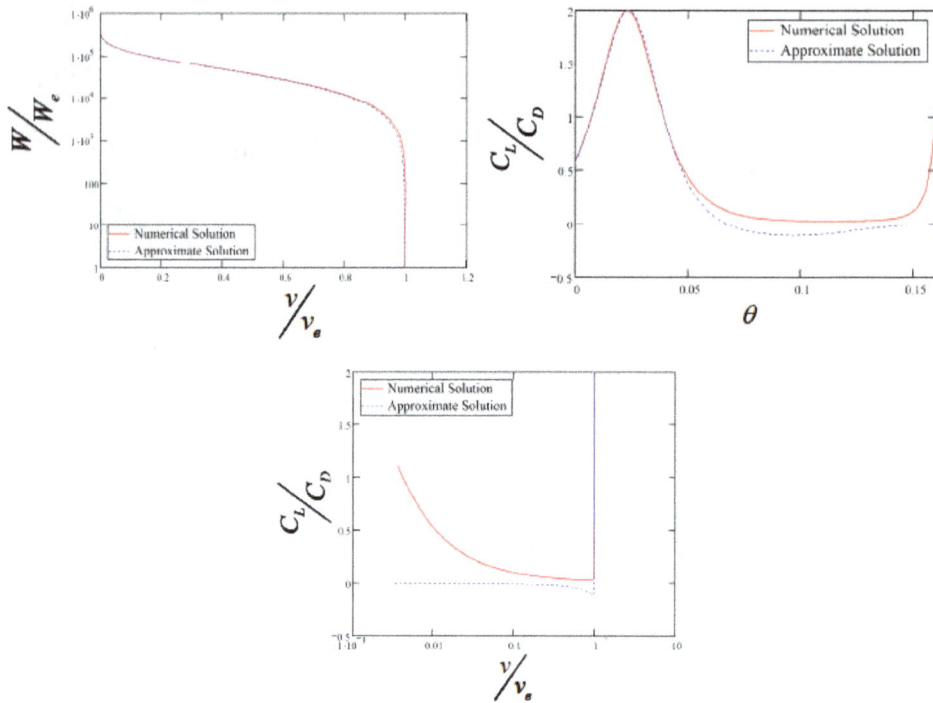

Chapter 11

1. Begin by expanding out the mean motion and simplifying:

$$\frac{2na^2\omega_\oplus\left(1-e^2\right)^{\frac{1}{2}}\cos i}{{}^I V^2} = \frac{2\left(\dfrac{\mu}{a}\right)^{\frac{1}{2}} a\omega_\oplus\left(1-e^2\right)^{\frac{1}{2}}\cos i}{{}^I V^2}$$

Next, approximate the velocity by its initial periapsis value using the vis-via equation

$${}^I V_{p_0} = \left[\mu\left(\frac{2}{r_{p_0}} - \frac{1}{a}\right)\right]^{\frac{1}{2}} = \left[\mu\left(\frac{2a - r_{p_0}}{r_{p_0}a}\right)\right]^{\frac{1}{2}} = \left(\frac{\mu}{a}\right)^{\frac{1}{2}}\left(\frac{(1+e)}{(1-e)}\right)^{\frac{1}{2}}$$

and replace *one* of the velocities in the equation. At the same time, replace the inclination with its initial value. This gives:

$$\frac{2na^2\omega_\oplus\left(1-e^2\right)^{1/2}\cos i}{{}^I V^2} \approx \frac{2\left(\dfrac{\mu}{a}\right)^{1/2} a\omega_\oplus\left[(1-e)(1+e)\right]^{1/2}\cos i_0}{{}^I V_{p_0}\left[\left(\dfrac{\mu}{a}\right)^{1/2}\left(\dfrac{1+e}{1-e}\right)^{1/2}\right]}$$

$$= \frac{2a\omega_\oplus\left(1-e\right)\cos i_0}{{}^I V_{p_0}}$$

$$= \frac{2r_{p0}\omega_\oplus\cos i_0}{{}^I V_{p_0}}$$

This is the desired result.

4. For small changes in the orbit, we can assume the denominator is almost a constant. (The only changing value is F, and it's a constant close to 1.) Thus, we can take the ratio between the "higher" orbit and the "lower" orbit to get:

$$\frac{\Delta t_{higher}}{\Delta t_{lower}} \approx \frac{e^{\beta\left(h_0+\beta^{-1}\right)}-1}{e^{\beta h_0}-1}$$

Realize $e^{\beta\left(h_0+\beta^{-1}\right)} > e^{\beta h_0} \gg 1$ and simplify:

$$\frac{\Delta t_{higher}}{\Delta t_{lower}} \approx \frac{e^{\beta\left(h_0+\beta^{-1}\right)}-1}{e^{\beta h_0}-1}$$

$$\approx \frac{e^{\beta\left(h_0+\beta^{-1}\right)}}{e^{\beta h_0}}$$

$$= \frac{e^{\beta h_0+1}}{e^{\beta h_0}}$$

$$= e$$

$$\approx 2.7$$

Chapter 12

1. After equating, the three component equations are:

$$^R\dot{V}\sin\gamma + {}^RV\dot{\gamma}\cos\gamma - \frac{{}^RV^2\cos^2\gamma}{r} = \frac{L}{m}\cos\sigma\cos\gamma - \frac{D}{m}\sin\gamma - g_r$$

$$^R\dot{V}\cos\gamma\cos\psi - {}^RV\dot{\gamma}\sin\gamma\cos\psi - {}^RV\dot{\psi}\cos\gamma\sin\psi$$
$$+ \frac{{}^RV^2}{r}\cos\gamma\cos\psi\left(\sin\gamma - \tan\phi\cos\gamma\sin\psi\right)$$
$$= -\frac{L}{m}\left(\cos\sigma\sin\gamma\cos\psi + \sin\sigma\sin\psi\right) - \frac{D}{m}\cos\gamma\cos\psi$$

$$^R\dot{V}\cos\gamma\sin\psi - {}^RV\dot{\gamma}\sin\gamma\sin\psi + {}^RV\dot{\psi}\cos\gamma\cos\psi$$
$$+ \frac{{}^RV^2}{r}\cos\gamma\left(\cos\gamma\cos^2\psi\tan\phi + \sin\gamma\sin\psi\right)$$
$$= \frac{L}{m}\left(-\cos\sigma\sin\gamma\sin\psi + \sin\sigma\cos\psi\right) - \frac{D}{m}\cos\gamma\sin\psi - g_\phi$$

The second and third component equations combine to gives one force equation

$$^{R}V\dot{\psi} = \frac{L\sin\sigma}{m\cos\gamma} - \frac{^{R}V^{2}}{r}\cos\gamma\cos\psi\tan\phi - g_{\phi}\frac{\cos\psi}{\cos\gamma}$$

and another equation we'll need:

$$^{R}\dot{V}\cos\gamma - {}^{R}V\dot{\gamma}\sin\gamma + \frac{^{R}V^{2}}{r}\cos\gamma\sin\gamma = -\frac{L}{m}\cos\gamma\sin\gamma - \frac{D}{m}\cos\gamma - g_{\phi}\sin\gamma$$

Combining this latter equation with the first component equation yields the other two force equations:

$$^{R}\dot{V} = -\frac{D}{m} - g_{r}\sin\gamma - g_{\phi}\sin\gamma\cos\gamma$$

$$^{R}V\dot{\gamma} = \frac{L}{m}\cos\sigma - g_{r}\cos\gamma + g_{\phi}\sin^{2}\gamma + \frac{^{R}V^{2}}{r}\cos\gamma$$

Bibliography

1. Allen, Julian H., and Eggers, A. J., Jr. "A Study of the Motion and Aerodynamic Heating of Ballistic Missiles Entering the Earth's Atmosphere at High Supersonic Speeds," NACA TR 1381, 1958.

2. Andersen, Benjamin M. and Whitmore, Stephen A. "Aerodynamic Control on a Lunar Return Capsule using Trim-Flaps," AIAA Paper No. 2007-1200, Presented at the 44th AIAA Aerospace Sciences Meeting and Exhibit, Reno NV, 9-12 January 2006.

3. "Apollo 10 Mission Report," *National Aeronautics and Space Administration,* Manned Spacecraft Center Report MSC-00126, August 1969.

4. "Apollo Command and Service Module System Specification (Block 1)," North *American Aviation, Inc.,* NASA Contractor Report NASA-CR-1167571, 1 October 1964.

5. Bate, Roger R. et al. Fundamentals of Astrodynamics. New York: Dover Publications, Inc., 1971.

6. Beyer, William H. CRC Standard Mathematical Tables, 27th Edition. Boca Raton: CRC Press, Inc., 1984.

7. Brace, F. C. "An Improved Chapman Theory for Studying Entry Into Planetary Atmospheres," Ph.D. Thesis, The University of Michigan, 1974.

8. Brink, Raymond W. Spherical Trigonometry. New York: Appleton-Century-Crofts, Inc., 1942.

9. Brogan, William L. Modern Control Theory, Third Edition. Upper Saddle River, NJ: Prentice Hall, 1991.

10. Bryson, Arthur E., Jr. and Ho, Yu-Chi. Applied Optimal Control – Optimization, Estimation, and Control. New York: Hemisphere Publishing Corporation, 1975.

11. Busemann, A., et al. "Optimum Maneuvers of Hypervelocity Vehicles," NASA Contractor Report CR-1078, June 1968.

12. Cameron, J.D.M. "Explicit Guidance Equations for Maneuvering Re-Entry Vehicles," TP3-3:30, General Electric Company, Philadelphia PA, 1973.

13. Chern, Jen-Shing and Vinh, Nguyeg Xuan. "Optimum Reentry Trajectories of a Lifting Vehicle," NASA Contractor Report 3236, 1980.

14. Chapman, D. R. "An Analysis of the Corridor and Guidance Requirements for Supercircular Entry into Planetary Atmospheres," NASA Technical Report R-55, 1960.

15. Chapman, Dean R. "An Approximate Analytical Method for Studying Entry into Planetary Atmospheres," NACA Technical Note 4276, 1958.

16. Chapman, D. R. "An Approximate Analytical Method for Studying Entry into Planetary Atmospheres," NASA Technical Report R-11, 1959.

17. Chapman, Dean R. "On the Corridor and Associated Trajectory Accuracy for Entry of Manned Spacecraft into Planetary Atmospheres," Proceedings of the Xth International Astronautical Congress (Springer-Verlag, Vienna, pp. 254-267), 1960.

18. Chapman, D. R. and Kapphahn, A. K. "Tables of Z Functions for Atmospheric Entry Analyses," NASA TR-R-106, 1961.

19. Coddington, Earl A. An Introduction to Ordinary Differential Equations. New York: Dover Publications, Inc., 1961.

20. Danby, J.M.A. Fundamentals of Celestial Mechanics, Second Edition. Richmond VA: Willmann-Bell, Inc., 1992.

21. DeRose, Charles E. "Trim Attitude, Lift and Drag of the Apollo Command Module with Offset Center-of-Gravity Positions at Mach Numbers to 29," NASA Technical Note TN D-5276, June 1969.

22. "Dynamic Pressure." *National Aeronautics and Space Administration.* Retrieved 24 October 2007, from the World Wide Web. http://www.grc.nasa.gov/WWW/K-12/airplane/dynpress.html

23. Eggers, Alfred J., Jr. "The Possibility of a Safe Landing," Space Technology. Ed. Howard Seifert. New York: John Wiley and Sons, Inc., 1959. 13.01-13.53.

24. Fitzpatrick, Philip M. Principles of Celestial Mechanics. New York: Academic Press, Inc., 1970.

25. García-Llama, Eduardo. "Analytic Guidance for the First Entry in a Skip Atmospheric Entry," AIAA Paper No. 2007-6897, Presented at the AIAA Atmospheric Flight Mechanics Conference and Exhibit, Hilton Head SC, 20-23 August 2007.

26. Gazley, C. "Deceleration and Heating of a Body Entering a Planetary Atmosphere from Space," The RAND Corporation Report P-955, February 1957.

27. Gell, David Allen. "Optimum Aerodynamic Maneuvers of Orbiting Vehicles," PhD Dissertation, The University of Michigan, 1983.

28. Gracey, C., et al. "Fixed-Trim Re-Entry Guidance Analysis," Journal of Guidance and Control, pp. 558-563, Volume 5, Number 6, 1982.

29. Graves, Claude A. and Harpold, Jon C. "Apollo Experience Report – Mission Planning for Apollo Entry," NASA Technical Note TN D-6725, March 1972.

30. Griffen, Michael D. and French, James R. Space Vehicle Design. Washington, DC: American Institute of Aeronautics and Astronautics, Inc., 1991.

31. Hankey, Wilbur L. Re-Entry Aerodynamics. Washington, DC: American Institute of Aeronautics and Astronautics, Inc., 1988.

32. Hicks, Kerry D. Simple Modeling Tools for Force Applications Reentry Vehicles. Edwards AFB, California: Phillips Laboratory, June 1993. (PL-TR-92-3033).

33. Jorris, Timothy R. "Common Aero Vehicle Autonomous Reentry Trajectory Optimization Satisfying Waypoint and No-Fly Zone Constraints," Doctoral Thesis, AFIT/DS/EN/0704. Graduate School of Engineering and Management, Air Force Institute of Technology (AU), Wright-Patterson Air Force Base OH, September 2007.

34. Kaya, Emre. "Crew Exploration Vehicle (CEV) Skip Entry Trajectory," Master's Thesis, AFIT/GSS/NEY/08-M06. Graduate School of Engineering and Management, Air Force Institute of Technology (AU), Wright-Patterson Air Force Base OH, March 2008.

35. King-Hele, Desmond. Theory of Satellite Orbits in an Atmosphere. London: Butterworth, 1964.

36. Lees, Lester., et al. "Use of Aerodynamic Lift During Entry into the Earth's Atmosphere," ARS Journal, pp. 633-641, September 1959.

37. Lees, L. and Probstein, R.F. "Hypersonic Viscous Flow Over a Flat Plate," Princeton University Aeronautical Engineering Laboratory, Report No. 195, 1952.

38. Li, Ting-Yi and Nagamatsu, H.T. "Shock Wave Effects on the Laminar Skin Friction of an Insulated Flat Plate at Hypersonic Speeds," GALCIT Memorandum, No. 9, 1952.

39. Loh, W. H. T. Dynamics and Thermodynamics of Planetary Entry. Englewood Cliffs, NJ: Prentice-Hall, Inc., 1963.

40. Longuski, James Michael and Vinh, Nguyen Xuan. "Analytic Theory of Orbit Contraction and Ballistic Entry into Planetary Atmospheres," JPL Publication 80-58, 1 September 1980.

41. Manders, R. "Apollo 11 Entry Postflight Analysis," NASA MSC Internal Note No. 70-FM-30 (MSC-01806), 20 February 1970.

42. McCuskey, S. W. Introduction to Celestial Mechanics. Reading, MA: Addison-Wesley Publishing Company, Inc., 1963.

43. Meyerott, Roland E. "Radiation Heat Transfer to Hypersonic Vehicles," Third Combustion and Propulsion Colloquium, pp. 431-447, London, England: Pergamon Press, 1958.

44. Putnam, Z. R., et al. "Entry System Options for Human Return from the Moon and Mars," AIAA Paper No. 2005-5915, Presented at the 2005 AIAA Atmospheric Flight Mechanics Conference, San Francisco CA, August 2005.

45. Putnam, Z. R., et al. "Improving Lunar Return Entry Footprints Using Enhanced Skip Trajectory Guidance," AIAA Paper No. 2006-7438, Presented at the AIAA Space 2006 Conference, San Jose CA, September 2006.

46. Regan, Frank J. Re-Entry Vehicle Dynamics. Washington, DC: American Institute of Aeronautics and Astronautics, Inc., 1984.

47. Regan, Frank J. and Anadakrishnan, Satya M. Dynamics of Atmospheric Re-Entry. Washington, DC: American Institute of Aeronautics and Astronautics, Inc., 1993.

48. Sänger, E. and Bredt, J. "A Rocket Drive for Long Range Bombers," Translation No. CGD-32, Technical Information Branch, Buaer Navy Department, August 1944.

49. Sellers, Jerry Jon. <u>Aerospace Science: The Exploration of Space.</u> Boston MA: The McGraw-Hill Companies, Inc., 2003.

50. Shkadov, L. M., <u>et al.</u> " Mechanics of Optimum Three-Dimensional Motion of Aircraft in the Atmosphere," <u>NASA Technical Translation F-777,</u> March 1975. (Originally published by Mashinostroyeniye Press, Moscow 1972.)

51. Szelc, D. "Apollo 10 Entry Postflight Analysis," <u>NASA MSC Internal Note No. 69-FM-283 (MSC-00126, Supplement 10),</u> 7 November 1969.

52. Sutton, George P. <u>Rocket Propulsion Elements: An Introduction to the Engineering of Rockets (Fifth Edition).</u> New York NY: John Wiley & Sons, 1986.

53. Tewari, Ashish. <u>Atmospheric and Space Flight Dynamics.</u> Boston MA: Birkhäuser Boston, 2007.

54. Undurti, Aditya. "Optimal Trajectories for Maneuvering Reentry Vehicles." MS Thesis, Draper Report No. T-1581, Massachusetts Institute of Technology (MIT), Department of Aeronautics and Astronautics, Cambridge MA, May 2007.

55. Undurti, Aditya, <u>et al.</u> "Optimal Trajectories for Maneuvering Re-entry Vehicles," <u>Paper No. AAS 07-306,</u> Presented at the 2007 AAS/AIAA Astrodynamics Specialist Conference, August 2007.

56. Vallado, David A. <u>Fundamentals of Astrodynamics and Applications.</u> New York: The McGraw-Hill Companies, Inc., 1997.

57. Vinh, Nguyen X., <u>Optimal Trajectories in Atmospheric Flight.</u> Amsterdam: Elsevier Scientific Publishing Company, 1981.

58. Vinh, Nguyen X., <u>et al.</u> <u>Hypersonic and Planetary Entry Flight Mechanics.</u> Ann Arbor: The University of Michigan Press, 1980.

59. Vinh, N. X. and Brace, F. C. "Qualitative and Quantitative Analysis of the Exact Atmospheric Entry Equations Using Chapman's Variables," <u>IAF Paper No. 74-010,</u> Presented at the XXVth Congress of the International Astronautical Federation, Amsterdam, The Netherlands, October 1974.

60. Vinh, N. X. and Chern, J. S. "Three-Dimensional Optimum Maneuvers of a Hypervelocity Vehicle," <u>IAF Paper No. 79-F-184,</u> presented at the XXXth Congress of the International Astronautical Federation, Munich, F.R.G., September 1979.

61. Vinh, N. X., et al. "Optimum Three-Dimensional Atmospheric Entry," Acta Astronautica, pp. 593-611, 1975.

62. Vinh, N. X., et al. "Flight with Lift Modulation inside a Planetary Atmosphere," AIAA Journal, pp. 1617-1623, November 1977.

63. Walter, Ulrich. Astronautics, Weinheim, Germany: Wiley-VCH Verlag GmbH & Co. KGaA, 2008.

64. Wang, Kenneth and Ting, Lu. "An Approximate Analytic Solution of Re-Entry Trajectory With Aerodynamic Forces," ARS Journal, pp. 565-566, June 1960.

65. Wiesel, William E. Modern Astrodynamics, First Edition, Printing 1.05. Beavercreek OH: Aphelion Press, 2003.

66. Wiesel, William E. Space Flight Dynamics, Second Edition. New York: The McGraw-Hill Companies, Inc., 1997.

67. Whitmore, Stephen A., et al. "Direct-Entry, Aerobraking, and Lifting Aerocapture for Human-Rated Lunar Return Vehicles," AIAA Paper No. 2006-1033, Presented at the 44th AIAA Aerospace Sciences Meeting and Exhibit, Reno NV, 9-12 January 2006.

68. Zipfel, Peter H. Modeling and Simulations of Aerospace Vehicle Dynamics, Second Edition. Reston VA: American Institute of Aeronautics and Astronautics, Inc., 2007.

Index

A

Acceleration, 65, *See also: Deceleration*

    centripetal, 39

    Coriolis, 39

    drag, 356–58, 372

    relative, 39

Aerodynamic heating, 215, *See also: Heating*

Altitude

    geocentric, 382, 385

    geodetic, 382–83

    non-dimensional, 69, 242–43, 294

Apoapsis, 13

Apollo

    Apollo 10 reentry, 375–79, 384–91, 393–99,

        401–7, 409–12, 414–15, 417, 419–21

    Apollo 11 reentry, 375, 379, 404

Approach conic, 222–25

    periapsis, 225–27

Argument of latitude, 348

Argument of periapsis, 11, 255

Ascending node, 11, 251, 255

Atmosphere

    1962 Standard, 400, 407–8

    exponential, 68–69, 400

    isothermal, 242, 400

B

Ballistic coefficient, 356

Ballistic entry

    shallow, 131–38, 140

    steep, 143, 160–62, 200–215, 216

Ballistic missile. *See Missile*

Bank angle, 44

    for planar entry, 59

    variable, 290–96, 315–41

C

Control

    optimal, 315, 330, 331

Crew Exploration Vehicle (CEV), 280–83

Cross range

    maximum, 319–39

D

Decay

    orbital, 369–71

Deceleration, 65–66, 71, 162–63, *See also:*
    Acceleration

Diveline guidance, 339–40

Drag, 43, 346, 356–58, 360–61, 364

    perturbation, 357, 364, 370, 360–61, 372

Drag polar, 291

    parabolic, 296–98

Dynamic pressure, 310

E

Eccentric anomaly, 13, 14, 357

451

U

Unified theory. *See Universal equations of motion*

Universal equations of motion

    application, 261–69, 272–80, 281, 280–83, 283–85, 300–304, 304–8, 308–10, 310–12, 315–16, 316–19, 319–39, 417–22

    development

        constant drag coefficient, 240–45

        variable drag coefficient, 290–96

    reduction to Loh's solutions, 255–61

    reduction to orbital equations, 245–55

V

Velocity

    circular, 15, 132

    finding from orbital elements, 21–23

    orbital, 12

    reference, 78

Vis-viva equation, 12

Vostok, 138

Y

Yaroshevskii, 131, 138, 140, *See also: Ballistic entry*

This page intentionally blank.